Gas Chromatography

Gas Chromatography

L. Szepesy, C.Sc.

Head of the Research Department of the Hungarian Oil and Gas Research Institute (MÁFKI) and Associate Professor at the Veszprém University of Chemical Engineering

English translation edited by

E. D. Morgan, B.Sc., M.A., D.Phil., F.R.I.C.

Formerly research chemist, Shell Chemical Co. Ltd, now Lecturer in Chemistry, University of Keele

ILIFFE BOOKS LTD
42 RUSSELL SQUARE
LONDON, W. C. 1

This book is the enlarged English version of the
"GÁZKROMATOGRÁFIA"
published by Műszaki Könyvkiadó, Budapest 1963

Revised by Prof. P. Benedek
Translated by F. Sós

592 01238 7

English edition first published in 1970 by Iliffe Books Ltd, in co-edition with Akadémiai Kiadó, Publishing House of the Hungarian Academy of Sciences, Budapest

Printed in Hungary

Contents

Preface

Development in the past 16 years has raised gas chromatography to the ranks of the most widely applied analytical methods both in laboratory and in industry. Though many instrumental analytical methods have undergone a similar spectacular development in recent years, the importance and wide-spread use of gas chromatography surpass that of any other method and appear to be quite unprecedented. The success of gas chromatography is due to the simplicity, rapidity and high sensitivity of the method which provides for efficient separation and opens possibilities for manifold applications. From permanent gases to high boiling point liquids or volatile solids, a great variety of organic and inorganic substances can be analyzed gas chromatographically. Because of the high sensitivity of the detectors, analysis requires only extremely small samples (of the order of micrograms) and impurities in ppm or even ppb concentrations can be detected and determined. Besides its application in analysis, gas chromatography has acquired growing importance in the preparation of pure substances, in the study of the chemical structures of compounds, in the investigation of the kinetics and mechanism of chemical reactions, including the determination of physico-chemical constants, and is also being used in automatic operation and process control.

Gas chromatography equipment is today an indispensable and fundamental tool in both research and industrial laboratories; without it the solution of certain analytical problems would be quite inconceivable in many fields. It is estimated that more than 60,000 gas chromatographs are now in use in the world. The wide-spread interest in gas chromatography is also reflected in the number of relevant publications: up to the end of 1967 more than 15,000 papers were published on this subject, of which more than 60% appeared in the last four years. Many excellent books on the subject have appeared in the past five or six years. Most of these books are, however, intended for specialists with some experience in gas chromatography and are therefore difficult to understand for beginners or for specialists of other fields who may wish to use the method.

The present book aims at presenting a general outline of the theoretical and practical aspects of gas chromatography, including the possibilities of its application and the trend of developments. By presenting a detailed survey of the relevant literature we wish to assist the reader in his eventual further search and to facilitate the acquisition of the necessary data. Theoretical questions will be dealt with only to the extent necessary for the explanation of chromatographic separation as a physical process and for the understanding of the fundamental correlations which are indispensable for the solution of practical problems. The description of the main parts of gas chromatography equipment should help the reader to select the appropriate apparatus and optimum operating conditions. The various types of columns and packings will be discussed in detail, for when faced with a given task of separation the choice of a suitable column and stationary phase is the paramount problem in practical gas chromatography. The description of special techniques should give some guidance for the solution of more complicated analytical tasks and of other problems. Beyond the discussion of qualitative and quantitative analytical methods a general survey of the analytical applications of gas chromatography will also be presented.

Preparative gas chromatography and process control are highly important special fields of application which together with some other special applications supplement the survey of the general scope of gas chromatography.

L. Sz.

1. General Survey

1.1 Classification of chromatographic methods

Essentially chromatography is a method of separation belonging to the class of diffusion mass transfer operations where the components of the mixture are separated as a result of their partition between two immiscible heterogeneous phases.

Chromatographic methods were first evolved for analytical purposes to study certain complex natural and synthetic substances which were incapable of examination by other methods. Besides their analytical applications, chromatographic methods have gained increasing importance, especially in the last twenty years, in the solution of other problems, such as the preparation of pure substances, determination of physico-chemical constants, reaction kinetic studies, investigation of molecular structures, and so on.

The development of chromatographic methods has followed a long and tortuous path; the work of many researchers, a number of discoveries and much development work were necessary to evolve the methods in use today. An essentially chromatographic method was used to clear sugar solutions as far back as the middle of the last century. Day[1] carried out the first laboratory chromatographic experiments at the turn of the century when passing crude oil through a column packed with finely powdered fuller's earth and collecting the various petroleum fractions, similar to the fractions obtained by distillation. Tswett[2] at the Institute of Botany of Warsaw University made some important contributions to the development of chromatography. In a paper published in 1906 he described the separation of the dye components of green leaves on a calcium carbonate adsorbent column. After rinsing with petrol ether the components of the dye mixture formed zones of different colours on the column: this observation gave the method its name (chroma = colour). Chromatographic methods had however to wait for their extensive study and application till the 1930's when following the work of Kuhn and Lederer and of the Hungarian scientists Zechmeister and Cholnoky more and more workers began to apply chromatography to the separation of dyes, sugars, amino acids, etc., that is, to substances which were difficult to separate by other methods.

The disclosure of new applications led to the development of new methods which in turn opened still newer fields of application. Today chromatographic methods are widely used in all branches of chemistry and of the allied sciences from the processing of micro-quantities up to commercial processes. In the course of time the meaning of the word "chromatography" has considerably expanded, and today the methods have but little to do with "separation according to colour". Though the original designation has only historical significance, it is still generally accepted and used for the characterization of a class of operations. As already indicated, chromatographic methods belong to the class of diffusion mass transfer operations. No special definition of chromatographic methods can be given on the basis of the physical principles of separation; the generally accepted definition refers only to the type of operation.

Chromatography is a collective name for all separation processes where the separation of the components is effected by their partition (differential sorption) between a stationary (fixed) phase with a large surface and a moving (percolating) phase which flows over the first.

Though there are several definitions in the literature, a strict and accurate definition seems to be unimportant and only certain characteristics will have to be fixed for the unequivocal treatment of chromatographic methods. Strictly speaking the above definition refers only to fixed bed procedures, but may be reasonably extended to moving bed processes where the two heterogeneous phases move in counter-current. From the point of view of the separation process the decisive factor is the relative flow between the phases.

Chromatographic methods may be classified according to various aspects. Unfortunately there is some considerable disagreement in the literature concerning the characteristics of various methods and their classification. One type of classification is based on the sorption process:[3] 1. adsorption chromatography, 2. partition (solution) chromatography and 3. ion exchange. A fourth group of chromatographic processes should also be mentioned here, namely electrochromatography where the separation of the substances is effected either by the electric mobility induced by an electric current which is imposed on the system, or by the overall effect of electric mobility and sorption processes. This classification leads however to a confusing picture, thus for example liquid–liquid, gas–liquid and the less well-known foam and emulsion chromatography would all belong to the class of partition chromatography. Neither does a classification based on various geometrical groupings (unidirectional flow through columns, strips or layers; bidirectional or transverse flow; radial flow from a centre) provide an adequate principle. For the unambiguous study of chromatographic processes a classification based on the state of the participating phases seems to be the most appropriate, as summed up in Table 1.1. Depending on the state of the moving phase a distinction is made between liquid and gas chromatography. As stationary phase, various solid adsorbents with large surfaces may be used (in the old nomenclature this would be called adsorption chromatography), or a liquid on a solid support may constitute the second phase. Separation is then the result of selective solubilities in the stationary liquid phase (called partition chromatography in the old

Table 1.1

Moving phase	*Stationary phase*	
	Solid support	*Liquid phase*
Liquid	Liquid–solid chromatography (LSC) Methods of implementation: 1. Column chromatography 2. Ion exchange chromatography 3. Paper chromatography 4. Thin layer chromatography	Liquid–liquid chromatography (LLC) Methods of implementation: 1. Column chromatography 2. Paper chromatography 3. Thin layer chromatography
Gas	Gas–solid chromatography (GSC) Methods of implementation: 1. Packed column 2. Open tubular (capillary) column	Gas–liquid chromatography (GLC) Methods of implementation: 1. Packed column 2. Open tubular (capillary) column

nomenclature). Classification may also be based on the nature of the stationary phase and on the nature of the sorption process, or on the method used for the preparation of the stationary phase. The most generally accepted classification is based on the method of implementation; column chromatography (packed and capillary columns), ion exchange chromatography, paper chromatography and thin layer chromatography.

In gas chromatography there are two basic types of column: namely the packed column and the open tubular (capillary or Golay) column. In gas–solid chromatography the column is packed with an adsorbent of small particle size or the inner surface of the capillary tube is coated with a thin adsorbent layer. In gas–liquid chromatography the liquid phase is fixed on the inert solid support in the packed column, while in the open tubular column the internal wall of the narrow tube is coated with the liquid phase. To ensure a uniform and thicker film the inside of the tube can be coated with a thin layer of some solid (support or adsorbent) substance prior to wetting. Packed capillary columns represent a transition between the packed and open tubular columns when the free space, which is not greater than a few times the diameter of the particles, is filled with a solid. The various types of columns and their fields of application will be dealt with in detail in the chapter devoted to the discussion of columns.

Practical implementation of chromatographic methods

In both liquid and gas chromatography three different techniques are used, namely: 1. frontal analysis; 2. displacement development and 3. elution processes.

1. In frontal analysis the mixture to be separated is fed at constant speed into the chromatographic column. On passing through the column the various components are sorbed selectively and to different degrees on the stationary phase, and thus move at different rates in the direction of flow. The least adsorbed component appears first ("breaks through") at the end of the column followed by the other components in the order of their sorption capacity. The emerging components can be detected by one or other of their physical properties with the help of various detectors. Figure 1.1 shows diagrammatically the separation obtained by the frontal method. In frontal chromatography only the component with lowest sorptivity leaves in the pure state, all other components are mixed with others with lower sorption power (overlapping).

2. In displacement chromatography the chromatogram is developed by the introduction of a substance of higher sorptivity than that of any of the components in the mixture to be separated and thus is capable of displacing the latter from the sorbent phase. As a result of

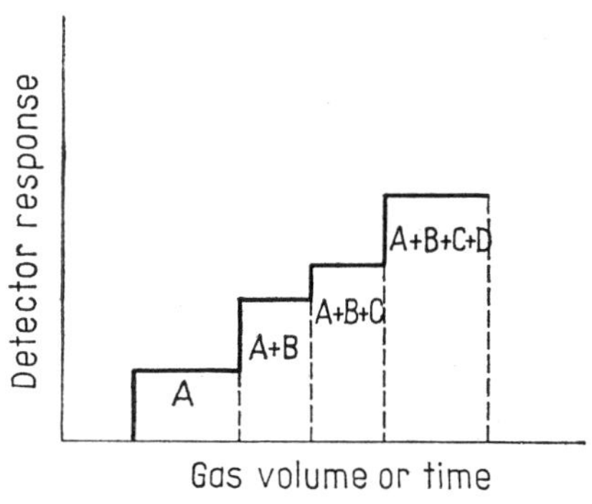

Fig. 1.1. Frontal chromatography

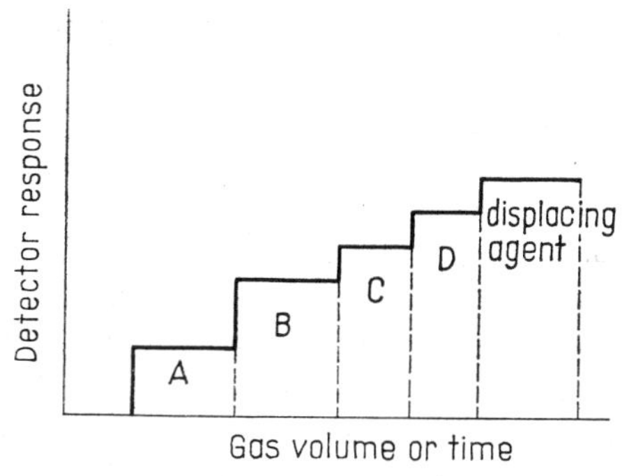

Fig. 1.2. Displacement chromatography

displacement the components of the sample under investigation take up an arrangement in the order of their binding power forming thereby distinct zones and emerging from the column in succession. Continuous analysis of the emerging material gives the consecutive steps shown in Fig. 1.2. The height of the steps can be used for the identification of the components, while the length of the step is proportional to the quantity of the component.

3. In elution chromatography the sample under investigation is swept through the column by elution with a fluid (eluent) or gas (carrier gas). The eluent passes through the column in a constant stream into which the sample to be analyzed is fed. The various

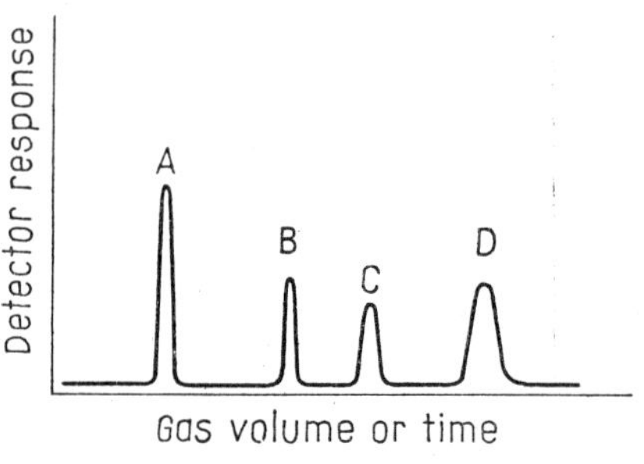

Fig. 1.3. Elution chromatography

components in the sample will travel, under the rinsing effect of the eluent, with different velocities through the column when the velocity will depend on the sorptivity of the component. Provided there is a sufficient difference between the partition coefficients, the components will leave the column after a certain distance completely separated. The chromatogram can be determined by a continuous detection of the emerging material stream; the form of such a chromatogram is shown in Fig. 1.3. The shape of the elution curves corresponds more or less to that of the Gaussian distribution curve. The various components can be identified from the volume of the carrier gas (eluent) which has flowed through the column till the component appears in its maximum concentration, or by measuring the time required for the component to appear. For quantitative evaluations the areas below the peaks are measured.

Let us now briefly compare the three different techniques:

Because of the overlapping mentioned, the frontal technique is not suitable for analytical purposes, but may be successfully applied to the measurement of phase equilibria (isotherms), for preparative separations, etc. The overwhelming majority of industrial chromatographic processes use frontal techniques.

The displacement technique is used in liquid chromatography for analytical purposes, while in gas chromatography this technique is mainly an auxiliary method for the preliminary concentration of components present in low concentrations. Displacement techniques may also be used successfully for preparative processes.

The frontal and displacement techniques have the common disadvantage that at the end of separation the column is still full of the sample or of the substance used for displacement and has to be regenerated before its next use.

In the case of elution only the eluent remains in the column at the end of the analytical process and the column is directly ready for the next analysis. The elution method has the disadvantage that the strongly bound components travel through the column fairly slowly, but by raising the temperature and the flow rate of the eluent the elution of the components can be considerably accelerated. The latest methods applied in gas chromatography, namely temperature and flow programming, offer possibilities of changing the temperature and the rate of flow of the carrier gas during analysis, thereby accelerating the elution of the high boiling, more strongly sorbed components in the sample.

The elution technique dominates gas chromatography and has an exclusive place in gas chromatographic analysis. We shall therefore restrict ourselves to a discussion of the theory and application of elution methods. For an excellent and detailed treatment of the frontal and displacement techniques we refer the reader to Schay's book.[4]

1.2 Development and applications

The adsorption of gases and vapours has been the subject of study for many years, and attempts at the application of selective adsorption to gas analysis date back to the 1920's. Adsorptive gas analysis however developed not on the line of classical chromatographic methods, but on the analogy with fractionated distillation. The first important works in this field were published by Peters and Weil,[5] and Eucken and Knick[6] on gas analysis by fractionated adsorption and desorption. Hesse *et al.*[7] were the first to apply in 1941 a gas flow in a column packed with adsorbent for the elution of the components. The first conscious application of the elution method is associated with the names of Damköhler and Theile[8] (1943), and later of Cremer *et al.*,[9]

Janák[10] and Ray.[11] The application of the frontal and displacement methods also began in 1943, following the work of Turner,[12] Tiselius[13] and Claesson.[14]

The present wide-spread use of gas chromatography is associated with the implementation of gas–liquid chromatography. When working out the method of liquid–liquid partition chromatography Martin and Synge[15] suggested as far back as 1941 the use of gas as the moving phase, nevertheless the first paper on the application of gas–liquid partition chromatography was published only in 1952. James and Martin[16] worked out the method for the microanalysis of fatty acids which soon gained wide-spread application and developed into one of the most generally and widely used analytical methods in research and industrial laboratories.

The history of gas chromatography can be divided broadly into four stages. In the first stage (between 1930 and 1940) gas chromatography passed through its very first applications to gas analysis problems and was combined with various adsorbents.

The second stage began in 1952 with the appearance of gas–liquid chromatography. In the years between 1952 and 1956 the solid foundations of gas chromatography were laid. The first apparatus for reproducible quantitative analyses was evolved and the fundamental problems of the application of the method were clarified. A number of research establishments became engaged in the development of gas chromatography and in the investigation of its application possibilities which brought about a greatly accelerated development in the following years.

The years between 1956 and 1962 may be considered as the third stage in the development of gas chromatography characterized by an extremely rapid development in both apparatus and methods. In this period high capacity, versatile laboratory gas chromatographs with exchangeable accessories were evolved and the first automatic preparative apparatus and process chromatographs appeared. Development of the apparatus was, of course, closely linked to developments in techniques, thus for instance the development of high sensitivity ionization detectors brought with it the evolution and application of capillary (open tubular) columns. Of the most important results achieved in this period the implementation of programmed temperature, pyrolysis gas chromatography and the application of microreactors in general, various pre-columns and the development of column arrangements, the evolution of secondary detectors or selective detectors should be mentioned.

The fourth stage began in 1962–63 and still continues. This last stage is characterized by a further development in apparatus and methods, the appearance of new, high-capacity equipment on the market and the consequent extremely rapid extension of the field of applications. Today gas chromatography is an indispensable method in every branch of chemistry. In addition to the study of natural and synthetic organic substances, gas chromatography has come to play an increasingly greater role in some hitherto less frequented fields of inorganic analysis. In other branches of chemistry, such as the chemistry of foodstuffs, fermentation, biochemistry, medical and forensic chemistry, etc. gas chromatographic analytical methods ensure new possibilities for the detection and control of substances. Beyond their importance in analytical chemistry gas chromatographic methods have gained an important role also in the study of the mechanism and kinetics of chemical reactions. A recent trend in development applies gas chromatography — eventually combined with other methods, e.g. spectrometry — to the determination of the structure of chemical compounds.

Another feature of developments in this last stage is the wide-spread investigation of the theoretical problems involved in gas chromatographic separation. In the preceding years research was mainly directed to the solution of problems raised by the practical application of the method and to finding new fields of application. Theoretical studies provided mainly working hypotheses, primarily for the qualitative study of the processes. In recent years research has been greatly intensified with respect to the study of the theoretical foundations of separation and of the influence of various interfering factors. Though these problems are far from being solved, the results obtained so far have already greatly contributed to the clarification of the separation process and of the effect of experimental conditions. These results opened a way to the improvement of the performance and efficiency of columns, to the evolution of new column types, to the reduction of the time needed for analysis — to mention only the most important achievements.

In connection with the improvement of column performance, work on the improvement in quality of solid supports, the appearance of new supports and adsorbents, the search for and application of new liquid phases should be mentioned. Detailed study of the interaction processes between the solute and the stationary phase provides the basis on which the final aim can be achieved, namely that the selection of column packing for any given task of separation should be possible without preliminary measurements. This selection is still quite empirical.

Considerable steps forward have been made in the qualitative and quantitative evaluation of the results of gas chromatographic analysis. In qualitative analysis, in addition to selective detectors, other chromatographic, spectrometric and chemical methods are used. The accuracy of quantitative analysis can be improved by the reliable operation of the analyzer and application of integrators for the measurement of the areas under the peaks.

The most recent trend of development is the ever growing application of computers. Computers coupled to gas chromatographs allow completely automatic operation of laboratory instruments, furnishing results faster and more precisely than before. On the other hand, computers make possible the storing and processing of gas chromatographic data for complete and rapid evaluation, including qualitative identification and quantitative calculations, printing an analysis report that contains the weight per cent composition of the sample components. These tasks were first solved by large time shared computers, recently, however, new small and relatively inexpensive computers have been developed for gas chromatographic data handling.

The question arises here, how can the rapid development and spreading of gas chromatography be explained, which is almost unique even in the field of instrumental analysis? The success of gas chromatography is due to the simplicity, high separation power and speed of the method, as well as to its wide scope of application. With gas chromatography such multicomponent and complex mixtures can be analyzed for which earlier the combination of several methods was necessary or which appeared insoluble by known analytical methods.

In general, gas chromatography is suitable for the analysis of substances the vapour pressures of which are at least 10 mm Hg at the temperature of the column. The maximum applicable column temperature is determined by the volatility and heat stability of the stationary phase, the stability and reactivity of the components in the sample and the structural material of the detector and apparatus.

Non-volatile or heat sensitive substances can be analyzed after pyrolytic degradation by analysis of their pyrolysis products. Commercially available apparatus can be used up to 250–300 °C, the latest types up to 400–500 °C. A special apparatus works at 600 °C and even higher temperatures, up to 1000 °C.

The advantages of gas chromatography

The advantages of the gas chromatographic method can be summed up as follows:

1. Compared with other instruments of similar performance (e.g. spectrometers), gas chromatographs are relatively cheap, their maintenance cost is low, they are easy to operate, their operation requires no special previous qualification.
2. The apparatus can be used for the solution of widely varying analytical tasks from the analysis of permanent gases to that of high boiling liquids or volatile solids.
3. Because of the low viscosity of the moving gas phase very long columns with high separation power can be used.
4. Because of the high diffusion rate in the gas phase the rate of mass transfer between the moving and stationary phases is high and analysis can be performed rapidly. The time required for one analysis is in general 10–20 min, with special apparatus the analysis of multicomponent systems can be performed in seconds.
5. For the detection of the separated components in the carrier gas various relatively simple and highly sensitive detectors with low time constants are available. Consequently very small samples will suffice; generally 1–3 ml of gas and 1–10 μl of liquid and often even smaller quantities will be enough.
6. The signal from the detector can be easily recorded. In the latest apparatus, recording and evaluation of the analytical results and storing of data are all automatic.
7. As a partition phase, various adsorbents and liquid phases can be used in the column. With the proper choice of the partition phase components with identical boiling points or of similar structure (e.g. optical isomers) can be separated.
8. Gas chromatography has opened new ways for the study of chemical processes and reactions and also for the determination of the chemical structure of compounds.

From the point of view of the construction and operation of the apparatus there is no difference whatsoever between the two types of gas chromatography, namely between gas–solid and gas–liquid chromatography. The difference between the two methods is in the method of separation; partition between the two phases is in the first case based on adsorption and in the second case on dissolution in the liquid.

Applications

Gas chromatography is primarily an analytical method, but may be used because of its above properties in many other fields with excellent results.

1. Analytical applications: direct analysis of gases, liquids, volatile solids; examination of non-volatile substances through the analysis of their pyrolysis products.
2. Preparative applications: for the preparation of pure substances or narrow fractions as standards or for further detailed investigations.
3. Continuous monitoring and automatic process control.
4. Study of the structure of chemical compounds.
5. Study of the mechanism and kinetics of chemical reactions.
6. Physico-chemical measurements: determination of isotherms, heat of solution, diffusion constants, activity coefficients.
7. Application of gas chromatographic techniques to other tasks, such as the determination of the specific surface of solids, plotting of distillation curves, and elementary analysis of organic substances.

The problems in 4, 5 and 6 are essentially special applications of analytical gas chromatography. The variety of techniques would require a theoretical discussion beyond the scope of the present book. We shall, however, deal briefly with the most important methods pertaining to item 7 as these raise new problems.

1.3 Apparatus and procedure

Every laboratory gas chromatograph consists essentially of six parts:

1. carrier gas system;
2. sampling device;
3. column;
4. thermostat;

5. detector;
6. recording and evaluating system.

The simplified principle and scheme of the gas chromatograph is shown in Fig. 1.4. To illustrate the process the part representing the column is divided into three sections, the area *c* stands for the small

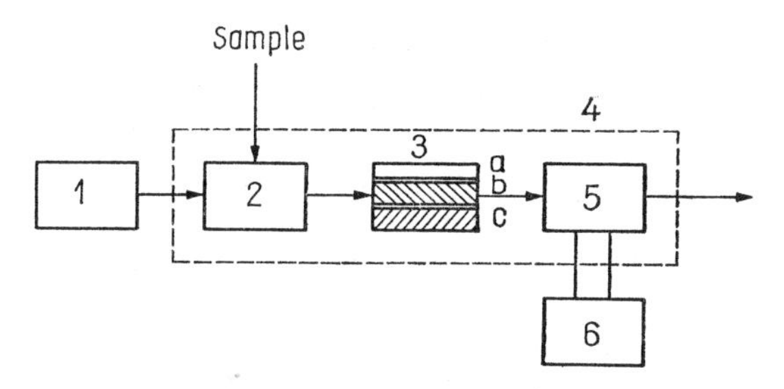

Fig. 1.4. Schematic diagram of a gas chromatographic apparatus. 1. Carrier gas system; 2. sample injector; 3. column; 4. thermostat; 5. detector; 6. recorder

particles of the solid support, covered with a thin layer of the non-volatile liquid phase *b*, while *a* represents the free space between the solid particles coated with the liquid through which the carrier gas flows.

The operation of the apparatus can be summed up briefly as follows: A gas stream, the so-called carrier gas, is fed into the column at a constant rate. The sample under investigation is injected into the carrier gas stream (in the case of liquid samples this is evaporated) and reaches the column together with the carrier gas. In the column the components of the sample travel in the direction of the gas flow with different velocities depending on their binding to the stationary phase. If the properties of this latter are suitable for the task in hand and the column is sufficiently long, the components of the sample will be completely separated and will appear in succession in the emerging carrier gas.

Figure 1.5 shows diagrammatically the separation of the components and the position of the zones of the latter at various times. By a zone we understand a section of the column in which the component or components are present in measurable quantities. When fed into the column the components in the same zone gradually separate as they travel along and emerge completely separated. The carrier gas emerging from the column is led through the detector which gives off signals proportional to the quantity of the components as they

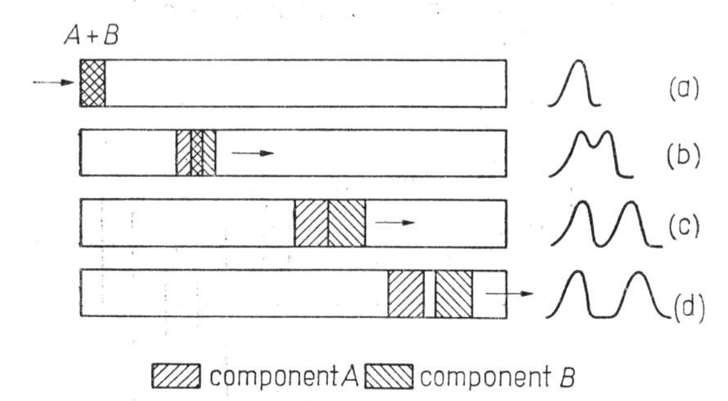

Fig. 1.5. Diagrammatic representation of the separation of two components

appear. By plotting or by electronically recording the signals from the detector a chromatogram characteristic of the actual separation is obtained.

The bell shaped curves on the right of Fig. 1.5 represent the detector signals corresponding to the position of the component zones should the end of the zone coincide with the end of the column. In case (a) the two components appear in the same peak, in case (b) a partial separation is effected, in case (c) the zones of the two components are just about to separate, while case (d) shows the chromatogram of two separated components leaving the column. The position of the peaks on the chromatogram serves for qualitative identification, while the height of the peaks or the areas below them are the bases for quantitative evaluations. The parts of the apparatus through which the sample travels are placed in a thermostat to ensure uniform temperature throughout the system.

General aspects of gas chromatographic separation

The two important characteristics of every chromatographic process are: the velocity at which the zone of the component travels in the column and the broadening of the zone during this progress.

Under given conditions, at constant temperature, pressure and carrier gas velocity (T = const, P = const, u = const), the migration rate of the zone is a simple function of the thermodynamic equilibrium of partition of the component between the moving and the stationary phase, i.e. of the partition coefficient, k. As k is temperature and pressure dependent, under any arbitrary condition the progression rate of the zone is determined by the value of k and the velocity of the carrier gas. Corresponding to its higher partition coefficient the more strongly bound (adsorbed or absorbed) component stays for a longer period of time in the stationary phase, thus its zone travels at a lower rate than the zone of the less readily sorbed component. If k is independent of the concentration (indicating a linear equilibrium correlation) the zone will travel at the same rate through the entire length of the column. In the idealized case the zone occupied by the component fed into the input end of the column will travel at a constant rate with unchanged width through the latter. This ideal case demands that the following conditions shall be fulfilled: 1. the flow rate should be the same in the entire cross-section of the column and constant in the entire column; 2. there should be no longitudinal diffusion; 3. equilibrium between the moving and stationary phases should be established instantaneously, that is to say, the mass transfer rate should be infinitely high.

However, in an actual column these conditions are far from being fulfilled, so that during its progress through the column the zone broadens (or spreads) and only the centre of the zone travels at a constant rate, depending on the partition coefficient k.

A fact often overlooked in the literature should be pointed out here, namely, that with a moving phase such an "ideal" column is impossible. When a carrier gas flows through the column, equilibrium cannot set in, not even in the case of very high mass transfer rates, simply because the gas phase moves. Equilibrium would be possible only for zero carrier gas rate. This question will be dealt with in detail in Chapter 3.

An improvement of column efficiency may be achieved by a study of the physical processes which induce broadening of the zones, i.e. by reducing the influence of these effects.

The chromatographic separation of two components requires the fulfilment of the basic condition that the zones of the components shall separate while travelling through the column, which means their progression rates, and hence their partition coefficients, shall be different. However, from the point of view of separation, the other factor, namely the broadening of the zone, is also extremely important and often an excessive broadening of the zones spoils or entirely prevents separation. The efficiency or performance of the column may quite often be judged from its effect on the broadening of the zone.

It follows then that to study the processes in the gas chromatographic column, on the one hand the equilibrium conditions and, on the other hand, the hydrodynamic and kinetic processes (diffusion and mass transfer rate) in the column have to be known. These factors will be examined in detail in connection with the theoretical problems and the operation of the column. We shall deal with the theory of separation only to a depth which is necessary for the reader to get a somewhat simplified but comprehensible physical picture of the separation process and of the factors affecting it.

1.4 Literature

The rate of development of gas chromatography can be fairly judged from the number of publications. Gas chromatography is one of the most active branches of analytical chemistry; papers on gas chromatography amount to about 10% of all analytical chemical publications and their number is exceeded only by the papers on optical methods. Between 1954 and 1962 the number of relevant publications rose exponentially. While in 1955 a total of 55 papers appeared on gas chromatography in international literature, the number of publications was over 1100 in 1960 and over 1800 in 1962. From 1963 onwards the rate of this growth decreased, since then the number of publications is estimated at yearly 1900–2000. According to certain estimates the present high number of publications will be maintained for some years, after which it will drop to a lower stable level, as was generally observed in the case of other analytical methods.

From the available data the number of publications on gas chromatography up to the end of 1968 may be estimated to be about 17,000. Considering the extraordinary variety of applications and the constantly growing number of gas chromatographs in use, the high number of publications will probably be maintained for some years. Even a superficial survey of the available material presents a serious problem and today a perceptible specialization appears within gas chromatography itself.

An overwhelming majority of the publications — about 70–75% — deals with the application of gas chromatography and only about 25–30% with the methodology and theoretical aspects. It is of course impossible to draw a sharp distinction here, as new applications involve new technical solutions and new generally valid observations may come to light.

To facilitate a survey of the literature on gas chromatography we shall attempt to give a brief characterization and summary of useful publications.

Books.[17–49] The first comprehensive works were published at the end of the 1950's and later in 1962–1963 several excellent books appeared about the same time. After a short interval again several books appeared or were in the press in 1965–1966. In addition to the earlier comprehensive works, books on certain aspects of gas chromatography, such as, for example, programmed temperature gas chromatography, open tubular columns, analysis of steroids, the study of metal chelates, etc. were published in recent years. Lectures delivered at various gas chromatographic courses and postgraduate training schools, partly of a general character dealing with the latest results in the field and partly on some narrower special problems, e.g. the investigation of steroids, were also published in book form.

As well as books dealing only with gas chromatography, many books on general chromatography also contain shorter or longer chapters on the subject.[50–53] Books on certain special subjects often include detailed summaries of gas chromatographic methods.[54–57]

Symposia.[58–76] A considerable and highly valuable part of gas chromatographic publications may be found among the material of various international symposia. The Gas Chromatography Discussion Group acting within the framework of the Institute of Petroleum has been arranging biannual symposia since 1956; these may be considered the most important gatherings of experts in this field.[58, 60, 63, 66, 70, 73] The Instrument Society of America has also organized biannual international symposia in the odd years between 1957 and 1963. The international symposia organized by Houston University under the title Advances in Gas Chromatography in the years 1963–1965 were on a high scientific level. The lectures delivered to the first two symposia were published in *Analytical Chemistry*, the material of the third symposium appeared in book form.[72] In the German Democratic Republic symposia on gas chromatography are organized roughly every second year since 1958,[61, 62, 65, 69, 71] in the Soviet Union also every second year since 1959.[74–76] Of course many other national or local symposia and conferences have been organized on the subject. Beside meetings devoted to gas chromatography alone, there are special gas chromatographic sections at the congresses and conferences on general analytical chemistry, instrumental analysis, separation methods, chromatography in general, on molecular structure studies and even on some special fields (e.g. testing of foodstuffs, etc.), where many important lectures on gas chromatography are delivered.

Reports and reviews. Publications on gas chromatography appear in their greatest number in the following journals: *Analytical Chemistry, Journal of Chromatography, Nature* and *Zeitschrift für analytische Chemie.* Recently new periodicals have also been published in this field, in 1963 the *Journal of Gas Chromatography*, in 1967 *Separation Science*, and in 1968 *Chromatographia* have appeared.

Of the reviews perhaps the most detailed are those published biannually since 1960 in the Fundamental Review number of *Analytical Chemistry.*[77–80] The Applied Reviews volumes published every odd year also by *Analytical Chemistry* give summaries from various fields and large numbers of references on the application of gas chromatography, mainly to the analysis of hydrocarbons, atmospheric contaminants, medical chemistry and to the fields of petroleum and related products, drugs and narcotics, solid and gaseous fuels, etc.

Reports on the symposia form a valuable part of the reviews, as they give a well arranged picture of the material, including discussions on topical problems and trends in the field.[81–84] Beside the international symposia, reports also appear on more limited, informal conferences of which those of the Gas Chromatography Discussion Group of the Institute of Petroleum are perhaps the most interesting.[85–87]

Bibliographies, abstracts of gas chromatographic data. To facilitate the survey of and selection from the great number of publications on gas chromatography, various abstracting and reference works have been published in recent years. Since 1963 each number of the *Journal of Chromatography* contains a reference section arranged according to certain main themes. The December numbers of the *Journal of Gas Chromatography*[88–91] contain a list of titles of papers on gas chromatography published in the same year, with a detailed index. The same journal publishes bibliographic compilations also on some restricted aspects of the subject, e.g. on programmed temperature gas chromatography.[92–94] *Gas Chromatography Abstracts* edited by the Gas Chromatography Discussion Group of the Institute of Petroleum and published since 1958 in a yearly volume (of four quarterly issues)[95–105] are a great help to the practising gas chromatographer. The excellent subject index from which rapid information may be gained deserves special notice. Because of lack of space we omit a detailed description of the indexing system, and will mention only the main headings: 1. general papers; 2. theory, definitions and retention data; 3. apparatus and methods; 4. carrier gas and column packing; 5. sample types; 6. applications and special separations; 7. auxiliary methods and techniques.

These main sections are subdivided into many sub-sections ensuring maximum cross-reference for a rapid assessment of the problem in hand.

The most detailed and rapid information on gas chromatographic publications can be gained from the weekly punched card information sheets of the Preston Technical Abstracts Co. (USA)[105] (909 Pitner Ave, Evanston, Ill. 60202), who worked out a rapid and efficient method for literature search.[106] The abstracts are also available in the form of microfilms.[107]

Bibliographic compilations have also been published in book form.[108–110]

Several publications contain detailed tables of gas chromatographic retention data. The *Journal of Chromatography* and the *Journal of Gas Chromatography* also publish compilations of retention data. Such compiled works were also published in book form.[111, 112]

Patents. Earlier USA gas chromatographic patents have been summarized [113, 114] and the *Journal of Gas Chromatography* regularly gives a list of new patents in this field.

Other publications on gas chromatography. The practising gas chromatographer may find some very valuable and detailed information on gas chromatographs and auxiliary accessories, as well as on applications in the regularly appearing pamphlets of the apparatus manufacturers. These publications give a number of practical examples and descriptions of new methods as well as detailed descriptions of new types of apparatus and accessories. Only some of these publications should be mentioned here; this list does not aim at completeness and neither reflects relative merits: Hewlett–Packard, F. and M. Scientific Division: *Facts and Methods*; — Applied Science Laboratories: *Gas-Chrom Newsletter*; — Perkin–Elmer: *Instrument News, GC Newsletter*; — Varian Aerograph: *Previews and Reviews, Research Notes*; — Pye Unicam: *Column*; — Carlo Erba: *Short Notes*, etc.

References

1. D. T. DAY, *Proc. Amer. Phil. Soc.*, **36**, 112 (1897)
2. M. TSWETT, *Ber. deut. botan. Ges.*, **24**, 316, 384 (1906)
3. H. G. CASSIDY, "*Fundamentals of Chromatography*", Interscience, New York (1957)
4. G. SCHAY, "*A gázkromatográfia alapjai*" (Fundamentals of gas chromatography), Akadémiai Kiadó, Budapest (1961)
5. K. PETERS and K. WEIL, *Z. angew. Chem.*, **43**, 608 (1930)
6. A. EUCKEN and H. KNICK, *Brennstoff-Chem.*, **17**, 241 (1936)
7. G. HESSE, H. EILBRECHT and F. REICHENEDER, *Annalen*, **546**, 251 (1941)
8. G. DAMKÖHLER and H. THEILE, *Angew. Chem.*, **56**, 353 (1943)
9. E. CREMER and F. PRIOR, *Z. Elektrochem.*, **55**, 66 (1951)
10. J. JANÁK, *Chem. Listy*, **47**, 464, 700, 817 (1953)
11. N. H. RAY, *J. Appl. Chem.*, **4**, 21, 82 (1953)
12. N. C. TURNER, *Petr. Ref.*, **22**, 140 (1943)
13. A. TISELIUS, *Arkiv. Kemi. Mineral. Geol.*, **16A**, No. 18 (1943)
14. S. CLAESSON, *Arkiv. Kemi. Mineral. Geol.*, **23A**, 1 (1946)
15. A. J. MARTIN and R. L. SYNGE, *Biochem. J.*, **35**, 1358 (1941)
16. A. T. JAMES and A. J. MARTIN, *Biochem. J.*, **50**, 679 (1952)
17. C. S. G. PHILLIPS, "*Gas Chromatography*", Butterworths, London (1956)
18. A. J. M. KEULEMANS, "*Gas Chromatography*", Reinhold Corp., New York (1957)
19. E. BAYER, "*Gas-Chromatographie*", Springer-Verl., Berlin—Göttingen—Heidelberg (1959). English Edition, Elsevier, New York (1961)
20. R. L. PECSOK (Ed.), "*Principles and Practice of Gas Chromatography*", J. Wiley, New York—London (1959)
21. R. KAISER, "*Gas-Chromatographie*", Akad. Verlag, Leipzig (1960)
22. R. KAISER, "*Chromatographie in der Gasphase*, I. Teil. Gas-Chromatographie", Bibliograph. Institut, Mannheim (1960)
23. D. AMBROSE and B. A. AMBROSE, "*Gas Chromatography*", G. Newnes Ltd., London (1961)
24. G. SCHAY, "*Theoretische Grundlagen der Gaschromatographie*", VEB Verlag der Wissenschaften, Berlin (1961)
25. H. A. SZYMANSKI (Ed.), "*Progress in Industrial Gas Chromatography*", Plenum Press, New York (1961)

26. A. A. ZHUHOVITSKII and N. M. TURKELTAUB. "*Gazovaya Khromatografiya*", Gostroptyehizdat, Moscow (1962)
27. E. BAYER, "*Gas-Chromatographie*", Springer Verlag, Berlin—Göttingen—Heidelberg, 2nd Ed. (1962)
28. R. KAISER, "*Chromatographie in der Gasphase*, II. Teil. Kapillar-Chromatographie", Bibliograph. Institut, Mannheim (1962)
29. R. KAISER, "*Chromatographie in der Gasphase*, III. Teil. Tabellen zur Gas-Chromatographie", Bibliograph. Institut, Mannheim (1962)
30. S. DAL NOGARE and R. S. JUVET, "*Gas-Liquid Chromatography. Theory and Practice*", Interscience, New York (1962)
31. A. B. LITTLEWOOD, "*Gas Chromatography: Principles, techniques, and applications*", Academic Press, London (1962)
32. J. H. PURNELL, "*Gas Chromatography*", J. Wiley, New York (1962)
33. J. H. KNOX, "*Gas Chromatography*", J. Wiley, New York (1962)
34. H. P. BURCHFIELD and E. E. STORRS, "*Biochemical Applications of Gas Chromatography*", Academic Press, New York (1962)
35. L. SZEPESY, "*Gázkromatográfia. Laboratóriumi és üzemi alkalmazások elvi és gyakorlati alapjai*" (Gas Chromatography. Theoretical and practical fundamentals of application in laboratory and industry), Műszaki Könyvkiadó, Budapest (1963)
36. R. KAISER, "*Gas Phase Chromatography* (English translation of Vols I–III of "Chromatographie in der Gasphase"), Butterworths, London (1963)
37. H. A. SZYMANSKI (Ed.), "*Lectures on Gas Chromatography — 1962*", Plenum Press, New York (1963)
38. P. G. JEFFERY and P. J. KIPPING, "*Gas Analysis by Gas Chromatography*", Pergamon Press, Oxford (1964)
39. H. SZYMANSKI (Ed.), "*Biomedical Applications of Gas Chromatography*", Plenum Press, New York (1964)
40. J. TRANCHANT (Ed.), "*Manual Pratique de Chromatographie en Phase Gaseuse*", Masson et Cie, Paris (1964)
41. L. R. MATTICK and H. A. SZYMANSKI (Eds.), "*Lectures on Gas Chromatography* — 1964", Plenum Press, New York (1965)
42. R. KAISER, "*Chromatographie in der Gasphase*", IV. Teil. Quantitative Auswertung, Bibliograph. Institut, Mannheim (1965)
43. L. S. ETTRE, "*Open Tubular Columns in Gas Chromatography*", Plenum Press, New York (1965)
44. R. W. MOSHIER and R. E. SIEVERS, "*Gas Chromatography of Metal Chelates*", Pergamon Press, New York (1965)
45. J. C. GIDDINGS, "*Dynamics of Chromatography*", Part I. Principles and Theory, Marcel Dekker, New York (1965)
46. M. B. LIPSETT, "*Gas Chromatography of Steroids in Biological Fluids*", Plenum Press, New York (1965)
47. W. E. HARRIS, "*Programmed Temperature Gas Chromatography*", J. Wiley, New York (1966)
48. E. LEIBNITZ and H. G. STRUPPE (Eds.), "*Handbuch der Gas Chromatographie*", Akademische Verlagsgesellschaft, Leipzig (1966)
49. M. TARAMASSO, "*Gas Chromatografia*", Franco Angelli, Milano (1966)

49a. L. S. ETTRE and A. ZLATKIS (Eds.), *The Practice of Gas Chromatography*, Interscience, New York (1967)
49b. J. KRUGERS (Ed.), "*Instrumentation in Gas Chromatography*", Centrex Publ. Co., Eindhoven (1968)
50. E. LEDERER and M. LEDERER (Eds.), "*Chromatography*", Elsevier, New York, 2nd Ed. (1957)
51. E. HEFTMANN (Ed.), "*Chromatography*", Reinhold Corp., New York (1961)
52. R. STOCK and C. B. RICE, "*Chromatographic Methods*", Reinhold Corp., New York (1963)
53. J. M. KOLTHOFF and P. J. ELVING (Eds.), "*Treatise on Analytical Chemistry*", Part I. Vol. 3. Interscience, New York (1961)
54. R. NEHER, "*Steroid Chromatography*", Elsevier, New York (1964)
55. C. T. BISHOP, in "*Advances in Carbohydrate Chemistry*", Vol. 19 (Ed. M. L. Wolfram), Academic Press, New York (1964)
56. E. C. HORNING, A. KARMEN and C. C. SWEELEY in "*Progress in the Chemistry of Fats and Other Lipids*" (Ed. L. T. Holman) Pergamon Press, New York (1964)
57. F. P. WOODFORD, in "*Fatty Acids: Their Chemistry, Properties, Production and Uses*", Part 3 (Ed. K. S. Markley) Interscience, New York (1964)
58. D. H. DESTY (Ed.), "*Vapour Phase Chromatography*", Proceedings of the Symposium held in London, May, 1956, Academic Press, New York, Butterworths Sci. Publ., London (1957)
59. V. J. COATES, H. J. NOEBELS and J. S. FAGERSON (Eds.), "*Gas Chromatography*". I. Symposium, East Lansing, Michigan, August, 1957. Academic Press, New York (1958)
60. D. H. DESTY (Ed.), "*Gas Chromatography*, 1958", Second Symposium, Amsterdam 1958. Academic Press, New York, Butterworths Sci. Publ. London (1958)
61. H. P. ANGELE (Ed.), "*Gas Chromatographie*, 1958", 1. Symposium in der DDR, Oktober 1958, Leipzig. Akademie-Verlag, Berlin (1959)
62. R. E. KAISER and H. G. STRUPPE (Eds.), "*Gas Chromatographie*, 1959". 2. Symposium in der DDR, Böhlen, Oktober 1959. Akademie-Verlag, Berlin (1960)
63. R. P. W. SCOTT (Ed.), "*Gas Chromatography*, 1960", Third Symposium, Edinburgh, June, 1960. Academic Press, New York, Butterworths, London (1960)
64. H. J. NOEBELS, R. F. WALL and N. BRENNER (Eds.), "*Gas Chromatography*", II. International Symposium, ISA, June, 1959. Academic Press, New York and London (1961)
65. H. P. ANGELE and H. G. STRUPPE (Eds.), "*Gas-Chromatographie*, 1961", 3. Symposium in der DDR, Mai 1961, Schkopau. Akademie-Verlag, Berlin (1961)
66. M. VON SWAAY (Ed.), "*Gas Chromatography*, 1962", Fourth Symposium, Hamburg, Butterworths, London (1962)
67. N. BRENNER, J. E. CALLEN and M. D. WEISS (Eds.), "*Gas Chromatography*", III. International Symposium (ISA) June, 1961. Academic Press, New York and London (1962)

68. L. FOWLER (Ed.), "*Gas Chromatography*", IV. International Symposium (ISA), June 1963. Academic Press, New York and London (1963)
69. H. P. ANGELE and H. G. STRUPPE (Eds.), "*Gas-Chromatographie*, 1963". 4. Symposium in der DDR, Mai, 1963, Leuna. Akademie-Verlag, Berlin (1964)
70. A. GOLDUP (Ed.), "*Gas Chromatography*, 1964". Fifth Symposium, Brighton, September, 1964. Butterworths, Institute of Petroleum, London (1965)
71. H. G. STRUPPE (Ed.), "*Gas-Chromatographie*, 1965". 5. Symposium in der DDR, Mai, 1965, Berlin. Akademie-Verlag, Berlin (1965)
71a. H. G. STRUPPE (Ed.), "*Gas-Chromatographie* 1968". 6. Symposium in der DDR, Mai, 1968, Berlin, Akademie-Verlag, Berlin (1968)
72. A. ZLATKIS and L. S. ETTRE (Eds.), "*Advances in Gas Chromatography*, 1965". Preston Techn. Abstr. Co., Evanston. Ill. (1966)
73. A. B. LITTLEWOOD (Ed.), "*Gas Chromatography*, 1966". Sixth Symposium, Rome, 1966. Institute of Petroleum, London (1967)
73a. C. L. A. HARBOURN and R. STOCK (Eds.), *7th International Symposium on Gas Chromatography and Its Exploitation*, Copenhagen, 1968. Preprints, Institute of Petroleum, London (1968)
74. A. A. ZHUKHOVITSKII *et al.* (Eds.), *Gazovaya Khromatografiya I.* (Trudi Pervoi Vsesoyuznoi Konferentsii) Akademia Nauk, Moscow (1960)
75. A. A. ZHUKHOVITSKII *et al.* (Eds.), *Gazovaya Khromatografiya II.* (Trudi Vtoroi Vsesoyuznoi Konferentsii) Nauka, Moscow (1964)
76. A. A. ZHUKHOVITSKII *et al.* (Eds.), *Gazovaya Khromatografiya III.* (Trudi Tretei Vsesoyuznoi Konferentsii) OKBA, Dzerzhinsk (1966)
77. S. DAL NOGARE, *Anal. Chem.*, **32,** (5) 19 R (1960)
78. S. DAL NOGARE and R. S. Juvet, *Anal. Chem.*, **34,** (5) 35 R (1962)
79. R. S. Juvet and S. DAL NOGARE, *Anal. Chem.*, **36,** (5) 36 R (1964)
80. S. DAL NOGARE and R. S. JUVET, *Anal Chem.*, **38,** (5) 61 R (1966)
81. S. DAL NOGARE, *J. Gas Chromatog.*, **2,** 189 (1964)
82. A. B. LITTLEWOOD, *J. Gas Chromatog.*, **2,** 349 (1964)
83. ANON, *Angew. Chem.*, **77,** 104 (1965)
84. L. S. ETTRE, *J. Gas Chromatog.*, **4,** 16 (1966)
85. A. B. LITTLEWOOD, *J. Gas Chromatog.*, **2,** 186 (1964)
86. A. B. LITTLEWOOD, *J. Gas Chromatog.*, **3,** 111 (1965)
87. A. B. LITTLEWOOD, *J. Gas Chromatog.*, **4,** 32 (1966)
88. S. T. PRESTON, G. HYDER and M. GILL, *J. Gas Chromatog.*, **2,** 391 (1964)
89. M. GILL and S. T. PRESTON, *J. Gas Chromatog.*, **2,** 391 (1964)
90. S. T. PRESTON and M. GILL, *J. Gas Chromatog.*, **3,** 399 (1965)
91. S. T. PRESTON and M. GILL, *J. Gas Chromatog.*, **4,** 435 (1966)
92. W. E. HARRIS and H. W. HABGOOD, *J. Gas Chromatog.*, **4,** 144 (1966)
93. W. E. HARRIS and H. W. HABGOOD, *J. Gas Chromatog.*, **4,** 168 (1966)
94. W. E. HARRIS and H. W. HABGOOD, *J. Gas Chromatog.*, **4,** 217 (1966)
95. C. E. H. KNAPMAN (Ed.), "*Gas Chromatography Abstracts*, 1958", Butterworths, London (1960)
96. C. E. H. KNAPMAN (Ed.), "*Gas Chromatography Abstracts*, 1959", Butterworths, London (1960)
97. C. E. H. KNAPMAN (Ed.), "*Gas Chromatography Abstracts*, 1960", Butterworths, London (1961)

98. C. E. H. Knapman (Ed.), "*Gas Chromatography Abstracts*, 1961", Butterworths, London (1962)
99. C. E. H. Knapman (Ed.), "*Gas Chromatography Abstracts*, 1962", Butterworths, London (1963)
100. C. E. H. Knapman (Ed.), "*Gas Chromatography Abstracts*, 1963", Butterworths, London (1964)
101. C. E. H. Knapman (Ed.), "*Gas Chromatography Abstracts*, 1964", Butterworths, London (1965)
102. C. E. H. Knapman (Ed.), "*Gas Chromatography Abstracts*, 1965", Butterworths, London (1966)
103. C. E. H. Knapman (Ed.), "*Gas Chromatography Abstracts*, 1966", Institute of Petroleum, London (1967)
104. C. E. H. Knapman (Ed.), "*Gas Chromatography Abstracts*, 1967", Institute of Petroleum, London (1968)
105. Card Abstracts. Abstracting Service, Preston Technical Abstracts Co., Evanston, Ill., USA
106. Termatrex Index — Gas Chromatography Literature, Preston Technical Abstracts Co., Evanston, Ill., USA
107. Microfilmed Gas Chromatography Abstracts, Preston Technical Abstracts Co., Evanston, Ill., USA
108. S. T. Preston and G. Hyder, "*A Comprehensive Bibliography and Index to the Literature on Gas Chromatography*", Preston Technical Abstracts Co., Evanston, Ill., USA (1964)
109. A. V. Signeur, "*Guide to Gas Chromatography Literature*", Plenum Press, New York (1964)
110. A. V. Signeur, "*Guide to Gas Chromatography Literature*", Supplement No. 1, Plenum Press, New York (1966)
110a. A. V. Signeur, "*Guide to Gas Chromatography Literature*", Vol. 2, Plenum Press, New York (1967)
111. J. S. Lewis, "*Compilation of Gas Chromatographic Data*", ASTM Publ. No. 343, Philadelphia (1963)
111a. O. E. Schupp and J. S. Lewis, "*Compilation of Gas Chromatographic Data DS* 25 *A*", ASTM Publ., Philadelphia (1968)
112. W. O. McReynolds, "*Gas Chromatographic Retention Data*", Preston Technical Abstracts Co., Evanston, Ill., USA (1966)
113. S. T. Preston and G. Hyder, *J. Gas Chromatog.*, **1**, 22 (1963)
114. S. T. Preston and G. Hyder, *J. Gas Chromatog.*, **1**, 24 (1963)

2. Fundamental Theory

It has already been pointed out in the general discussion of gas chromatography that the rate at which the component zones travel in the column depends on the equilibrium conditions and the flow rate of the carrier gas, while the broadening of the zones is a function of the flow conditions in the column, of longitudinal diffusion and of mass transfer rate. Before going into the details of gas chromatographic theories it is helpful to survey briefly these elementary processes in their relation to the operating conditions of the gas chromatographic column.

2.1 Hydrodynamics of the column

Let us first consider the flow in the column. Investigation of flow conditions dominates both the theoretical and practical study of column operation at the moment. It should be noted here that, but for a few exceptions, earlier work more or less omitted the study of the effect of flow conditions. There is a considerable difference between flow conditions in the two basic types of gas chromatographic columns, namely in open tubular or capillary, and packed columns which necessitates their separate treatment.

Open tubular columns

Flow of the fluid in the open tube is characterized by the dimensionless Reynolds number, *Re*:

$$Re = \frac{du\rho}{\mu} = \frac{dG}{\mu} \qquad 2.1$$

where d is the diameter of the tube (cm), u the linear velocity of the gas (cm/sec), ρ the density of the gas (g/cm^3), G the mass velocity (g/cm^2, sec) and μ the absolute viscosity (g/cm, sec). Depending on the Reynolds number Re two types of flow are distinguished, namely laminar, or viscous, and turbulent flow. When $Re < 2000$, flow is laminar, when $Re > 4000$, flow is turbulent. Between the two regions

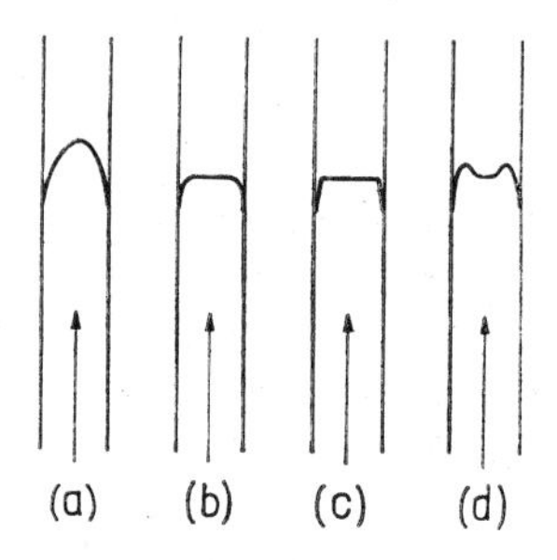

Fig. 2.1. The different flow profiles. (a) Laminar flow; (b) turbulent flow; (c) plug flow; (d) flow in a packed tube

there is an intermediate region. Laminar flow is characterized by the formation of layers with different velocities after a certain length (about 50 d). Velocity is lowest near the wall (on the wall itself it is, of course, zero) and highest in the centre of the tube. Layers with different velocities do not mix, but travel parallel to each other in the direction of flow. The characteristic parabolic velocity distribution of laminar flow is shown in Fig. 2.1(a). The maximum velocity in the middle of the tube is twice the average velocity.

When the flow rate is increased eddies cause the internal currents to mix leading to a certain equalization of flow rates. The characteristic rates of turbulent flow are shown in Fig. 2.1(b).

In open tubular gas chromatographic columns, flow rates are relatively very low and the columns operate always under laminar conditions. This is illustrated by the following few examples corresponding to the most frequently encountered flow rates:

$$d = 0.03 \text{ cm}, \quad T = 23\,°\text{C}, \quad \mu_{H_2} = 88.2\ \mu\text{P}, \quad \mu_{N_2} = 176.5\ \mu\text{P}$$

	$V = 2$ cm^3/min	$V = 10$ cm^3/min
H_2 carrier gas	$Re = 1.3$	$Re = 6.5$
N_2 carrier gas	$Re = 9.3$	$Re = 46.7$

Packed columns

There is a substantial difference between the flow conditions in packed and in open tubular columns. The irregularly shaped multidirectional

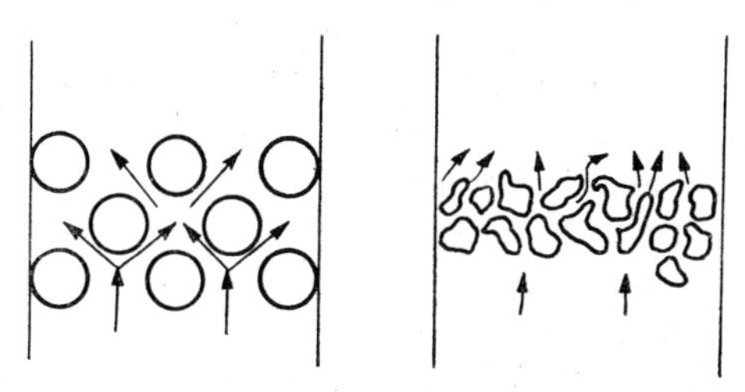

Fig. 2.2. Fluid flow changes direction as a result of packing

channels between the particles of the packing cause a lateral deflection and mixing of the flowing layers which is essentially a random process. This is demonstrated in Fig. 2.2 for bidimensional beds of uniform spheres and irregularly shaped particles. In the case of irregular hollow particles this effect is, of course, far more complex than for regular spheres. Flow in a packed bed may be described by the modified Reynolds number which contains as a characteristic geometric dimension the diameter, d_p (cm), of the particles:

$$Re_m = \frac{d_p G}{\mu} \tag{2.2}$$

In the region of laminar flow $Re_m < 10$, in the intermediate region $10 < Re_m < 200$ and in the region of turbulent flow Re_m is over 200.

In the practical application of packed gas chromatographic columns $Re_m < 10$, so that the column operates always in the laminar region. As an example the calculated values for the most frequently occurring conditions can be given:

$$d = 0.3\ \text{cm}, \quad d_p = 0.02\ \text{cm}, \quad T = 23\ °\text{C}$$

	$V = 25\ \text{cm}^3/\text{min}$	$V = 50\ \text{cm}^3/\text{min}$
H_2 carrier gas	$Re_m = 0.11$	$Re_m = 0.22$
N_2 carrier gas	$Re_m = 0.78$	$Re_m = 1.56$

Examination of flow conditions in the packed bed shows a clear, distinct velocity distribution in every channel, while the velocity at the surface of the particles is zero. As a result of this retarding effect in a given cross-section of the packed column, the rate differences are balanced and consequently rate distribution will not be parabolic, even in the case of laminar flow, but will be similar to the ideal plug flow, as illustrated in Fig. 2.1(c). In their investigation of packed reactors Schwartz and Smith[1] have shown that a rate distribution like that in Fig. 2.1(d) will develop in the packed bed because of the lower density of the bed in the vicinity of the wall compared to that in the centre of the tube. Maximum velocity develops at a distance of 1–2 particle diameters from the wall and in the case of an appropriate packing might be 100% higher than the velocity in the middle of the tube. This effect is the more significant the smaller the tube-diameter : particle-size ratio. With increasing tube diameter obviously other effects will also appear because of radial particle size distribution due to inadequate packing and consequent fluctuations in bed density.

The following parameters may serve for the description of the packed bed: a) particle diameter of the packing; b) shape factor of the particles; c) dimension and distribution of pores; d) surface roughness; e) porosity of the bed or void volume. The dimensions of the particles are characterized by the average particle diameter which in the case of a non-spherical packing may involve a rather considerable error. As a matter of fact, different mesh fractions are used corresponding to a certain particle size distribution. Deviation from the spherical shape is accounted for by the shape factor which cannot be calculated directly and is usually determined from the measurement of the pressure drop. Pore size and pore distribution can be determined with the porosimeter or calculated from the measurement of physical adsorption. In the case of laminar flow surface roughness has no effect on flow conditions, but influences the density of the packed bed: from smooth surfaced particles, beds of higher density can be prepared.

Porosity or void volume is one of the most important parameters of packed beds. The void volume, ε, is the ratio of the empty spaces between the particles to the volume of the packing (cm^3/cm^3). Void volume depends on the size, shape and surface roughness of the particles and on the method of packing. For the densest packing of spheres with identical diameters $\varepsilon = 0.26$, for the random distribution in a packed bed ε is generally 0.3–0.45; in gas chromatography an ε value of 0.38 is considered a good average. True flow rate in the packed bed can be calculated when the void volume is known, as flow takes place in the void between the particles. The true linear velocity of the gas is:

$$u_t = \frac{u}{\varepsilon}$$

where u_t is the true linear velocity in a packed bed (cm/sec); u is linear velocity calculated for the empty tube cross-section (cm/sec); and ε is void volume of the packing (cm^3/cm^3).

Irregularities in the packing cause changes in radial and axial directions resulting in fluctuations of the flow rate and a corresponding distortion of the velocity profile. The effect of the latter will be investigated in connection with the rate theory.

Pressure drop in the column

Flow in both the open tubular and packed column is the result of a pressure difference between the inlet and outlet ends. When the outlet pressure is fixed, the necessary inlet pressure will depend on the resistance of the column, that is, on the pressure drop.

In the case of laminar flow in a tube the pressure drop can be calculated from the Hagen–Poiseuille equation:

$$\Delta p = 32 \frac{L u \mu}{g d} \qquad 2.3$$

where g is the gravitation constant (cm/sec^2); and L is the length of the tube (cm).

A number of formulae are given in the literature for the calculation of the pressure drop in a packed bed. For the examination of the pressure drop on adsorbers Ergun's equation[2] was found to be applicable in practice, as it is valid for every flow range:

$$\Delta p = \frac{L}{g} \left[150 \frac{(1-\varepsilon)^2 u \mu}{\varepsilon^3 d_p^2} - 1.75 \frac{1-\varepsilon}{\varepsilon^3} \frac{G u}{d_p} \right] \qquad 2.4$$

For laminar flows and low gas rates, such as occur in gas chromatography, the second term is very small and negligible compared with the first term.

In the study of gas chromatographic flow conditions another important factor, namely the compressibility of the gas, also has to

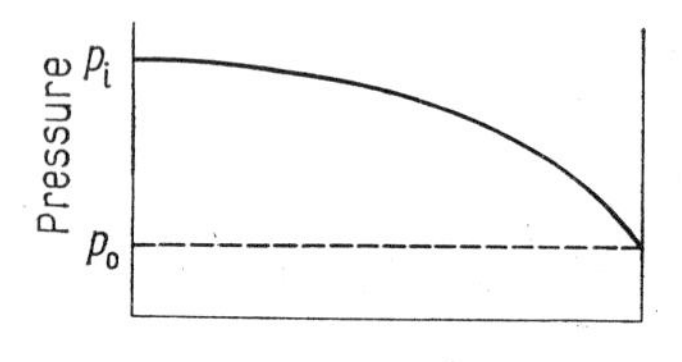

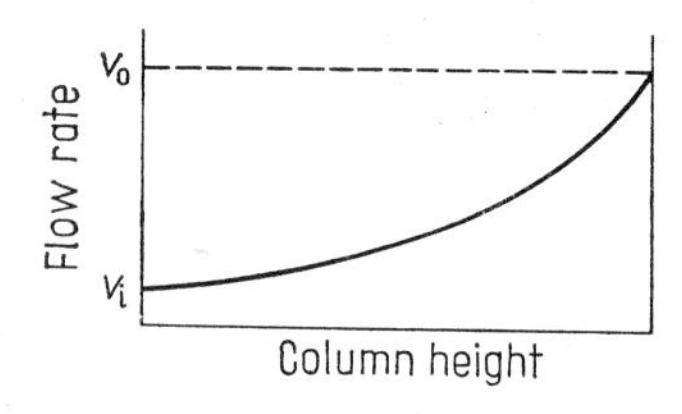

Fig. 2.3. Pressure and flow rate changes in a column

be accounted for. As a result of the pressure drop in the column, gas volume, and consequently also flow rate, undergo significant changes. The changes of pressure and flow rate in the column are illustrated in Fig. 2.3. The simplest way to account for the compressibility of the gas is to calculate the average pressure from the inlet and outlet pressures and to assume that the entire column operates at this average pressure. This, of course, is only an approximation, though quite satisfactory for practical purposes.

The average mean pressure $\bar{p}$ (integral mean value) is according to James and Martin:[3]

$$\bar{p} = \frac{2}{3}\left[\frac{p_i^3 - p_o^3}{p_i^2 - p_o^2}\right] \qquad 2.5$$

where p_i and p_o stand for the inlet and outlet pressures, respectively.

Rearrangement of the equation leads to:

$$\frac{\bar{p}}{p_o} = \frac{2}{3}\left[\frac{(p_i/p_o)^3 - 1}{(p_i/p_o)^2 - 1}\right] \qquad 2.6$$

The flow rate of the gas is usually measured after the gas has emerged from the column. As the pressure is inversely proportional to the volume:

$$\frac{\bar{p}}{p_0} = \frac{v_0}{\bar{v}} \quad \text{or} \quad \bar{v} = v_0 \frac{p_0}{\bar{p}}$$

where v_0 and $\bar{v}$ are the outlet and average volume rates in the column, respectively.

The pressure correction factor introduced by James and Martin is:

$$\frac{p_0}{\bar{p}} = j = \frac{3}{2}\left[\frac{(p_i/p_0)^2 - 1}{(p_i/p_0)^3 - 1}\right] \qquad 2.7$$

Thus the average volume rate and the average linear rate in the column are:

$$\bar{v} = jv_0 \quad \text{and} \quad \bar{u} = ju_0 \qquad 2.8$$

In the packed column of average size most frequently used in gas chromatography the usual pressure drop is 0.5–2 atm, in capillary columns it varies between 2 and 6 atm. Pressure correction (Eq. 2.7) can be satisfactorily used up to about 10 atm pressure drop and is thus adequate for the solution of practical problems.

The permeability of the column, or more accurately its specific permeability coefficient K, is the most important hydrodynamic parameter of the column which gives an indication of the ease of flow through the column. Its dimension is cm^2. K should not be confused with the porosity of the packing which represents the free space. To the calculation of pressure drop in the packed bed in cases of laminar flow Darcy's law

$$\Delta p = \frac{\mu \bar{u} L}{gK} \qquad 2.9$$

may be applied. Usually there are no permeability data available for gas chromatographic packings, and therefore the above correlation is not often used. As it appears from the equation, the permeability of the packing can be calculated from the measurement of the pressure drop. Combining Eq. 2.9 with Ergun's equation Reisch *et al.*[4] arrived at the following equation for the permeability:

$$K = \frac{d_p^2}{150} \frac{\varepsilon^3}{(1 - \varepsilon)^2} \qquad 2.10a$$

It appears from Eq. 2.10 that permeability is independent of the operation conditions (T, p, u) and of the dimensions (length, diameter) of the column, but is proportional to the square of the particle diameter and the porosity of the packing. In the case of open tubular columns permeability is a simple function of the diameter of the tube:

$$K = \frac{r^2}{8} \qquad 2.10b$$

where r is the radius of the tube.

Permeability plays an important role in the study of column operation and in the comparison of columns of different types.

2.2 Fundamental laws of diffusion

Diffusion is a process occurring in every system with concentration differences and aims at the elimination of these differences. The fundamental definition of diffusion is given by Fick's first law on the correlation between mass flow and concentration gradient:

$$\frac{\partial N}{\partial t} = -D\frac{\partial n}{\partial l} \qquad 2.11$$

where N is the number of molecules passing through the unit surface, t is the time, D the diffusion coefficient, n the concentration, l the length. The negative sign indicates that diffusion takes place in the direction of lower concentrations.

According to Eq. 2.11 the quantity of material diffusing through a unit surface in unit time is proportional to the concentration gradient in the direction of diffusion.

Fick's second law reflects the principle of the conservation of mass for diffusion processes. As in the gas chromatographic column only longitudinal diffusion is subjected to examination, the one-dimensional form of Fick's second law will be used for the description of the diffusion process:

$$\frac{\partial n}{\partial t} = D\frac{\partial^2 n}{\partial l^2} \qquad 2.12$$

A particular solution of Eq. 2.12 is:

$$n = \frac{A}{\sqrt{t}} e^{-l^2/4Dt} \qquad 2.13$$

where A is an arbitrary constant.

When in a system the total quantity of diffusing material, m, is given, we obtain for the constant A:

$$A = \frac{m}{2\sqrt{\pi D}} \qquad 2.14$$

and the solution of the differential equation will be:

$$n = \frac{m}{2\sqrt{\pi D_t}} e^{-l^2/4Dt} \qquad 2.15$$

which is the equation of a Gaussian curve. The Gaussian curve, also called the bell curve or error function, was first applied in probability calculation, but is generally characteristic of random processes occurring in nature.

The characteristics of the Gaussian curve are the maximum and the width of the curve. The width of the curve may be given by the distance between the inflection points which, in our case, would be $2\sqrt{2Dt}$, thus the width of the curve depends on the product of D and t. Figure 2.4 shows several Gaussian curves. The similarity between the broadening and shape of the diffusion zone and the gas chromatographic elution peak are of considerable help in the study of the broadening process of the gas chromatographic zone. In gas chromatography, zone broadening is always of diffusional nature (though, of course, not entirely the result of diffusion effects) and may be described by a single effective diffusion coefficient. From the analogy with probability calculation the width of the Gaussian curve is usually called the standard deviation, σ, that is to say:

$$\sigma = \sqrt{2Dt} \qquad 2.16$$

The square of standard deviation is called variancy which will appear in the theoretical studies of gas chromatography

$$\sigma^2 = 2Dt \qquad 2.17$$

This correlation is also called Einstein's equation, as Einstein was the first to point out the random nature of diffusion processes.

As already apparent from the correlation quoted above, the width of the diffusion curve is determined not only by time, but also by the

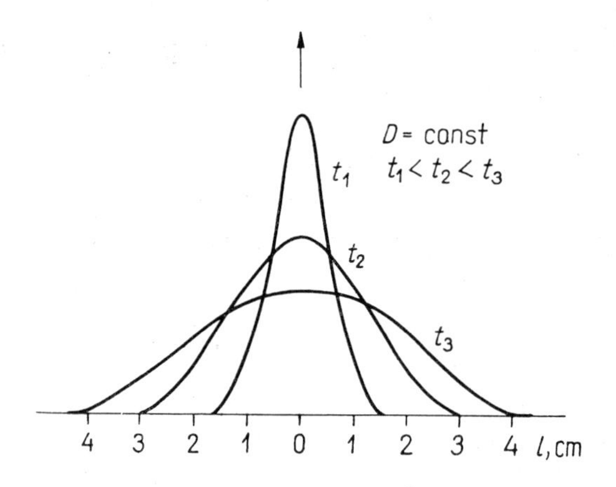

Fig. 2.4. Gaussian curves

value of the diffusion coefficient, D. The laws and findings discussed so far are of general validity and are applicable to both gases and liquids, but because of the essential difference between the diffusion of gases and liquids it is better to discuss the two separately.

Diffusion in gases

For ideal gases the diffusion coefficient can be derived from the kinetic gas theory:

$$D = \frac{1}{3} \Lambda \bar{u} \qquad 2.18$$

where Λ is the mean free path, and $\bar{u}$ the mean velocity of the molecule, thus the dimension of D is length^2/time. The diffusion coefficient defined by Eq. 2.18 refers to the diffusion of a single gas. The equation for the counter-diffusion of a binary mixture of molecules A and B of different sizes is:

$$D_{AB} = \frac{1}{3} \frac{n_A \Lambda_A \bar{u}_A + n_B \Lambda_B \bar{u}_B}{n_A + n_B} \quad 2.19$$

where n_A and n_B are the numbers of moles of A and B in the unit volume, Λ_A and Λ_B are the corresponding mean free paths and $\bar{u}_A$ and $\bar{u}_B$ the mean velocities. In the case of counter-diffusion the diffusion of both components can be expressed with the same diffusion coefficient D, or:

$$D_{AB} = D_{BA}$$

In the following the indices referring to the components will be omitted and the symbol D_g will be used for gas diffusion.

There are several empirical equations in the literature for the diffusion coefficient, of which the theoretically corroborated correlation of Gilliland[5] is the most generally employed:

$$D_g = 0.0043 \frac{T^{3/2}}{P(V_A^{1/3} + V_B^{1/3})^2} \sqrt{\frac{1}{M_A} + \frac{1}{M_B}} \quad 2.20$$

where T is the absolute temperature, °K, P is the total pressure in atm, M_A and M_B are the molecular weights and V_A and V_B the molecular volumes of the gases at normal boiling points in cm³/g-mole.

The error in the calculation of D_g is between 10 and 40%, which is due mainly to the deviation of real molecules from the rigid, spherical shapes assumed in kinetic gas theory.

The value of D_g is practically independent of the concentration and inversely proportional to the pressure. It increases with temperature according to the 1.5th (or, in the opinion of some authors, to the 2nd) power of T and is also a function of the molecular weights and molecular volumes of the carrier gas and of the diffusing substance. The value of D_g varies in general between 10^{-1} and 1 cm²/sec. The diffusion coefficient D_g, is obviously characteristic of any given binary gas mixture, thus there may be considerable fluctuations in its value for different carrier gases. From the measurements of Clarke and Ubbelohde[6] the diffusion coefficients of, for example, n-heptane in different carrier gases at 30 °C and 1 atm pressure are:

	H_2	N_2	Ar
D_g cm²/sec	0.283	0.074	0.0658

In the free or molecular diffusion discussed so far diffusion rate is determined by collision of the molecules. If, however, diffusion takes place in a medium in which the voids are smaller than the mean free path, diffusion will occur according to some other mechanism and

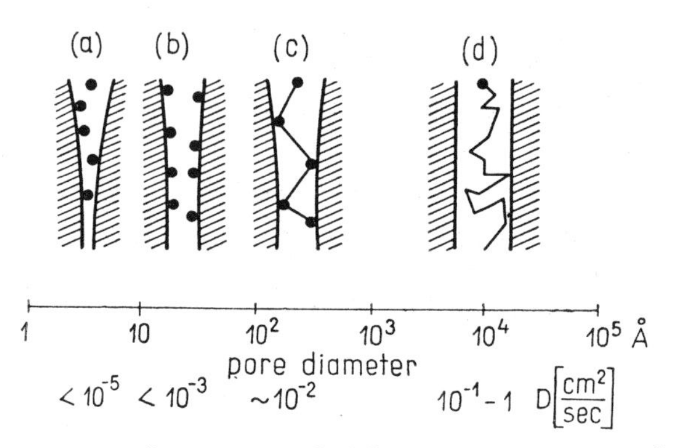

Fig. 2.5. Different types of diffusion on porous solids.[7] *(a) Solid diffusion; (b) surface or Volmer diffusion; (c) Knudsen diffusion; (d) free or molecular diffusion*

Eq. 2.20 is inadequate[7] for the calculation of D_g. In a bed packed with a porous solid the diffusion coefficient D_g will follow other laws. The diagram of the four possible types of diffusion is shown in Fig. 2.5. From Fig. 2.5 the diffusion processes in the packed bed can be summed up as follows: the diffusion process is molecular in the space between particles and in pores with diameters greater than 0.1 μm; the diffusion constant here has the highest value. From the point of view of longitudinal diffusion in the gas chromatographic column the D_g value for molecular diffusion will be the decisive factor. In pores with diameters below 0.1 μm collisions with the wall are more frequent than intermolecular collisions; this type of diffusion is called Knudsen or capillary diffusion. While, according to Volmer in smaller pores the adsorbed molecules diffuse on the pore walls towards less densely coated areas, in pores with diameters of the order of 0.001 μm (10 Å), where the adsorption forces of the neighbouring walls overlap, a so-called solid diffusion takes place. The average D_g values characteristic of the different types of diffusion are shown in Fig. 2.5. In gas chromatography the three latter types of diffusion occur only in gas–solid chromatography on microporous adsorbents (active charcoal, molecular sieves, etc.), though Knudsen diffusion may also take place on supports commonly used in gas–liquid chromatography. As the D_g values in these diffusion processes are several powers of ten lower than in molecular diffusion, they do not interfere with the longitudinal diffusion process, but may greatly affect the mass transfer rate which is again an important factor in the broadening of the zone.

Diffusion in liquids

The theory of diffusion in liquids has not been elucidated to the same degree as that in gases. The accuracy of the empirical equation for the calculation of the diffusion coefficient of liquids, D_l, is considerably lower than that for gas diffusion and may involve errors of several hundred per cent. There are few reliable values for liquid diffusion in the literature and none for the special liquids used in gas–liquid chromatography.

In general the value of D_l is between 10^{-4} and 10^{-5} cm^2/sec, thus four to five orders of magnitude lower than the values of D_g. Accordingly, the longitudinal diffusion in the liquid may be neglected from the point of view of zone broadening. The diffusion rate in liquids and the value of D_l are important in considering mass transfer in gas–liquid chromatography, as under certain conditions the diffusion of the liquid will determine the rate of mass transfer (see Chapter 3).

2.3 Phase equilibria

In gas chromatography separation is the result of partition between the moving gas phase and the stationary solid or liquid phase. For the study of the separation process the physical chemistry of phase equilibria provides the necessary starting point.

Partition of a component between two phases is characterized by the partition coefficient, k, which is the ratio of the concentrations in the two phases:

$$k = \frac{C_1}{C_2} \qquad 2.21$$

When both phases are ideal mixtures, k will be independent of the concentration. An ideal mixture may be defined by Nernst's law according to which the chemical potential of every component can be written as:

$$\mu = \mu^0 + RT \ln C \qquad 2.22$$

where μ^0 is independent of the concentration and is a function of the temperature only.

In equilibrium the chemical potentials of the same component are the same in both phases, $\mu_1 = \mu_2$, thus we may write:

$$\mu_1^0 + RT \ln C_1 = \mu_2^0 + RT \ln C_2 \qquad 2.23$$

It follows from Eqs 2.21 and 2.23 that

$$\frac{C_1}{C_2} = \exp[(\mu_1^0 - \mu_2^0)RT] = k \qquad 2.24$$

The right hand side of Eq. 2.24 is independent of the concentration, so the partition coefficient, k, is really independent of concentration.

From Eq. 2.22 other correlations may also be deduced for ideal solutions, such as e.g. Raoult's law. According to Raoult's law partial pressure in the gas phase, p_i, of any component in the ideal solution is proportional to its mole fraction in the liquid phase, x_i. The proportionality factor, p_i^0, is the vapour pressure of the component at the given temperature:

$$p_i = p_i^0 x_i^0 \qquad 2.25$$

In practice the systems under investigation cannot be considered ideal. The equilibrium conditions of real systems can be approached by two methods:

1. By the partition function in its concentration dependence. This method usually involves the plotting of sorption isotherms and will be discussed in detail in connection with adsorption equilibria.

2. The second method is based on the concept of activity. Deviation from ideal behaviour is accounted for by the activity coefficient. For a real mixture Eq. 2.22 is modified as follows:

$$\mu = \mu^0 + RT \ln \gamma C \qquad 2.26$$

where γ is the activity coefficient, and γC the activity. The activity coefficient may be either greater (positive deviation) or smaller (negative deviation) than 1 and is a function of the concentration.

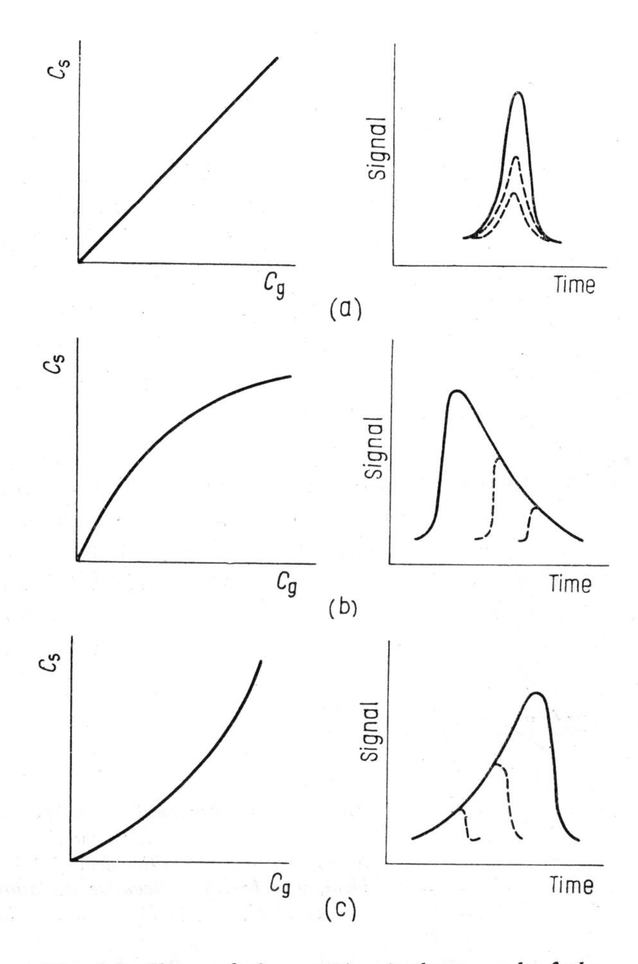

Fig. 2.6. Shape of the partition isotherms and of the corresponding elution peaks. (a) Linear isotherm — symmetric elution peak (position of peak maximum constant); (b) concave isotherm — distorted peak (tailing) (position of peak maximum depends on the quantity of the sample); (c) convex isotherm — distorted peak (leading) (position of peak maximum depends on the quantity of the sample)

The partition coefficient expressed in terms of activities will be:

$$k_{act} = \frac{\gamma_1 C_1}{\gamma_2 C_2} \qquad 2.27$$

and is independent of the concentration.

In gas chromatographic processes depending on the state of the stationary phase either adsorption (gas–solid chromatography, GSC) or solution (gas–liquid chromatography, GLC) takes place.

To describe the concentration dependence of the partition equilibrium the partition isotherm is generally used which indicates the concentration changes in the two phases at a constant temperature. At the low concentrations occurring in gas chromatography the three types of partition isotherms shown in Fig. 2.6 are found.

2.3.1 Solution equilibrium

In ideal solutions the partition of the solute between the two phases is, according to Eq. 2.24, independent of the concentration, or in other words, a linear partition isotherm is obtained. In non-ideal solutions the partition isotherm deviates from linearity. In Fig. 2.7 curve I

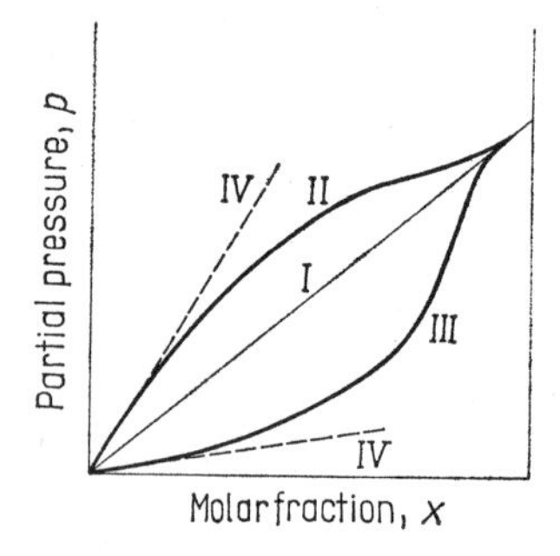

Fig. 2.7. Diagrammatic representation of solution equilibria: I — Ideal solution, Raoult's law, $\gamma = 1$; II — Positive deviation from the ideal, $\gamma > 1$; III — Negative deviation from the ideal, $\gamma < 1$; IV — Henry's law

illustrates the linear correlation according to Raoult's law for ideal solutions. Curve II is characteristic of a positive, curve III of a negative deviation from the ideal. At the low concentrations encountered in gas chromatography the constancy of the partition coefficient may often be valid also for non-ideal systems. To this case Henry's law may be applied according to which the partial pressure of a component (p_i) over its dilute solution is proportional to its mole fraction

$$p_i = H_i x_i \qquad 2.28$$

Equation 2.28 differs from Eq. 2.25 (Raoult's law) in that the proportionality factor is not the vapour pressure of the pure component, but the empirical Henry constant H. The dashed line in Fig. 2.7 illustrates Henry's law.

2.3.2 Adsorption equilibrium

Adsorption is essentially the condensation and binding of gases and vapours on the surface of solids in contact with the former. This binding is the result of the net fields of the elements constituting the surface of the solid and of their interaction with the particles reaching the surface. For any given pair of adsorbent–adsorbate the quantity bound by adsorption is a function of the equilibrium gas pressure and of temperature:

$$a = f(p, t) \qquad 2.29$$

where a is the adsorbed quantity in ml/g or g/g, p the equilibrium pressure and t the temperature in °C.

From this function with three variables a more convenient relationship is derived when one of the variables is kept constant. In this way we arrive at the adsorption isotherm:

$$a = f(p) \qquad t = \text{const} \qquad 2.30$$

or at the adsorption isobar:

$$a = f(t) \qquad p = \text{const} \qquad 2.31$$

and at the adsorption isoster:

$$p = f(t) \qquad a = \text{const} \qquad 2.32$$

Adsorption equilibrium is usually characterized by the adsorption isotherm.

For the explicit description of the isothermic function of adsorption equilibrium several theories may be found in the literature. In the following the most frequently applied isotherm equations will be briefly discussed.

The Freundlich equation[8]

$$a = kp^{n} \qquad 2.33$$

where both k and n are constants.

This was first a purely empirical correlation, but it was proved later that it may be theoretically corroborated.

For linear isotherms $n = 1$, for concave isotherms $n<1$ and for convex isotherms $n > 1$.

Logarithmic plotting of the equation leads to a straight line (log a *vs.* log p) from which the constants can be determined.

The Langmuir equation[8]

$$a = \frac{a_m bp}{1 + bp} \qquad 2.34$$

where a_m is the gas volume necessary for the monomolecular coating of the adsorbent, and b is a constant characteristic of the adsorption strength.

Langmuir's equation gives the first theoretically founded description of the adsorption of gases and vapours which can be applied in the pressure range from zero up to saturation. The validity of the equation can be tested from its linearized form. The Langmuir equation is valid when the plotting of the measured data as p/a *vs.* p gives a straight line. The slope of the straight line is $1/a_m$, the intercept of the ordinate $1/a_m b$, thus the two constants in the equation can be determined from the straight line. Langmuir's equation is derived by assuming monomolecular adsorption and a homogeneous surface. Because of the inhomogeneity of the surface of actual adsorbents the equation can be applied only after certain modifications, especially to the description of adsorptions at low pressures.

The Sips equation[9]

$$a = \frac{a_m b' p^{n}}{1 + b' p^{n}} \qquad 2.35$$

This differs from Langmuir's equation in the "constant" b characteristic of the adsorption strength which becomes here a power function of the pressure:

$$b = b' p^{n-1} \qquad 2.36$$

Substitution of Eq. 2.36 into Eq. 2.34 leads to Sips's equation which can be used satisfactorily for the description of adsorption on inhomogeneous surfaces in the entire pressure range including low pressures.[10]

2.3.3 Separation factor

The partition coefficient and the correlations for the description of phase equilibria which have been discussed above depict the partition of one component between two phases. The separation of the components in a sample depends on the relative values of these partition coefficients. To characterize the ease or difficulty with which two components may be separated the relative volatility or separation factor, α, is introduced which can be expressed as the ratio of the partition coefficients of the components:

$$\alpha = \frac{k_a}{k_b} = \frac{C_{a1} C_{b2}}{C_{a2} C_{b1}} \qquad 2.37$$

where C_{a1} and C_{a2} are the concentrations of the component a, and C_{b1} and C_{b2} the concentrations of the component b in phases 1 and 2.

In gas–liquid chromatography, as in distillation theory, the ratio of the products of the activity coefficients and of the vapour pressures is used for the description of relative volatility:

$$\alpha = \frac{\gamma_b p_b^0}{\gamma_a p_a^0} \qquad 2.38$$

where p_a^0 and p_b^0 are the vapour pressures of the two components at the temperature investigated.

To describe the equilibria conditions of binary mixtures in distillation processes a well-known diagram is used which shows the correlation between the mole fractions of the more volatile component in the vapour and liquid phases. If in Eq. 2.37 the concentrations are written in the form of mole fractions so that for the more volatile component the mole fractions should be y and x and for the heavier component 1 and $1-x$, then the following explicit equation is obtained for the equilibrium curve:

$$y = \frac{\alpha x}{1 + (\alpha - 1) x} \qquad 2.39$$

Equation 2.39 is illustrated in Fig. 2.8. For the equation to be valid α must be constant. For systems for which Raoult's or Henry's law is valid α is constant. If α changes with concentration, the equilibrium curve can be plotted from experimental data.

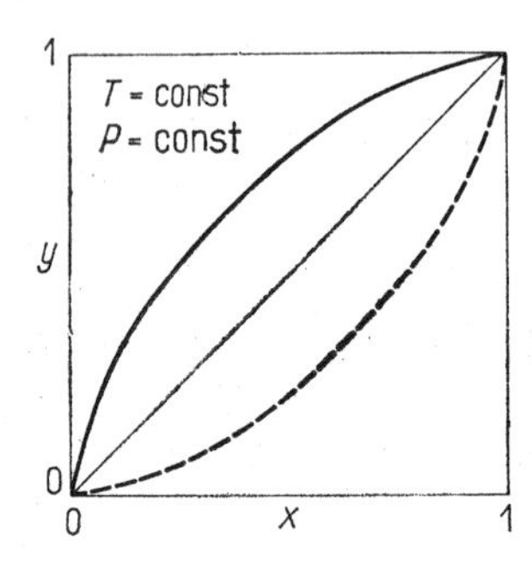

Fig. 2.8. Equilibrium curves of a binary mixture

The equilibrium curve may also be used for the characterization of the solution and adsorption equilibria. These deviate from the distillation equilibrium curve because for absorption and adsorption the correlation refers to constant pressure and temperature (P, $T =$ const). In a given system α will slightly decrease with increasing pressure, but this effect is in general negligible. The value of α decreases with increasing temperature, that is to say, separation deteriorates at higher temperatures.

If $\alpha = 1$ the equilibrium curve will coincide with the diagonal of the diagram, i.e. it will not be possible to separate the two components. The greater the value of α above 1 the easier will be the separation of the two components. The value of the separation factor refers, of course, to a given system, i.e. to a certain stationary phase. Separation may be varied between wide limits by changing the nature of the stationary phase, i.e. by applying different adsorbents or liquid phases. Depending on the nature of the stationary phase the order of sorptivity may be reversed. The continuous curve in Fig. 2.8 illustrates the equilibrium of an ethylene–ethane mixture on active charcoal adsorbent, and the dashed curve in the same figure represents the same equilibrium on silica gel. While on non-polar active charcoal the degree of adsorption depends on the molecular weight, on polar silica gel molecular structure is the decisive factor and ethylene with its unsaturated bond will be the more readily adsorbed substance.

The main advantage of gas chromatography and of chromatographic methods in general over separation by distillation is that in chromatography the number of degrees of freedom is a higher one, as P and T are both independent variables. While in distillation a given separation factor belongs to each given separation, in chromatography the separation factor may be varied between wide limits by the choice of the stationary phase. The possibilities are particularly numerous in gas chromatography, where many solid adsorbents and liquid phases of different properties are available for the solution of a great variety of separation problems.

2.4 Symbols and definitions

The spreading of gas chromatographic methods necessitated the unambiguous definition of concepts and terms used for the description of the methods and evaluation of the results. The initiative in this field was taken by the Gas Chromatography Symposium in London in 1956 where a uniform nomenclature for gas chromatographic methods (gas–solid chromatography and gas–liquid chromatography) was accepted and suggestions were put forward for the definition of several important terms and factors. The International Union of Pure and Applied Chemistry (IUPAC) has set up an international committee to co-ordinate the suggestions made by the working committees in the USA[11] and Great Britain[12] and to evolve a gas chromatography nomenclature. The suggestions of the international committee concerning the chromatogram and process parameters were universally accepted.[13] These terms refer to elution gas chromatography and primarily to gas–liquid chromatography.

In the field of gas–solid chromatography further suggestions were asked for certain terms, but these questions are still open, mainly because the practical and theoretical problems of gas–solid chromatography have become the subject of thorough research only in recent years.

The suggested definitions and symbols have been complemented by others in the course of years, while some have more or less lost their practical importance, but the fundamental terms remained, of course, unchanged.

The chromatogram. By the chromatogram we understand the signal from the detector plotted against time or carrier gas volume. The idealized chromatogram of a single component measured with a differential or integral detector (see below) is shown in Fig. 2.9.

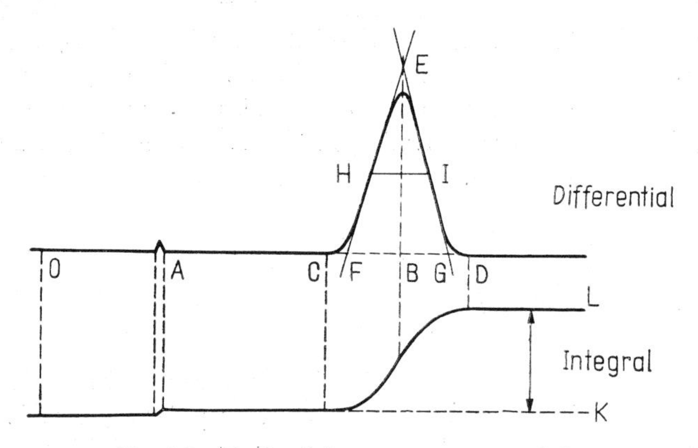

Fig. 2.9. Idealized chromatograms recorded by differential and integral detectors

The following definitions refer either to diagrams obtained directly with differential detectors or to differential curves plotted from chromatograms which were obtained with integral detectors. The base line is the section of the chromatogram which is recorded when only the carrier gas emerges from the column. The peak is that section of the chromatogram which is signalled by the detector on the appearance of the component in the carrier gas (on imperfect separation two or more components of the sample may appear in the same peak). The peak base $\overline{CD}$ is the distance between the extreme limits of the peak, measured on the base line. Under the area of the peak we understand the area bordered by the curve and the base. Peak height $\overline{BE}$ is the distance between the peak maximum and the base, measured parallel to the axis which records the signal from the detector. The peak width $\overline{FG}$ is the distance on the base between the tangents of the inflection points of the curve. The curve intersects the straight line parallel to the base at a distance $\overline{HJ}$ which is the width of the curve at half peak height.

In chromatograms recorded with integral detectors, when sensing the components of the sample the instrument plots an S-shaped curve and the base line is shifted. This shifting is called a step and the distance between the base line before and after the step is the step height.

Retention parameters. The retention volume, V_R, is the uncorrected carrier gas volume necessary for the elution of any given component:

$$V_R = t_R v_0 \tag{2.40}$$

where t_R is the retention time, i.e. the period of time between the injection of the sample and the appearance of peak maximum and v_0 is the volume rate of the carrier gas at the pressure and temperature of the outlet end of the column. In Fig. 2.9, t_R corresponds to the distance $\overline{OB}$, but in the following definitions the horizontal axis is assumed to be the volume of the carrier gas.

Gas hold-up, V_M, is the uncorrected retention volume of some non-adsorbed sample:

$$V_M = t_0 v_0 \tag{2.41}$$

and is the carrier gas volume necessary for carrying such a sample from the sample injector to the detector when the gas volume is measured at the outlet pressure. This includes the void volume of the column, the volumes of the injector and of the detector, and can be determined easily for any column by the elution of a substance whose partition coefficient is relatively small compared to that of the other components. Usually nitrogen, air or one of the rare gases is used for this purpose. The small quantity of air which generally reaches the column through the injector may also be utilized for the determination of gas hold-up. In the case of flame ionization detectors feeding of methane with the sample is recommended.

The void volume of open tubular columns can be calculated from the geometrical dimensions of the column.

From the gas hold-up multiplied by the pressure drop correction factor (see Section 2.1), j, the contribution of the column to V_M or the free gas volume of the latter (V_G) is obtained:

$$V_G = jV_M \tag{2.42}$$

Thus if the gas volume of the column is known (it can be calculated from the known void volume) the volumes of the sample injector and detector can be determined.

The ratio of the free gas volume of the column and the volume of the liquid phase in the column (V_L) is one of the important parameters of gas–liquid chromatographic columns; it is called the phase ratio and is given the symbol β:

$$\frac{V_G}{V_L} = \beta \tag{2.43}$$

The retention volume defined by Eq. 2.40 disregards the rate changes due to the compressibility of the carrier gas. If V_R is multiplied by the pressure drop correction factor, j, the so-called corrected retention volume V_R^0 is obtained:

$$V_R^0 = jV_R = j\overline{OB} \tag{2.44}$$

This quantity is also a function of the free column volume, the volumes of the sample injector and the detector.

The adjusted retention volume, V_R' may be calculated from V_R and the gas hold-up of the column, i.e. the retention volume for the inert peak:

$$V_R' = V_R - V_M = \overline{AB} \tag{2.45}$$

If this value is corrected for the pressure drop the net retention volume, V_N, is obtained:

$$V_N = jV_R' = V_R^0 - V_G = j\overline{AB} \tag{2.46}$$

The net retention volume depends on the quantity of liquid in the column, which necessitates the introduction of a further term, namely, the specific retention volume, V_g, which is the retention volume calculated for 0 °C and refers to one gram of the liquid phase:

$$V_g = \frac{V_N 273}{w_L T} \tag{2.47}$$

where w_L is the mass of the liquid phase in the column ($V_L \rho_L$) and T is the absolute temperature (in °K).

The thermodynamic definition of the partition coefficient, k, which characterizes phase equilibria, was given in Section 2.3. The definition of the partition coefficient for the gas chromatographic column is:

$$k = \frac{\text{mass of the component in the stationary phase}}{\text{mass of the component in the moving phase}}$$

$$\times \frac{\text{volume of the stationary phase}}{\text{volume of the moving phase}}$$

It is assumed that under the conditions of gas chromatography the partition coefficient is concentration independent.

For the correlation between the retention volume and the partition coefficient it may be deduced that:

$$V_R^0 = V_G + k V_L \tag{2.48}$$

or from Eq. 2.46:

$$V_N = k V_L \tag{2.49}$$

These correlations stand for the relationship between the thermodynamic quantity, k, and the column parameters. Equation 2.48 shows that the retention volume is the sum of two terms, the first of which is the gas volume of the column, i.e. the gas volume necessary for the elution of the sample provided no gas is sorbed by the stationary phase ($k = 0$) and the second is the volume necessary for the elution of the solute when repeated phase exchange takes place. It also appears from this equation that the higher the partition coefficient the greater will be the retention volume.

From Eqs 2.46, 2.47 and 2.48 k may be expressed as:

$$k = \frac{V_N}{V_L} = \frac{V_N \rho_L}{w_L} = \frac{V_g T \rho_L}{273} \tag{2.50}$$

To indicate the relative position of the component peak to the inert gas peak the so-called capacity ratio, k', or partition ratio is usually employed; this is the ratio of the partition coefficient and the phase ratio as defined by Eq. 2.43:

$$k' = \frac{k V_L}{V_G} = \frac{k}{\beta} \tag{2.51}$$

In practice the value of k' is between 2 and 200 for packed columns and between 0.2 and 20 for open tubular columns.

These correlations refer to retention volumes with respect to the relative position of a single component peak to the inert gas peak. In practice the measurement of retention times is quite often preferred and in many publications the location of the peak is given by the retention time rather than by the retention volume. Obviously quite analogous equations may also be written for the retention time.

If the appearance or retention time of the inert component is given the symbol t_0 and that of the component the symbol t_R then:

$$t_R = t_0 \left(1 + k \frac{V_L}{V_G}\right) = t_0 (1 + k') \qquad 2.52$$

Retention time like retention volume is the sum of two terms. The first term stands for the period of time necessary for a non-sorptive component ($k = 0$) to travel through the free gas volume of the column. The second term is the retention time due to sorption on the stationary phase. The latter depends on the partition coefficient and the phase ratio, that is to say, on the partition ratio k'.

The net retention time, t'_R, i.e. the period of time during which the component resides in the stationary phase is:

$$t'_R = t_R - t_0 \qquad 2.53$$

From Eqs 2.52 and 2.53, the capacity ratio, k', may be expressed in terms of the retention times:

$$k' = \frac{t'_R}{t_0} \qquad 2.54$$

that is to say k' represents the ratio of retention times in the two phases. Thus k' may be directly determined from the chromatogram with the help of Eq. 2.54. This is illustrated by Fig. 2.10 which is a chromatogram plotted as a time function. The travelling times of the components through a given column depend on their partition coefficients, but on average each component stays for the same time in the gas phase. Retention time in the gas phase is determined by the free gas volume of the column and the average gas flow rate, that is to say, corresponds to the period of time necessary for the travelling of a non-sorptive ($k = 0$) component through the column.

In chromatographic methods the retention of the component is usually characterized by the retardation factor R_F which according to definition is the ratio of the travelling rates of the component zone and of the eluent zone. With the symbols accepted in gas chromatography R_F may be written as:

$$R_F = \frac{V_G}{V_G + k\, V_L} = \frac{t_0}{t_0 + t'_R} = \frac{1}{1 + k'} \qquad 2.55$$

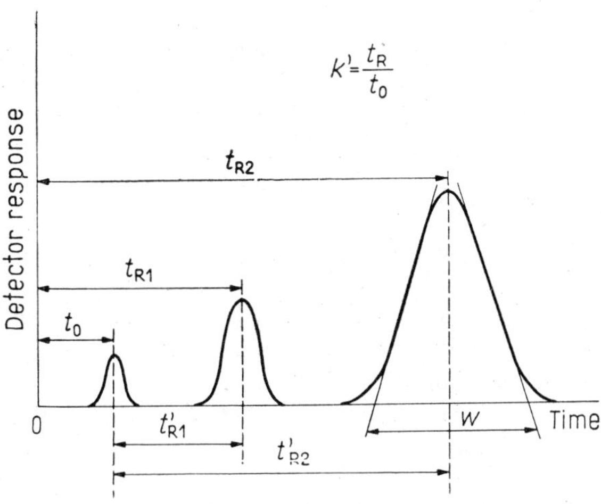

Fig. 2.10. Determination of k′ from the chromatogram

In addition to the fundamental definition in gas chromatography R_F may be given three other physical definitions: 1. the fraction of solute in the gas phase; 2. the fraction of the retention time of the solute in the gas phase; 3. the probability of a soluble molecule being in the gas phase. It should be noted however that in gas chromatography R_F is only a rarely employed process parameter.

In the foregoing paragraphs parameters referring to the peak of one component or to the relative position of the latter to the inert gas phase were subjected to examination. However, it is possible to give the retention parameters with reference to the parameters determined under identical conditions of another component, e.g. of a standard substance. Relative retention r_{12} defined in this way is an analogous concept to relative volatility, α, and may be expressed in various ways:

$$r_{12} = \frac{V_{g2}}{V_{g1}} = \frac{V_{N2}}{V_{N1}} = \frac{t'_{R2}}{t'_{R1}} = \frac{k'_2}{k'_1} = \frac{k_2}{k_1} = \alpha \qquad 2.56$$

The most reliable determination of relative retention is obtained from a chromatogram which contains the peaks of both components. This determination as illustrated in Fig. 2.10 consists in the measuring of the corresponding distances. Thus for a given stationary phase relative retention is independent of the column and is a thermodynamic quantity characteristic of the degree to which the two components can be separated. Relative retention is used on the one hand for the characterization of column operation and represents on the other hand a method for the qualitative identification of the components.

A further fact should also be pointed out here in connection with retention data. The above correlations are valid for constant partition coefficients, that is to say for linear isotherms. In the case of non-linear isotherms the size of the sample also affects retention data; when the isotherm is concave, retention time and retention volume will decrease with increasing sample size and, in the case of convex isotherms, these parameters will increase (see Fig. 2.6). In these cases retention data can be interpreted only at very low concentrations, that is to say, at the initial linear section of the isotherm.

This question will be discussed in detail in the section dealing with the effect of sample size.

References

1. C. E. Schwartz and J. M. Smith, *Ind. Eng. Chem.*, **45,** 1209 (1953)
2. S. Ergun, *Chem. Eng. Progr.*, **48,** 89 (1952)
3. A. T. James and A. J. P. Martin, *Biochem. J.*, **50,** 679 (1952)
4. J. C. Reisch, C. H. Robinson and T. D. Wheelock, see Chapter 1, ref. 67, p. 91
5. E. R. Gilliland, *Ind. Eng. Chem.*, **26,** 681 (1934)
6. J. K. Clarke and A. R. Ubbelohde, *J. Chem. Soc.*, 2050 (1957)
7. W. Peters and H. Jüntgen, *Brennstoff-Chemie*, **46,** (2) 56 (1965)
8. S. Brunauer, "*The Adsorption of Gases and Vapours*", Princeton Univ. Press, Princeton, Vol. 1 (1945)
9. J. Sips, *J. Chem. Phys.*, **16,** No. 5. 490 (1948)
10. L. Szepesy, V. Illés and L. Fáy, *Acta Chim. Hung.*, **37,** 71 (1963)
11. H. W. Johnson and F. H. Stross, *Anal. Chem.*, **30,** 1586 (1958)
12. D. A. Ambrose, A. J. M. Keulemans and J. H. Purnell, *Anal. Chem.*, **30,** 1582 (1958)
13. D. Ambrose *et al.*, Chapter 1, ref. 63, p. 423

3. Theory of Gas Chromatography

3.1 General problems

In the first stage of the development of gas chromatography attention centred not so much on the study of theoretical fundaments, but more on the solution of concrete analytical tasks. Gas chromatography, like other new methods, developed first on empirical lines. This may be quite satisfactory for a while, but eventually it will become necessary to find some kind of classification and evaluation for the collected mass of data and experience. Only theory can provide the system needed for the treatment and evaluation of the experimental data and theory will light the path to further extension and understanding.

There are several hundred publications on the theoretical aspects of gas chromatographic separation and the number of theoretical publications is still on the rise, nevertheless the problems involved are far from being clarified. As already pointed out in Chapter 1 the processes in gas chromatographic columns can be interpreted from an investigation of the migration rate of the component zones and of zone spreading. The fundamental conditions for the separation of two components are, on the one hand, different migration velocities of the two component zones in the column and, on the other hand, a spreading of the zones during progress which will not impede their separation. The migration velocities of the zones depend on the equilibria conditions, i.e. on the partition coefficient, k, and the flow rate of the carrier gas, while zone spreading is a function of column geometry and of the elementary processes of flow, diffusion and mass transfer.

It appears immediately even from this sketchy presentation of column processes that fundamentally the theoretical problems involved in gas chromatography may be classified into two groups. One group of problems centres around the choice of the stationary phase and is based on the interaction between the stationary phase and the components, i.e. on understanding of the equilibria conditions. Gas chromatography involves the study of solution and adsorption equilibria.

The theoretical calculation of solution equilibria and even more of adsorption equilibria is one of the classic, but yet unsolved, problems of physical chemistry, so the selection of the stationary phase is still essentially an empirical procedure. The general aspects of stationary phase selection will be discussed later in detail in the chapter on columns.

To the other group of theoretical problems belong the elementary processes associated with the dynamics of gas chromatography which affect zone spreading and determine column efficiency. Gas chromatography theories deal with the study of these dynamic effects. The fundamental theory of gas chromatography describes the development and spreading of the zones in their dependence on the flow, diffusion and mass transfer conditions.

The development and shape of the zone may be comparatively simply described by general differential equations. The main problem is to find the mathematical solution of these equations. Every theory is founded on the adequate description and characterization of the system under investigation. It may perhaps come as a surprise to those less versed in physical chemistry and chemical engineering that our knowledge is still inadequate for the description of both the geometry of the system and the fundamental processes. For an adequate characterization of a packed bed it would be necessary to know the geometry of the voids, i.e. the dimensions and directions of the channels between the particles. Such a description of a packed bed is one of the unsolved problems of chemical engineering. In gas–liquid chromatography it is impossible to determine the distribution of the liquid phase either for open tubular columns or packed columns, as this distribution depends on the wetting of the support, on capillary condensation, solid–liquid interaction, etc. Only rather inaccurate, approximate methods are known for the description of fundamental flow, diffusion and sorption processes.

Another aspect of the problems is the treatment of hydrodynamic, diffusion and mass transfer processes and of their interaction. As these processes in themselves are far from being clarified, their simultaneous treatment and description can also be only approximate. To obtain differential equations of any use extensive simplifications often bordering on compromise have to be applied to the choice of the initial and limiting conditions. These problems can, of course, be approached in different ways which explain the many theoretical publications and gas chromatographic "theories". It is obvious that for the description of a given process several theories cannot be valid. A theory is adequate if by it the process can be described quantitatively from the fundamental parameters of the system which are determined by independent methods and from operational conditions, without requiring experimental measurements. The description of gas chromatography can be considered only qualitative so far. This raises the question whether it is absolutely necessary for the practising gas chromatographer to be acquainted with the theoretical problems. It is undoubtedly possible to perform gas chromatographic analyses without knowledge of the theory, but the choice of an appropriate column and stationary phase and of the optimum operation conditions requires, for the solution of a new separation problem, acquaintance with the process and the effects of interfering factors. Through a theoretical study of column processes it became possible to achieve considerable improvements in column performance, to reduce the time required for analysis and to work out new column types and analytical methods. A classical example here is the open tubular column coated inside with the liquid phase, the applicability of which has been demonstrated by Golay on purely theoretical basis at a time when no such column was yet in operation.

The discussion of the theoretical fundamentals of gas chromatography in the previous chapter included the basic laws of hydrodynamics and diffusion. Before proceeding to the description of the various theories it seems expedient to give a brief survey of the third effect governing zone spreading, namely of the kinetics of mass transfer. Mass transfer rate is characteristic of the degree of deviation from equilibrium. This effect is often called in the literature "non-equilibrium". Deviation from equilibrium is characteristic of all cases of mass transfer between a stationary and a moving phase. Let us now examine this effect for a single component in connection with the process taking place in a gas chromatographic column.

Part of the sample fed into the inlet end of a column is rapidly adsorbed while another part remains in the gas phase. If carrier gas flow were to cease equilibrium would be established between the two phases. The carrier gas stream however sweeps the solute from the gas phase and transports it to the next section of the column where it is again adsorbed, when liberated from the stationary phase it will continue its way through the column. Flow in the column hinders the molecule in the gas phase in once again reaching the stationary phase in the same cross-section of the column, thus flow continuously upsets, while mass transfer tends to establish a state of equilibrium. As a result of these two opposed effects a certain state of non-equilibrium develops at every

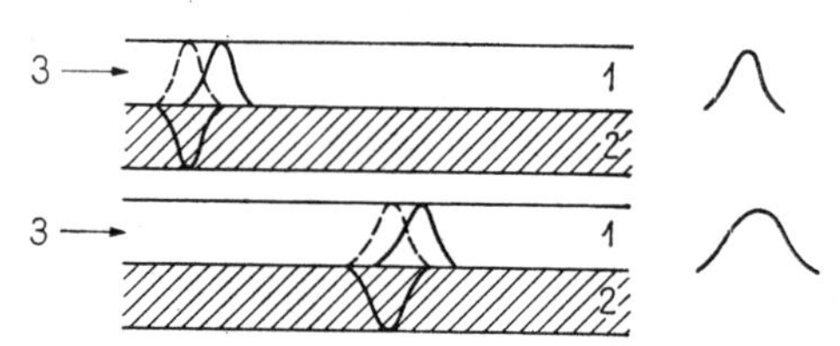

Fig. 3.1. Broadening of the component zone due to non-equilibrium. 1. Gas phase; 2. stationary phase; 3. carrier gas inlet

point of the zone. Flow rate tends to extend the zone, while rapid mass transfer reduces the width of the zone. In Fig. 3.1 deviation from equilibrium is shown by the changes in concentrations in the two phases. The dashed line among the curves for the gas phase represents the concentration changes corresponding to equilibrium. As a consequence of flow the actual concentration change runs ahead of the equilibrium position, that is to say in the gas phase at the beginning of the zone actual concentration is higher than the equilibrum concentration and at the end of the zone it is lower. Horizontal shift between the two concentration profiles is constant in the entire zone and is directly proportional to the flow rate of the carrier gas and inversely proportional to the rate of mass transfer. The degree of deviation from equilibrium is given by the vertical distance between the two profiles which is the smallest at the centre of the zone and the greatest in the region of sharp concentration changes.

The rate of migration of the zone is determined by the partition coefficient, k, which is the ratio of the concentrations in the stationary and gas phases. The zone of a readily soluble component (high k_2) travels at a lower rate than the zone of a poorly soluble component (low k_1). According to the concentration distribution as shown in Fig. 3.1 the k value for a given system will be approached only in the centre of the zone, so only this will travel at a constant rate depending on k. At the beginning of the zone, concentration (or mass ratio) in the gas phase will be higher than the equilibrium concentration, i.e. the condition $k_a < k$ will be valid (where k_a is the apparent partition coefficient) and therefore the leading part of the zone will travel at a higher rate than the centre of the zone. In the trailing portion of the zone, on the other hand, concentration in the gas phase is lower than the equilibrium concentration and the migration rate will therefore be lower, resulting in a gradually increasing distance between the two ends of the zone, i.e. in zone spreading.

It should be noted here that this scheme is valid only for linear isotherms (k = const). If $k \neq$ const, i.e. the isotherms are convex or concave, the migration rate will further depend on the concentration range or size of the sample, while, on the other hand, instead of the Gaussian curve the distorted elution curves shown in Fig. 2.6 are obtained.

The study of mass transfer rate is impeded by the effect of flow distribution and molecular diffusion on the deviation from the equilibrium state. Depending on the type of column and on the stationary phase (adsorbent or liquid coated solid support), mass transfer may proceed according to different mechanisms. In a given system the rate-controlling process may vary depending on operation conditions. We shall come back to the study of mass transfer rate in Section 3.2.2.

In connection with gas chromatography theories the following general statements may be postulated: The process taking place in the gas chromatography column is so complex that our present knowledge is insufficient for its accurate description. For want of the structural description of the system the following steps also lack foundation. As the accurate physical model of the system was unknown the exact description of models involving various simplifications and neglections was attempted. The various theoretical approximations have some common features, notably the assumption that deviation from equilibrium is small, retention time is long compared to the time required for equilibration and a Gaussian distribution may be assumed for the description of concentration distribution. The first gas chromatography theory was the plate theory evolved by Martin and Synge[1] and applied in the early stage of gas chromatography. Though the plate theory may today be considered obsolete, it is indispensable for the understanding of the development of theories and of column performance.

To classify theoretical approximations van Deemter *et al.*[2] and later Keulemans[3] introduced some new concepts on the basis of the interfering factors described earlier, namely equilibrium conditions, and flow and diffusion processes.

Depending on equilibrium conditions, i.e. the shape of the isotherm, a distinction may be made between a) linear and b) non-linear chromatography.

From the point of view of mass transfer processes the conditions in the column may be classified into a) ideal and b) non-ideal conditions. The criteria for ideal and non-ideal systems were given in Chapter 1.

By accounting simultaneously for the equilibrium conditions and mass transfer, chromatographic processes can be classified into four groups:

I. Linear — ideal chromatography.
II. Non-linear — ideal chromatography.
III. Linear — non-ideal chromatography.
IV. Non-linear — non-ideal chromatography.

This classification is obviously of no practical importance. In the case of gas flow an ideal system is even theoretically inconceivable. Theoretical gas chromatographic investigations are limited to the description of the linear — non-ideal system which is characteristic of gas–liquid chromatography. In the case of a non-linear isotherm, such as generally characterizes gas–solid chromatography, the mathematical treatment of the process is extremely difficult and no suitable correlation is known to describe the process.

The following theories refer mainly to systems of the third group.

3.2 Theory of packed columns

Description of gas chromatography processes can be approached by two methods: 1. the so-called plate theory and 2. the rate theory.

3.2.1 Plate theory

The general method for the description of multistage counter-diffusion processes is based on the introduction of the theoretical plate concept also called the equilibrium stage. Introduced into chromatography by Martin and Synge[1] the method was improved by Mayer and Tompkins[4] to determine the number of theoretical plates necessary for a given separation.

According to the plate theory a chromatographic column consists of a series of equilibrium stages. By equilibrium stage, or more commonly theoretical plate, we understand that section of the column in which the two phases reach equilibrium. The gas phase is transported by the carrier gas to the next column section where a new equilibrium, corresponding to the concentration conditions in this section, is established. This repeated equilibration results with respect to one component in the spreading of the component zone. In the case of two or more components repeated phase exchange brings about the separation of the components.

In the calculation of diffusion, mass transfer processes may, strictly speaking, be applied only to processes in which contact is stepwise. In a distillation column the phases separate after each plate, i.e. equilibration and a new phase exchange take place on the next plate under the effect of a new contact. In the case of continuous contact, as in packed columns, no equilibration can occur as the driving force of the mass transfer process is the deviation from equilibrium. For the characterization of packed columns the term "transfer unit" is used instead of "theoretical plate" to describe that section of the column in which the concentration change is equal to the mean driving force in the same section, that is to say to the average difference between the equilibrium and actual concentrations. The height equivalent to one theoretical plate (HETP) and the height of a transfer unit (HTU) will be practically identical in those cases in which the system passes through a great number of equilibrium steps and the concentration change in each equilibrium stage is small. From this aspect the use of HETP in gas chromatography would be quite acceptable, but as it will be shown later, this concept is for other reasons unsuited for the characterization of the separation process.

According to the physical model in the construction of the plate theory[1,4] a stepwise contact was assumed instead of continuous flow and therefore a binomial distribution, i.e. a non-continuous elution curve was obtained.

Glueckauf[5] improved the plate theory by assuming continuous flow which is valid for very low flow rates only when an equilibration of the two phases in the section corresponding to the theoretical plate may be assumed. Without going into details the following final correlations were obtained by Glueckauf for a column with N theoretical plates:

$$x_N = \frac{e^{-v} v^N}{N!} \qquad 3.1$$

where $x_N = x_{G_N}/x_{G_0}$, the quotient of the concentration of the gas phase on the Nth plate and on the first plate, v is the quantity of carrier gas which has flowed through the system divided by the effective plate volume.

Plotting from this equation x_N *vs.* v for various numbers of theoretical plates a group of curves is obtained in which every elution curve has one maximum and two inflection points. The peak (maximum) of the curve is at $v = N$ and for this value Eq. 3.1 is:

$$x_N\,(\max) = \frac{e^{-N} N^N}{N!} \qquad 3.2$$

assuming continuous flow results in a Poisson distribution for the elution curve. In the case of high plate numbers both the binomial and the Poisson distribution approximates the Gaussian curve, but Klinkenberg and Sjenitzer[6] have demonstrated the different peak width of the two models. They also maintain that the Gaussian curve may be approximated by assuming different mechanisms which however does not prove the validity of the assumed mechanism.

From the plate theory it is possible to give an approximate description of peak broadening during passage through the column. According to Glueckauf's deduction, peak width is proportional to the square root of the theoretical plate number.

Using the same symbols as in the chromatogram in Fig. 2.10, t_R time (or the corresponding distance), necessary for reaching the peak maximum, and w, the peak width (the distance on the base line between the tangents at the inflection points), can be expressed from Glueckauf's theoretical consideration[7] with the number of theoretical plates:

$$t_R = CN \qquad 3.3$$

$$w = C\,4\sqrt{N} \qquad 3.4$$

where C is a proportionality factor. Elimination of C from the two equations and rearrangement lead to the number of theoretical plates:

$$N = 16 \left(\frac{t_R}{w} \right)^2 \qquad 3.5$$

When the number of theoretical plates and the length of the column, L, are known, the value of the height equivalent to a theoretical plate (HETP) may be expressed as:

$$\text{HETP} = \frac{L}{N} = \frac{L}{16} \left(\frac{w}{t_R} \right)^2 \qquad 3.6$$

The plate theory gives an approximate description of the zone spreading for one component and with its help the "number of theoretical plates" characteristic of column efficiency can be calculated, from which and the column length the HETP value is obtained. The plate theory however disregards the effect of flow rate and the kinetics of the mass transfer process and reveals nothing about the factors influencing HETP value. Thus according to the plate theory, HETP continuously decreases with decreasing flow rate which disagrees with experience as the function HETP *vs.* flow rate always has a minimum.

Halász has dealt in detail with the criticism of the theoretical plate concept.[8] In the first stage of the development of gas chromatography by describing an analogous process plate theory he provided a work hypothesis without having a real physical content. The "theoretical plate" defined in gas chromatography and the theoretical plate in distillation denote entirely different and unrelatable concepts. In distillation (and in other counter-current mass transfer operations) the number of theoretical plates indicates the ease of separation of two components and represents that number of ideal or equilibrium stages on the equilibrium curve of the binary mixture which is necessary for a given concentration change; HETP is the column height producing a concentration change which would correspond to one equilibrium stage.

In gas chromatography the number of theoretical plates refers to the peak broadening of a single component during its passage through the column. It appears from Eq. 3.6 that in a given column (L = const) HETP is in fact the peak broadening relative to the retention time, consequently the N value represents that fraction of the column which is occupied by the relative peak broadening. The number of theoretical plates and the HETP value calculated from it refer to a given component, for another component the parameters of the elution peak are different, thus another N, i.e. HETP value is obtained. For components with higher sorptivity (high k value), i.e. with longer retention time, a greater number of theoretical plates and accordingly lower HETP value is obtained.

In gas chromatography, since N refers to a single component only, it is obviously not suitable to describe the separation of two components or the separating power of the column. In earlier books [9,10] and papers[11] on gas chromatography a comparison is drawn between distillation and gas chromatography theoretical plate numbers necessary for the performance of a given separation task. It was generally found that in the case of packed columns the necessary number of theoretical plates was ten times greater in gas chromatography than in distillation. With the introduction of new types, e.g. open tubular columns, it became apparent that no such simple relationship can be postulated. It is clear from what has been said so far that such a comparison has no sense, as the two different types of physical models are not comparable. In gas chromatography the "plate number" is insufficient for the characterization of the separating power of the column and serves only as an indication of peak broadening, that is to say of column efficiency and only for the comparison of columns of identical types. We shall come back to this problem in a discussion of column operation.

From the interpretation of the number of theoretical plates it appears that in gas chromatography the use of this concept and that of HETP may be misleading and may lead to incorrect analogies. According to recent opinions it would be more correct to replace the theoretical plate number by *efficiency* and plate height by *relative peak broadening.* As however earlier communications mention exclusively the old concept their use in this book is also unavoidable. In the following, "plate height" will be given the symbol H as is usual in recent gas chromatographic literature.

3.2.2 Rate theories

a) Gas–liquid chromatography (GLC)

Rate theory covers the study of processes in the column, the influence of flow conditions, diffusion and resistance to mass transfer on peak broadening. Contrary to the plate theory it indicates also the significant dependence of zone broadening, i.e. of the value of H, on flow rate and includes the effect of column and operation parameters.

There is an extensive literature dealing with the development of the rate theory in which the works of Lapidus and Amundson,[12] Glueckauf,[13] Klinkenberg and Sjenitzer[6] and of van Deemter, Zuiderweg and Klinkenberg[2] are the most important.

Van Deemter, Zuiderweg and Klinkenberg's equation, generally accepted for the description of gas chromatography, was evolved from the deductions of Lapidus and Amundson by extension of Glueckauf's theory.

The van Deemter equation

According to van Deemter *et al.* the diffusion process in gas chromatography can be described with the differential equations of the material balance of the gas and liquid phases. The solution of the two differential equations for one peak gives a Gaussian curve.

Between the H value which is characteristic of the kinetics of the separation process and the factors influencing mass transfer the following correlation is valid:

$$H = 2\lambda\, d_p + \frac{2\gamma\, D_g}{u} + \frac{8}{\pi^2} \frac{k'\, d_f^2}{(1+k')^2\, D_l} u \qquad 3.7$$

where λ is a factor characteristic of the packing, d_p is the particle diameter of the packing, γ is a correction factor for the unevenness of the spaces between the particles, D_g and D_l are the diffusion coefficients of the component in the gas and liquid phase respectively, u is the linear gas velocity, k' the capacity ratio (see Eq. 2.51) and d_f is the effective thickness of the liquid film on the support.

Corresponding to the factors affecting mass transfer there are three terms in the van Deemter equation. The first term accounts for the geometry of the packing, the second for the longitudinal diffusion in the gas phase and the third for the resistance to the mass transfer process. If we examine the criteria of ideal chromatography we find that the terms in Eq. 3.7 contain the deviations from these criteria, that is to say, the factors which cause these deviations. The usual simplified form of Eq. 3.7 is:

$$H = \mathbf{A} + \mathbf{B}/u + \mathbf{C}u \qquad 3.8$$

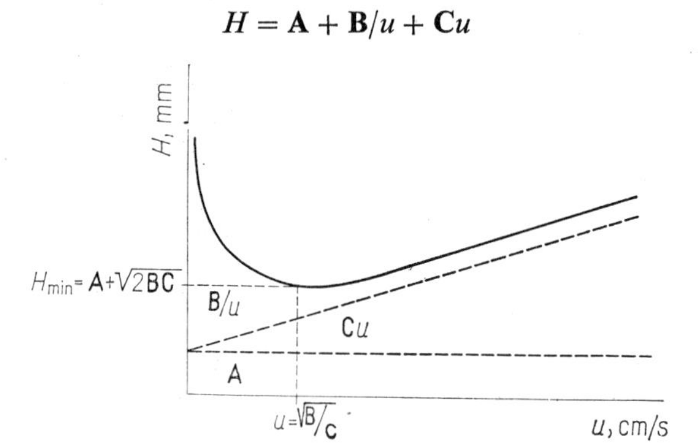

Fig. 3.2. Changes in H vs. linear gas velocity

The dependence of H on flow rate can be studied from the simplified form of the van Deemter equation. Figure 3.2 shows the changes in H *vs.* linear gas velocity. Equation 3.8 is that of a hyperbola with its minimum at $u = \sqrt{\mathbf{B}/\mathbf{C}}$ velocity, and the minimum H value $H_{min} = \mathbf{A} + 2\sqrt{\mathbf{BC}}$. It appears from Fig. 3.2 that gas velocity has a decisive influence on column efficiency; the column will operate most efficiently at a given gas velocity.

The constants **A, B** and **C** in the van Deemter equation can be determined by plotting the values of H calculated from experimental data according to Fig. 3.2. Their determination is easiest by graphic representation. The area under the curve can be divided into three regions corresponding to the three terms in Eq. 3.8 (see Fig. 3.2). The first term represents the zone distortion due to the inhomogeneity of the packing and is independent of the flow rate of the gas.

At gas velocities lower than the optimum, H is inversely proportional to gas velocity in accordance with the second term in the equation. At low flow rates the effect of molecular diffusion is greater and very high H values may result.

The resistance to mass transfer represented by the third factor is directly proportional to gas velocity. At optimum gas velocity, i.e.

at the minimum of the H *vs.u* curve, the effects of molecular diffusion and mass transfer resistance are equal and the values of the terms are the same. With increasing flow rate the effect of mass transfer resistance on H increases, above optimum gas velocities mass transfer is the controlling factor.

The absolute value of optimum gas velocity depends on mass transfer resistance, if this is low, high velocities may be applied. It should be noted here that the plotted experimental data give in general flatter curves and consequently in practice, the optimum rate can be approached over a fairly wide velocity range.

The constants can be more accurately determined by gradual approximation. **B** may be determined from the curve $H-\mathbf{C}u$ *vs.* $1/u$ and **C** from $H-\mathbf{B}u$ *vs.* u, when the slope of the straight lines will give the required values. The most reliable method for the determination of the constants is the method of least squares.

When examining the effect of flow rate the compressibility of the flowing gas has to be considered. Owing to the pressure drop in the column, flow rate will change and will always be higher at the outlet than at the inlet from which it follows that it is impossible to realize an optimum gas velocity in the entire length of the column. The operation of the column will approach the optimum conditions more closely the nearer the value of the ratio of inlet pressure to outlet pressure approaches unity. If this ratio is in practice below two, the effect of flow rate may be neglected.

Let us now see the physical meaning of the three terms in the van Deemter equation and their effect on the process.

The first term, $\mathbf{A} = 2\,\lambda\,d_p$, is characteristic of the solid particles and is independent of the nature of the liquid phase and the solute and also of operation conditions. This term is called by the authors the "eddy diffusional" term as it accounts for the irregular channels of varying dimensions between the particles, which result in a rate distribution similar to that encountered in turbulent flow. This term is rather unfortunate, as the flow rates applied in gas chromatography are in the range of laminar flow, so eddies characteristic of turbulent flow are hard to imagine. It seems more appropriate to accept Golay's suggestion[12] and to call this effect the "multipath" effect.

The term **A** in van Deemter's equation and in the following modified rate equations has been the subject of much discussion. In term **A**, λ is a complex structural factor characteristic of the packing for which widely contradictory data are found in the literature, including differences of several orders of magnitude, zero and negative values.[15,16] λ depends also on the particle size; and experimental data have shown it to be a function of the gas velocity also. In the case of uniform, regular particles the value of λ is approximately 1 and may be taken practically as 1. With decreasing particle diameter, d_p, the value of **A** diminishes. According to some recent views, where packing is adequate, **A** can be neglected, its presence points to some external factor (dead space).

The second term in Eq. 3.7, $\mathbf{B}/u = 2\ \gamma\ D_g/u$ indicates the effect of molecular diffusion on zone spreading. It was shown in paragraph 2.2 that in the gas phase spreading due to diffusion is proportional to the residence time in the gas phase, thus inversely proportional to the flow rate. In the same paragraph the magnitude of the diffusion coefficient D_g and its dependence on operation conditions were described. For the value of the correction factor γ, 0.5–1 was obtained from the measured data. γ is an empirical factor which accounts for the effect of narrowings and bends of the channels between the particles on free diffusion. Knox and McLaren[17] have recently worked out an approximate calculation method for the determination of γ. The calculated γ value of 0.63 for glass beads is in very good agreement with the experimentally measured $\gamma = 0.60$. Because of the lack of an appropriate structural model the discrepancy between the calculated and measured values is considerably greater in the case of irregularly shaped, porous particles. In the rate theories the term **B** is perhaps the best clarified, though it should be mentioned here that from the point of view of zone spreading, only at very low gas velocities will the effect of molecular diffusion be significant and under practical conditions the contribution of the term **B** to the value of H is very small.

The third term in Eq. 3.7 includes the resistance to mass transfer in the liquid phase. It should be pointed out however that originally van Deemter *et al.* had taken into consideration also a second mass transfer term for the gas phase, C_g which they believed to be negligibly small compared to the resistance in the liquid phase. According to Eq. 2.51:

$$k' = \frac{kV_L}{V_G}$$

the value of the capacity ratio, k', depends on the partition coefficient, k, and the phase ratio of the column. In general k' is greater than 1, thus with the increase of k' the coefficient **C** and accordingly the value of H will diminish. The effective liquid film thickness, d_f, appears as a squared term in the numerator on the right side of Eq. 3.7; the value of H will decrease with decreasing d_f. Thus through k' and d_f the quantity of liquid exerts two opposite effects. The value of the diffusion coefficient D_l depends mainly on the viscosity of the liquid. The greatest uncertainty in the term **C** is introduced by the supposition of a uniform liquid film and a corresponding film thickness d_f.

The equation of van Deemter *et al.* stimulated considerable progress in the theoretical study of gas chromatography columns and of the effects of operation conditions. By plotting, according to Eq. 3.8, the H values calculated from the recorded chromatogram, the **A**, **B** and **C** coefficients of the equation can be determined. Thus it is possible to investigate starting from the experimental data the influence of the various parameters on column operation. Van Deemter's equation was the subject of examination by a number of authors; the papers by Keulemans and Kwantes,[18] Cheshire and Scott,[19] de Wet and Pretorius,[20] Bohemen and Purnell,[15] Littlewood,[21] Scott[22] and Purnell[23] should be mentioned here as some of the most important in this respect. Experimental investigation of the equation indicates a qualitative agreement between the results and the equation, but rather large deviations were obtained between the numerical values of the various factors and the influence of operation conditions. These deviations are due partly to the determination of the H value and the inaccuracies in the determination of the coefficients **A**, **B** and **C**, as well as to the fact that effects outside the column were neglected and are partly the results of the approximate character of Eq. 3.7. The factors γ, λ and d_f in the equation are empirical constants which can be determined only from an adequate number of experimental data and are only characteristic of a given column.

Parallel to the application of van Deemter's equation attempts were made to improve the theory and to modify the equation. These modifications aim at a more thorough study of the mass transfer process and at the addition of new terms to the van Deemter equation on the one hand, and, on the other hand, at the improvement of the physical model and the study of the interaction of the elementary processes.

Modification of the van Deemter equation

One of the essential modifications of the equation is based on a more accurate description of the mass transfer process. Resistance to mass transfer consists not only of the diffusion resistance in the liquid film but diffusion resistance in the gas phase also contributes to the overall effect. The pattern is similar to the "two films" theory of diffusional mass transfer processes. Van Deemter supplemented his equation with a term accounting for mass transfer resistance in the gas phase as far back as 1957;[24] the application of this modified equation is discussed in the papers by Glueckauf, Golay and Purnell.[25] The usual form of this modified equation is:

$$H = \mathbf{A} + \mathbf{B}/u + \mathbf{C}_l\, u + \mathbf{C}_g\, u \qquad 3.9$$

where the first three terms are the same as in Eq. 3.7 and $\mathbf{C}_g$ is the mass transfer resistance in the gas phase. Ayers and Lloyd[26] have demonstrated that for low liquid coatings, i.e. thin liquid films, mass transfer resistance depends mainly on the resistance in the gas phase, that is to say, the term containing $\mathbf{C}_g$ is significant.

To provide a more detailed description of the process in the gas phase, Jones[27] added further terms to the equation:

$$H = \mathbf{A} + \mathbf{B}/u + \mathbf{C}_l\, u + \mathbf{C}_g\, u + \mathbf{C}_1\, u + \mathbf{C}_2\, u \qquad 3.10$$

In Eq. 3.10 the coefficient $\mathbf{C}_1$ accounts for the velocity distribution due to retarded gas flow in the layers which are in contact with the solid surface. The coefficient $\mathbf{C}_2$ is a correlation term accounting for the interaction of the two types of resistances in the gas phase. Kieselbach[28, 29] presents a detailed experimental study of the equation. $\mathbf{C}_g$ is in fact the resistance to diffusion from the bulk of the gas phase to the liquid phase on the support surface and is significant mainly for readily soluble components which are eluted at a later stage.

The term $\mathbf{C}_g$ is

$$\mathbf{C}_g\, u = c_a \frac{k'^2}{(1+k')^2} \frac{d_g^2}{D_g}\, u \qquad 3.11$$

where c_a is a proportionality factor, d_g is the gas diffusional path length, k' is the capacity ratio, and D_g is the diffusion coefficient of the component in the gas phase.

C_1 is the resistance to diffusion through the stagnant gas film on the boundary of the liquid phase. This will be significant mainly in the case of rapidly eluted, poorly bound components and its value depends greatly on the particle size:

$$C_1 u = \frac{c_b d_p^2}{D_g} u \qquad 3.12$$

where c_b is a proportionality factor (~ 1).

There are no experimental proofs to justify the sixth term in the equation.

Improvement of the rate theory was attempted also from another angle, namely by the stochastic theories based on the classic "random walk" problem. The first steps in this direction were made by Giddings *et al.*[30-32] and Beynon *et al.*[33] Instead of describing a great number of trials and modifications we shall restrict ourselves to their main characteristics and to the correlations deemed at present the most adequate. Giddings[34-37] approached the description of the process from the diffusional character of zone broadening by accounting for the effect of local non-equilibrium. He studied mass transfer in the gas phase[38] and the effect of liquid distribution on the H value[36, 39] with the help of general non-equilibrium theory. He paid attention to the fact that the processes in the gas phase cannot be independent of each other[40] and will mutually reduce each other's effect on the value of H. According to his coupling theory the term **A**, characteristic of flow and the effect of mass transfer resistance in the gas phase should be treated together.

Written in its simplified form the equation is:

$$H = \frac{1}{1/\mathbf{A} + 1/\mathbf{C}_1 u} + \frac{\mathbf{B}}{u} \mathbf{C}_l u + \mathbf{C}_g u + H_e \qquad 3.13$$

The first term in Eq. 3.13 is the result of merging "eddy diffusion" with the velocity distribution term of Eq. 3.10. At the velocities usual in gas chromatography this term is generally reduced to the term C_1 in Jones's equation, but at high flow rates it will correspond to Eq. 3.12.

The second, third and fourth terms are essentially the same as the corresponding terms in Eq. 3.10. The last term, H_e, is characteristic of the equipment, as unequal packing at the end of the column and other dead spaces in the equipment also cause zone broadening.

Starting from the general non-equilibrium theory the effect of mass transfer processes and the dependence of mass transfer resistances on the structural parameters of the column were subjected to a detailed study. Giddings[39] gives for different conditions eight possible correlations with which to calculate gas diffusion resistance and another eight correlations for $\mathbf{C}_l$.

Study of the diffusion resistance in the liquid phase brought about noteworthy progress, as the earlier assumed uniform liquid film is, in reality, non-existent. At the contact points of the particles, as well as in the voids and pores, the liquid forms islets of different thicknesses. Since $\mathbf{C}_l$ is proportional to the square of the thickness of the liquid phase, distribution has a very considerable effect. With the introduction of the configurational factor Giddings[41] gives a general method applicable to the description of islet type coats.

Littlewood[42] has published some interesting statements on the nature of the term $\mathbf{C}_g$. He demonstrated both experimentally and theoretically that uneven flow is the main source of zone spreading in the packed column. In certain parts of the cross-section the component zone is able to travel more rapidly because of a) uneven flow and b) uneven liquid distribution. In preparative columns of 3–10 cm diameter this effect is generally known, but according to Littlewood in 4–6 mm diameter analytical columns too, these are the main contributions to the value of the term $\mathbf{C}_g$.

Detailed study of the general non-equilibrium theory resulted in further additions to Eq. 3.13.[16, 43]

Depending on the nature of the packing and of the flow there are five possible types of mechanisms whose sum may be written instead of the first term in Eq. 3.13:

$$H = \sum_{i=1}^{5} \frac{1}{1/\mathbf{A} + 1/\mathbf{C}_1 u} + \frac{\mathbf{B}}{u} + \mathbf{C}_l u + \mathbf{C}_g u + H_e \qquad 3.14$$

The five terms of the sum account for: 1. processes taking place through the channels between the particles; 2. processes through the particles; 3. processes between uneven flow channels; 4. processes between inhomogeneous regions and 5. processes extending to the entire column.

It is more expedient to study the reduced forms of the equations which are arrived at by the definition of two dimensionless parameters:[44]

Reduced plate height:

$$h = \frac{H}{d_p} \qquad 3.15$$

Reduced linear velocity:

$$\nu = \frac{ud_p}{D_g} \qquad 3.16$$

Knox has dealt in detail with the reduced equations and found experimental proof of the modified theory.[45]

Influence of the pressure drop in the column. In the above correlations the compressibility of the gas stream was not taken into consideration. In columns where pressure drop is not negligible the changes in gas velocity must also be considered, as gas expansion also causes zone spreading. Giddings *et al.*[46] were the first to point out the necessity of applying a pressure correction. De Ford *et al.*[47, 48] in their study of the efficiency of packed columns have demonstrated the importance of pressure correction. In their equation the term **A** is considered negligible, but for the sake of completeness it will be shown in brackets.

$$H = (\mathbf{A}) + \frac{\mathbf{B}^0}{p_0 u_0} + (\mathbf{C}_g^0 + \mathbf{C}_1^0)\, p_0 u_0 f + \mathbf{C}_l^0 u_0 j \qquad 3.17$$

where p_0 is the outlet pressure, u_0 the outlet gas velocity, j the pressure correction factor of James and Martin and f is the pressure correction:

$$f = \frac{p_i (p_0 + 1) j^2}{2} \qquad 3.18$$

The correction factor f is usually about one and with the exception of accurate theoretical work can be neglected. The coefficients $\mathbf{B}^0$, $\mathbf{C}_g^0$, $\mathbf{C}_1^0$ and $\mathbf{C}_l^0$ can be determined by measuring the H values for different outlet pressures and outlet velocities.

Sternberg and Poulson[49] investigated the correlation between pressure drop and column efficiency from the point of view of choosing gas velocity. Sternberg[50] published a method of calculation by which the effect of the pressure drop can be determined. There are several publications on the effect of pressure on the efficiency and separation performance of the column.

Though recent theories on the processes occurring in the gas chromatographic column and on the effect of operation conditions represent a considerable step forward, the theoretical description of the process is far from being solved. The most favourable agreement between the theoretically calculated and experimental data is obtained for microbead packings when the difference between the calculated and measured H values is 20–50%.[39] This difference of generally 200–300% or more in the case of commonly used irregular porous packings is due mainly to the lack of an adequate description of the structure of the system, as already mentioned in Chapter 1.

The statistical foundations of the rate theories

The rate theories have the common feature that the H value which is characteristic of zone spreading is made up from the contribution of several elementary processes. Actual concentration distribution can be described by the Gaussian curve (see paragraph 2.2) the width of which is characteristic of the width of the zone and may be defined with the standard deviation, σ. For random processes σ is according to the random walk theory[51]

$$\sigma = l\sqrt{N} \qquad 3.19$$

where l is the length of the elementary steps and N the number of steps. If the elementary processes are considered to be independent and random processes, every elementary process contributes to the width of the Gaussian curve with a σ^2 variance.

$$\sigma^2 = \Sigma\sigma_i^2 \qquad 3.20$$

This simple correlation is valid only if there is no interaction between the processes which induce zone spreading, that is to say, if their effects can in fact be considered to be independent.

In the majority of papers dealing with the elaboration and development of rate theories the effect of elementary processes is treated according to Eq. 3.20, but this assumed "independent" contribution to the main process is only a rough approximation. The interaction between the elementary processes had been pointed out some time ago.[27, 35, 38] By applying an appropriate correlation coefficient[27] it is possible to form the sum of the variances, but there is no way to calculate this coefficient directly.

The H value characteristic of column efficiency may be expressed with the help of the variance:[27]

$$H = \frac{\sigma^2}{L} \tag{3.21}$$

which is the variance corresponding to unit column length.

Another expression frequently used is[48]

$$H = \frac{d(\sigma^2)}{dL} \tag{3.22}$$

i.e. the rate of variance increase during progress through the column.

If H is constant, variance will increase linearly with the length of the column, while the width of the peak (σ; standard deviation) is proportional to the square root of column length.

b) Gas–solid chromatography (GSC)

So far the process has been described and the theoretical problems studied from the aspect of gas–liquid chromatographic conditions. Considerably less work has been devoted to the study of the theory of gas–solid chromatography and only in recent years has attention turned to the application of solid adsorbents and the description of separation by adsorption.

Before embarking on the description of the process a few deviations from the concepts used so far should be pointed out.

In GSC the phase ratio defined by Eq. 2.43 is

$$\beta = \frac{V_G}{V_A} \tag{3.23}$$

where V_G is the same as the gas volume in Eq. 2.43 and V_A is the true adsorbent volume. By true volume we understand the weight of the adsorbent divided by its true density. As $V_A > V_L$, lower β and therefore (see Eq. 2.51) higher k' values occur in GSC than in GLC.

In the equations used for the description of the retention volumes the volume of the liquid phase should be substituted by the true volume of the adsorbent.

The rate theory for the description of gas–solid chromatography contains essentially the same terms as already discussed for GLC with the important difference that instead of the $\mathbf{C}_l$ term which characterizes mass transfer in the liquid phase the equation contains a $\mathbf{C}_k$ term characteristic of adsorption kinetics. Thus all the equations of the preceding section can be applied to gas–solid chromatography, provided the term $\mathbf{C}_k$ is included. Giddings has dealt in detail with the theoretical investigation of gas–solid chromatography and the term $\mathbf{C}_k$.[35, 39, 52] His equation for gas–solid chromatography is:

$$H = \sum_{i=1}^{5} \frac{1}{1/\mathbf{A} + 1/\mathbf{C}_1 u} + \frac{\mathbf{B}}{u} + \mathbf{C}_k u + \text{other terms} \qquad 3.24$$

Applying a reaction equation of the first order to the description of adsorption kinetics on homogeneous surfaces a relatively simple expression is obtained for $\mathbf{C}_k$. In the case of adsorbents with inhomogeneous surfaces usually encountered in practice this value has to be multiplied by a heterogeneity factor which depends on the energy distribution of surface sites. In his study of the general non-equilibrium theory Giddings quoted several expressions for the calculation of $\mathbf{C}_k$.[39] Theoretical considerations have indicated that primarily on homogeneous surfaces $\mathbf{C}_k$ is of a lower order than the $\mathbf{C}_l$ of GLC which points to a highly efficient applicability of GSC. So far no appropriate experimental data are available for the study of $\mathbf{C}_k$.

One of the main problems of the applicability of GSC are the asymmetric tailing peaks instead of the usual symmetric peaks of GLC. Earlier, tailing was attributed to the non-linear isotherm, but later it was demonstrated that even in the case of linear equilibrium correlation tailing may occur for kinetic reasons because of the difference between the adsorption and desorption rates at adsorption sites with different energies.

Everett has studied in detail the processes on solid surfaces.[53] The theoretical investigation of gas–solid chromatography involves the almost insoluble fundamental difficulty of finding an accurate mathematical description for the adsorption process even in the simplest case of a perfectly homogeneous surface, so that a similar solution for the adsorbents with heterogeneous surfaces with which we have to deal in reality is inconceivable at the present stage of our knowledge. The main obstacle which has to be overcome here is the lack of an adequate description of adsorbent structure and of the distribution and dimensions of the pores.

From the practical point of view the theoretical study — and to a certain degree the applicability — of gas–solid chromatography is hindered mainly by imperfect reproducibility in the manufacture of adsorbents, so that there are differences not only between the adsorbents of different origins, but even deviations between different batches from the same manufacturer.

In recent years many special adsorbents have been evolved for gas chromatographic purposes and interest in gas–solid chromatography is growing, so that further progress may be expected in this field too in the near future with respect to both applications and the elucidation of the theoretical problems involved.

3.3 Theory of open tubular columns

The invention of open tubular, or as they were first called, capillary columns is linked with the name of Golay.[54,55] He arrived at the application of tubes without packing but with liquid coated walls on purely theoretical grounds, through the investigation of zone spreading due to the particles of the packing and of the reduction of the pressure drop in the column. Examination of the diffusion of a flowing gas in an empty tube led Golay in 1957 to the theory of the open tubular (unpacked) columns, a theory still generally accepted. His deductions are based on the following suppositions: 1. that the internal wall of the tube is coated with a liquid film of uniform thickness; 2. there is laminar gas flow in the column. For the term H a correlation similar to the van Deemter equation is obtained

$$H = \frac{2D_g}{u} + \frac{2}{3}\frac{k'}{(1+k')^2}\frac{d_f^2}{D_l}u + \frac{1+6k'+11k'^2}{24(1+k')^2}\frac{r^2}{D_g}u \qquad 3.25$$

where r is the radius of the capillary tube and the other symbols are the same as in the previous expressions.

The simplified form of the equation is:

$$H = \frac{\mathbf{B}}{u} + \mathbf{C}_l u + \mathbf{C}_g u \qquad 3.26$$

Comparison of Eqs 3.8 and 3.26 indicates the similarities and differences. The term **A** of van Deemter's equation is missing as in this case there is no packing. The term $\mathbf{C}_1 u$ corresponds to the third term in Eq. 3.8, but for the proportionality coefficient which instead of being $\frac{8}{\pi^2}$ is here $\frac{2}{3}$ (later the factor 2/3 was used also in the description of packed columns). The term $\mathbf{C}_g u$ is new and represents the diffusion resistance in the gas phase. The modification of the van Deemter equation was investigated by using Golay's equation and the interpretation of the gas diffusion term which led to the extended Eq. 3.9. Golay also pointed out the difference between the mechanisms of the processes in the packed and open tubular columns.[14] He showed that mass transfer in the gas phase plays a far greater role in packed columns than in the open tubular column because of the irregular channels in the packing. The process i.e. the value of H in the open tubular column is primarily controlled by the diffusion resistance of the liquid film. Several authors including Scott and Hazeldean,[56] Desty and Goldup,[57] Halász and Schreyer,[58] Desty, Goldup and Swanton[59] have tried to find an experimental proof of Golay's equation and studied the performance of open tubular columns.

Golay's equation is generally accepted for the description of the operation of open tubular columns. Giddings[39,60] has studied the effect of non-uniform liquid films on mass transfer factors, but the differences can be considered only as approximations.

In the open tubular columns coated with an adsorbent layer which have been evolved lately, gas–solid chromatography takes place. For GSC Eq. 3.26 is modified as follows:

$$H = \frac{\mathbf{B}}{u} + \mathbf{C}_k u + \mathbf{C}_g u \qquad 3.27$$

where $\mathbf{C}_k$ is a term characteristic of adsorption kinetics.

3.4 Efficiency and operation of columns

A gas chromatography column has to meet several and somewhat contradictory demands such as high separation power, high speed of operation and high capacity. Any one of these properties can be improved only at the expense of the other two. Capacity, that is the quantity of the sample which can be injected, will be discussed in detail in the section on preparative columns. Beside the separating power of the column and the speed of analysis the conditions of analysis must also be considered which means the inlet pressure and temperature shall be within the limits generally used in commercial apparatus.[61]

The separating power of the column may be characterized partly by the broadening of the individual component peaks and partly by the separation of the peaks. The following parameters are examined in connection with the investigation of column operation:

column efficiency;
separating power of the column;
number of plates, i.e. column length necessary for a given separation;
operation parameters;
minimum analysis time.

3.4.1 Column efficiency

Number of theoretical plates

The efficiency of the column is characterized by the broadening of the peak which for a given column can be expressed by the number of theoretical plates. The inaccuracy and misleading nature of the expression has already been mentioned in connection with the plate theory, but in the literature the concept of plate height is generally established for the characterization of the width of the Gaussian curve. The number of theoretical plates can be calculated from the elution peak data on the chromatogram. For the determination of the plate number various correlations can be deduced from the retention data (V_R, V'_R, i.e. t_R, t'_R) and from some dimension characteristic of peak width. In practice the number of plates is determined by measuring the appropriate distances on the chromatogram. The characteristic data of the elution peak are illustrated in Fig. 3.3.

James and Martin[62] have suggested two methods for the determination of the number of plates, one method is that of standard deviation, the other the measurement of the area below the peak. Both methods are slow and cumbersome, so that two other methods have since been established in chromatographic practice.

One of these methods uses the correlation implicit in Eq. 3.5 which written for the distances measured on the chromatogram will be:

$$N = 16\left(\frac{x}{w}\right)^2 \qquad 3.28$$

where x is the distance corresponding to the peak maximum measured

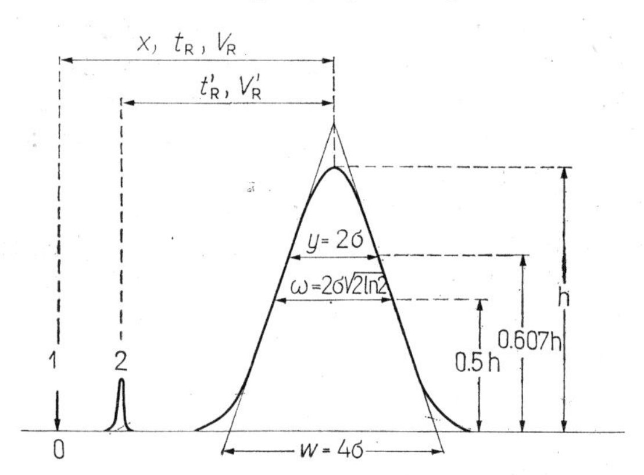

Fig. 3.3. The characteristic data of the elution peak. 1. Injection; 2. inert (air) peak

from the point of injection and w is the distance on the base line between the tangents of the peak. The International Nomenclature Committee of IUPAC has suggested this formula for general use.[63]

The other generally used expression is:

$$N = 5.54\left(\frac{x}{\omega}\right)^2 \qquad 3.29$$

where ω is the peak width at half peak height (the multiplication factor is obtained from $\omega^2 = 5.54\ \sigma^2$ where $5.54 = 8\ \ln 2$).

The determination of N includes the assumption that the signal from the detector changes linearly with the concentration, otherwise the calculated value of N cannot be correlated to column efficiency. We should like to mention here a fact often neglected in the literature, namely that the determination of N — especially in the case of peaks that are not perfectly symmetrical — is rather inaccurate, with 10–20% and often higher deviations. The efficiency of the column and the value of N depend on column operating conditions so that these too must be accurately defined when efficiency is determined. When comparing columns of the same type it is advisable to give the value of N for one metre length of the column. For packed columns the number of plates for one metre may vary between 500 and 5000. The number of theoretical plates of an open tubular column may be calculated in a similar manner, when in general a very high number is obtained which however is not a good indication of the separating power of the column.

Separation factor

Purnell introduced a new concept to describe column efficiency which is more appropriate than the number of theoretical plates[64, 65] and may be used satisfactorily for open tubular columns. The equation defining the separation factor S contains instead of the apparent, the reduced retention data, i.e. those measured from the air peak:

$$S = 16\left(\frac{V'_R}{w}\right)^2 = 16\left(\frac{t'_R}{w}\right)^2 \qquad 3.30$$

Effective number of plates

To characterize open tubular columns Desty *et al.*[59] have introduced the effective number of plates N' which may also be applied to packed columns. The difference between the two plate numbers is that the determination of the effective plate number contains retention data measured not from the sample injection but from the air peak (V'_R or t'_R), that is, the same as Purnell's separation factor S:

$$N' = 16\left(\frac{V'_R}{w}\right)^2 = 16\left(\frac{t'_R}{w}\right)^2 \qquad 3.31$$

The correlation between the two types of plate numbers is:

$$N' = N\left(\frac{k'}{1+k'}\right)^2 \qquad 3.32$$

Therefore N and N' (S) differ only by a constant factor which is characteristic of the given column. The plate height corresponding to the effective plate number is:

$$H' = H \left(\frac{1 + k'}{k'} \right)^2 \tag{3.33}$$

The values N' and S are particularly appropriate for the comparison of different types of columns, i.e. of packed and open tubular columns, as a considerably higher number of plates is obtained from Eq. 3.28 for open tubular columns than for packed columns when both are used for the solution of the same separation problem.

Separation number (n_{sep})

Column efficiency may be characterized by the possible number of peaks between two n-paraffin peaks with consecutive carbon numbers and calculated from the following correlation:[78]

$$n_{sep} = \frac{(t_1 - t_2)}{\omega_1 + \omega_2} - 1 \tag{3.33a}$$

where t is the retention time, ω the peak width measured at half peak height and 1 and 2 stand for the two consecutive n-paraffins (e.g. n-hexane and n-heptane). The separation number may be expediently used particularly for the characterization of capillary columns or when programmed temperature or pressure is applied.

Rohrschneider has presented a table on the correlation between the equations used for the characterization of column efficiency.[66]

3.4.2 Separating power of the column

Resolution power of the column

The efficiency of a column is given by the broadening of the peak of a single component or by the possible number of peaks between the peaks of two standard components. For the description of the separation of two neighbouring components the separation of the two peaks has to be determined. In general, separation of the successive peaks may be characterized by the degree of contamination with the other component. Glueckauf subjected to theoretical examination the dependance of contamination on plate number and on relative volatility,[7] α. The simplest treatment of this correlation is on the basis of the nomogram in Fig. 3.4 which presents the relationship between the number of theoretical plates, N, and the factor Φ for

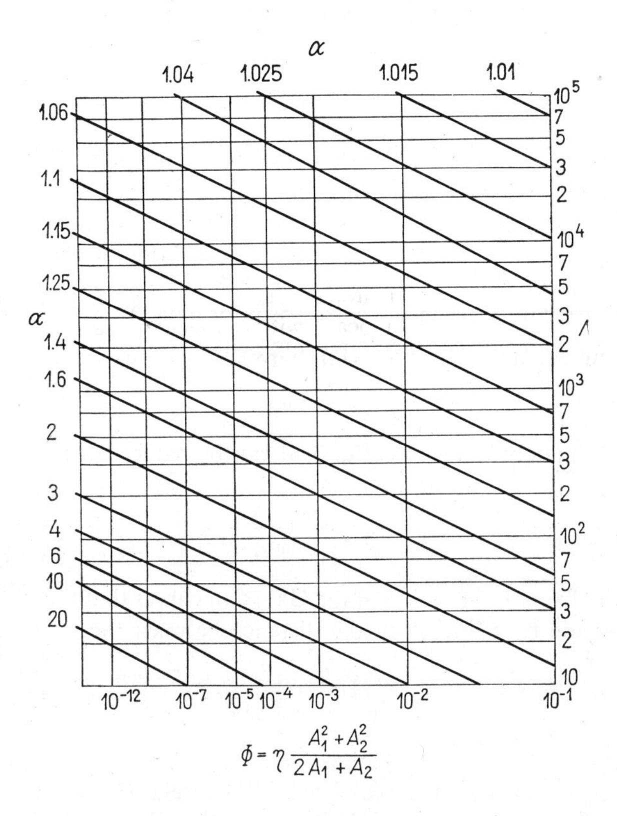

Fig. 3.4. Correlation between the number of theoretical plates and the degree of separation for various values of relative volatility[7]

cases of different relative volatilities. In the factor Φ is the weight fraction of contamination in the separated component, and A_1 and A_2 are the ratios of the areas under the peaks of the two components to the total area.

Resolution (R_s)

To characterize the separation of two components a simple method was suggested by the Nomenclature Committee of IUPAC[67] using the retention times and the peak broadenings:

$$R_s = 2\frac{t_{R2} - t_{R1}}{w_1 + w_2} \qquad 3.34$$

where t_R is time (or the corresponding distance) between sample injection and the peak maximum, w the distance on the base line between the tangents at the inflection points. Index 1 refers to the more weakly, index 2 to the more strongly sorbed component. There are also other correlations for the value of R_s in the literature which often cause confusion in the comparison of data.[68]

For the width of two neighbouring peaks

$$w_1 = w_2 \qquad 3.35$$

is approximately valid, thus Eq. 3.34 can be reduced to:

$$R_s = \frac{t_{R2} - t_{R1}}{w_2} \qquad 3.36$$

It follows from the equation of the Gaussian curves that the two peaks touch at the base line when the distance between the peak maxima is 4 σ, or

$$t_{R2} - t_{R1} = w_2 \qquad 3.37$$

Thus a separation by a distance of 4 σ is given by the value $R_s = 1$. However, in this case the peaks will still contain about two per cent of contamination. (Theoretically the separation of the Gaussian curves is never complete.) It may be considered a satisfactory separation from the analytical aspect when the distance between the peak maxima is 6 σ or $R_s = 1.5$. The equations given in the literature for the determination of the number of plates which are necessary for a given separation may differ according to the value of R_s chosen.

For the more general description of the resolution power Karger[69] introduced into the equation the $w_1/w_2 = n$ "peak ratio" characteristic of the ratio of peak widths and, depending on the relative position of the peaks, derived various equations for the description of the resolution and separation power.

The resolution power may also be given in Kováts retention index units to be discussed later (Chapter 6)[70]

$$R_s = \frac{I_2 - I_1}{\omega_I f} \tag{3.37a}$$

where I_1 and I_2 are the retention indices of the two components to be separated, ω_I is the peak width at half peak height in retention index units, $f = 1.699$ is a correction factor ($4\sigma = w = 1.699\omega$).

ω_I may be calculated from the peak data by means of the following correlation:

$$I = \frac{100 \log \left(\dfrac{2t' + \omega}{2t' - \omega}\right)}{\log t'_{n+I} - \log t'_n} \tag{3.37b}$$

The value ω_I is also appropriate for the characterization of column efficiency.[66]

Separation factor

Applied to two components Purnell's separation factor characterizes separation. Assuming a resolution $R_s = 1.5$ between the two peaks it follows from Eq. 3.30 and the definition of the relative volatility (Eq. 2.56) that

$$S = 36 \left(\frac{\alpha}{\alpha - 1}\right)^2 \tag{3.38}$$

The value S as defined by Eq. 3.38 is a thermodynamic quantity characteristic of separation and independent of the column. The actual S value for a given column can be determined from the measured chromatogram with the help of Eq. 3.30.

3.4.3 Required plate number and column length

There are several, more or less similar correlations in the literature for the determination of the number of plates necessary for separation.

This necessary number of plates is obtained from the equation defining theoretical plates and the equations for Purnell's S factor (Eqs 3.30 and 3.38) for the separation[71] ($R_s = 1.5$):

$$N_{ne} = 36\left(\frac{\alpha}{\alpha - 1}\right)^2\left[1 + 2\left(\frac{V_g}{V_{R2}}\right) + \left(\frac{V_G}{V_{R2}}\right)^2\right] \quad 3.39$$

As

$$\frac{V_{R2}}{V_G} = k'_2 \quad 3.40$$

Eq. 3.39 may also be written with the capacity ratio k' of the more readily sorbed component:

$$N_{ne} = 36\left(\frac{\alpha}{\alpha - 1}\right)^2\left(\frac{1 + k'}{k'}\right)^2 \quad 3.41$$

In general, without restricting the value of R_s the necessary number of plates will be:

$$N_{ne} = 16R_s^2\left(\frac{\alpha}{\alpha - 1}\right)^2\left(\frac{1 + k'}{k'}\right)^2 \quad 3.42$$

From Eq. 3.32 the necessary effective number of plates, N'_{ne}, will be:

$$N'_{ne} = 16R_s^2\left(\frac{\alpha}{\alpha - 1}\right)^2 \quad 3.43$$

Equation 3.42 written with the phase ratio, β, and the partition coefficient, k, on the basis of Eq. 2.51, will be:

$$N_{ne} = 16R_s^2\left(\frac{\alpha}{\alpha - 1}\right)^2\left(\frac{\beta}{k_2} + 1\right)^2 \quad 3.44$$

where k_2 is the partition coefficient of the more readily sorbed component.

Some important conclusions can be drawn from Eqs 3.42 and 3.44 on the relationship between the parameters of the system under investigation and the number of plates necessary for separation. It appears from the equations that required plate number depends on the relative volatility of the components, and the partition ratio characteristic of the column, that is to say, on the values of the phase ratio β and of the partition coefficient k.

The following table shows the values of the second term in Eq. 3.42 for various k' values:

k'	0.25	0.5	1	5	10	20	50
$\left(\frac{1+k'}{k'}\right)^2$	25	9	4	1.44	1.21	1.11	1.04

It appears from the table that if $k' > 5$ the necessary number of plates is essentially determined by the relative volatility, α, and the column itself has but little influence. In cases when $k' < 5$, on the other hand, the column parameters play an important role in the determination of the plate number. The table also indicates that if $k' > 20$ the theoretical and effective numbers of plates will be practically equal:

$$N \approx N' \qquad 3.45$$

From Eq. 3.44 further conclusions can be drawn on the influence of the phase ratio, β, and the partition coefficient, k. Figure 3.5 shows changes in the number of plates which is necessary for separation *vs.* the values of β and k according to Kieselbach.[29] If k is small, that is to say, in the case of poorly soluble, rapidly eluted components, a high number of plates will be necessary and this will greatly increase with increasing values of β. The value of the second term in Eq. 3.44 is determined by the ratio β/k, but with increasing values of k the the role of β will diminish, while at very high values of k, i.e. in the case of readily soluble components, the second term in the equation will be negligible. From Fig. 3.5 a clear picture is obtained of the difference in the operation of packed and open tubular columns. While in packed columns the value of β is around 10, in open tubular columns $\beta > 100$ which means that for the separation of poorly adsorbed components a very high number of plates is necessary in the open tubular column.

The column length necessary for the desired separation will be according to the correlation:

$$L_{ne} = N_{ne} H \qquad 3.46$$

$$L_{ne} = 16 R_s^2 H \left(\frac{\alpha}{\alpha - 1}\right)^2 \left(\frac{1+k'}{k'}\right)^2 \qquad 3.47$$

This equation may be used for the calculation of column length if the value of H which is characteristic of peak broadening is known. As the value of H is different for every component, the value for the more readily sorbed of the two components under investigation should be

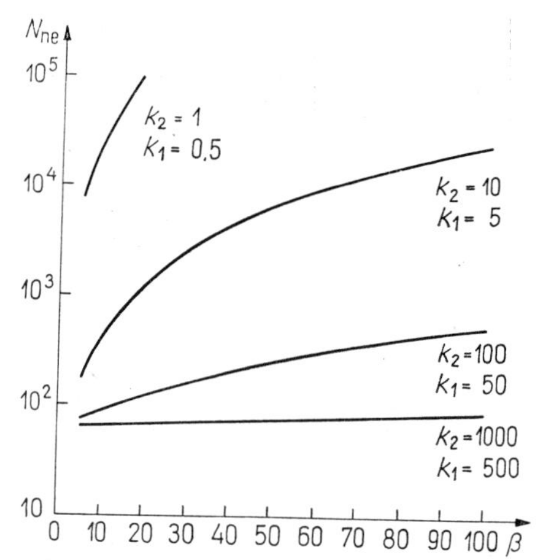

Fig. 3.5. The number of theoretical plates necessary for separation vs. the phase ratio, β, for different partition coefficients (α = 2, const)[29]

known. As already mentioned in connection with theoretical investigations, our present knowledge is yet insufficient for the calculation of the value of H from independent data, so at present Eq. 3.47 cannot be used directly for the determination of the necessary column length.

3.4.4 Operation parameters of the column

Performance index

When evolving the theory of the open tubular column Golay introduced the concept of the performance index,[54, 55] originally to control the flow conditions in such a column:

$$PI = \frac{\Delta t^4 \, t_0 \, \Delta p}{t^3 (t - 0.94\, t_0)} \qquad 3.48$$

where Δt is the width of the peak under investigation at half peak height, t the retention time of the component, t_0 the retention time of the air peak; Δp the pressure drop in the column. The dimension of PI is in poise.

Another way to describe the performance index with the column parameters is:

$$PI = 30.7\, H^2 \frac{\mu}{K} \frac{1 + k'}{k' + 1/16} \qquad 3.49$$

where μ is the viscosity of the carrier gas, K the permeability of the column and k the capacity ratio.

According to Golay for an ideal column $PI = 0.1$ poise. In the older packed columns PI was equal to 1000 poise or four powers of ten worse. Golay gave for open tubular columns values between 0.2 and 10, while in the recent highly efficient packed columns the value of PI is between 10 and 20. The value of PI is characteristic of the flow in the column and provides a relationship between elution time of the component and pressure drop. The performance index is of help in the improvement of column operation and in the selection of the optimum carrier gas rate. In the comparison of columns of different types the performance index is usually mentioned.

Performance parameter

Halász *et al.*[61] suggested the performance parameter PP instead of the performance index for the characterization of column operation. The performance parameter is obtained by merging the performance index and the N_{ne} value which is characteristic of the separating power of the column, but so far no unambiguous opinion has developed with respect to its application.

The values of both PI and PP are functions of k' and both pass through a minimum at around $k' = 3$.

Effective number of plates per unit time

Desty *et al.*[59] characterized column operation with the rate of development of the effective number of plates:

$$\frac{N'}{t_R} = \frac{\bar{u}}{H'(1 + k')} \qquad 3.50$$

or the same written for the number of theoretical plates:

$$\frac{N}{t_R} = \frac{\bar{u}k'^2}{H(1 + k')^3} \qquad 3.51$$

where $\bar{u}$ is the average linear gas velocity and the other symbols are the same as before. This accounts for characteristic column parameters and the time factor and offers a way of comparing columns of different types.[72]

3.4.5 Minimum operation time

In addition to the maximum separation power the other important demand which the gas chromatography column has to meet is the least possible time required for separation. This is an important factor even in the case of laboratory analyses, but is of particular importance for process chromatographs. There are accordingly many publications dealing with theoretical and experimental investigation of the minimum time required for analysis.[57, 71, 73, 74, 75]

The retention time of any given component:

$$t = \frac{L(1 + k')}{\bar{u}} = \frac{NH}{\bar{u}}(1 + k') \qquad 3.52$$

In the investigation of analysis time t refers to the more readily sorbed component.

Analysis time can be reduced firstly by applying higher gas rates, so that in the case of rapid analyses higher gas rates may occur than the rate for optimum value of H.

To determine the time required for analysis the number of plates, N, is substituted by the necessary number of plates N_{ne}, for a resolution R_s in Eq. 3.52 on the basis of Eq. 3.42:

$$t_{ne} = 16R_s^2 \frac{H}{\bar{u}} \left(\frac{\alpha}{\alpha - 1}\right)^2 \frac{(1 + k')^3}{k'^2} \qquad 3.53$$

The analysis time which is necessary for a resolution R_s is essentially the minimum analysis time.

It appears from Eq. 3.53 that analysis time depends on the relative volatility, α, which indicates the ease or difficulty of separation, on the partition coefficient k', which is a characteristic of the system, and on the value of $H/\bar{u}$.

$H/\bar{u}$ may be expressed from the van Deemter equation:[2]

$$\frac{H}{\bar{u}} = \frac{\mathbf{A}}{\bar{u}} + \frac{\mathbf{B}}{\bar{u}^2} + \mathbf{C}_l + \mathbf{C}_g \qquad 3.54$$

In high speed analysis the $\bar{u}$ values are high, so that the first two terms on the right side of the equation may be neglected:

$$\frac{H}{\bar{u}} = \mathbf{C}_l + \mathbf{C}_g \qquad 3.55$$

By substituting Eqs 3.42 and 3.55 into Eq. 3.53 we obtain:

$$t_{\mathrm{ne}} = N_{\mathrm{ne}}(\mathbf{C}_l + \mathbf{C}_g)(1 + k') \qquad 3.56$$

which shows that the time requirement of analysis is determined by the number of plates necessary for separation, by k' and by mass transfer resistance.

Reduction in mass transfer resistance may be achieved by increasing the gas velocity on the one hand and on the other hand by the production of a uniform, thin liquid film. In packed columns the optimum particle size is about 100 mesh (150 μm) and the optimum liquid phase loading (depending on the specific surface of the support) is 0.2 to 5 w% of the support.

Karger and Cook[76] and later Guiochon[77] when attempting to find optimal conditions for column operation have made a detailed study of the influence of various parameters on resolution and analysis time.

In columns of small diameter which are packed with small uniform particles and have low liquid loading an appropriate choice of the liquid phase and of operation conditions may result in a tenth of the usual 10–30 min analysis time. With special apparatus the analysis of multicomponent mixtures may be performed within seconds. The factors affecting column operation will be studied more thoroughly in the chapter on columns.

References

1. A. J. P. Martin and R. L. M. Synge, *Biochem. J.*, **35,** 1358 (1941)
2. J. J. van Deemter, F. J. Zuiderweg and A. Klinkenberg, *Chem.Eng. Sci.*, **5,** 271 (1956)
3. A. J. M. Keulemans, Chapter 1, ref. 18, p. 96
4. S. W. Mayer and E. R. Tompkins, *J. Am. Chem. Soc.*, **69,** 2866 (1947)
5. E. Glueckauf, *Trans. Farad. Soc.*, **51,** 34 (1955)
6. A. Klinkenberg and F. Sjenitzer, *Chem. Eng. Sci.*, **5,** 258 (1956)
7. E. Glueckauf, *Trans. Farad. Soc.*, **51,** 1540 (1955)
8. I. Halász, *J. Gas Chromatog.*, **4,** 8 (1966)
9. A. B. Littlewood, "*Gas Chromatography: Principles, techniques, and applications*", Academic Press, London (1962)
10. J. H. Purnell, "*Gas Chromatography*", J. Wiley, New York (1962)
11. E. F. Herington, Chapter 1, ref. 58, p. 30
12. L. Lapidus and N. R. Ammundson, *J. Phys. Chem.*, **56,** 984 (1952)
13. E. Glueckauf, "*Principles of Operation of Ion Exchange Columns*", Soc. of Chem. Ind., London (1955)
14. M. J. E. Golay, Chapter 1, ref. 63, p. 139
15. J. Bohemen and J. H. Purnell, Chapter 1, ref. 60, p. 6
16. J. C. Giddings, Chapter 1, ref. 70, p. 3
17. J. H. Knox and L. McLaren, *Anal. Chem.*, **36,** 1477 (1964)
18. A. J. Keulemans and A. Kwantes, Chapter, 1, ref. 58, p. 15
19. J. D. Cheshire and R. P. W. Scott, *J. Inst. Petroleum*, **44,** 74 (1958)
20. W. J. De Wet and V. Pretorius, *Anal. Chem.*, **30,** 325 (1958)
21. A. B. Littlewood, Chapter 1, ref. 58, p. 23
22. R. P. W. Scott, Chapter 1, ref. 58, p. 189
23. H. J. Purnell, *Ann. N. Y. Acad. Sci.*, **72.,** 592 (1959)
24. J. J. van Deemter, "*2nd Informal Symp. G. C. Disc. Group*", Cambridge, Sept. 1957
25. E. Glueckauf, M. J. E. Golay and J. H. Purnell, *Ann. N. Y. Acad. Sci.*, **72,** 612 (1959)
26. R. J. Loyd, B. O. Ayers and F. W. Karasek, *Anal. Chem.*, **32,** 698 (1960)
27. W. L. Jones, *Anal. Chem.*, **33,** 829 (1961)
28. R. Kieselbach, *Anal. Chem.*, **33,** 806 (1961)
29. R. Kieselbach, Chapter 1, ref. 67, p. 139
30. J. C. Giddings and H. Eyring, *J. Phys. Chem.*, **59,** 416 (1955)
31. J. C. Giddings, *J. Chem. Phys.*, **26,** 169 (1957)
32. J. C. Giddings, *J. Chromatog.*, **2,** 44 (1959)
33. J. H. Beynon, S. Clough, D. A. Crooks and G. R. Lester, *Trans. Farad. Soc.*, **54,** 705 (1958)
34. J. C. Giddings, *J. Chem. Phys.*, **31,** 1462 (1959)
35. J. C. Giddings, *J. Chromatog.*, **3,** 443 (1960)
36. J. C. Giddings, *J. Chromatog.*, **5,** 46 (1961)
37. J. C. Giddings, *J. Chromatog.*, **5,** 61 (1961)
38. J. C. Giddings, *Anal. Chem.*, **34,** 1186 (1962)
39. J. C. Giddings, *Anal. Chem.*, **35,** 439 (1963)

40. J. C. GIDDINGS and R. A. ROBINSON, *Anal. Chem.*, **34,** 885 (1962)
41. J. C. GIDDINGS, *J. Phys. Chem.*, **68,** 184 (1964)
42. A. B. LITTLEWOOD, Chapter 1, ref. 70, p. 77.
43. J. C. GIDDINGS, *Anal. Chem.*, **35,** 2215 (1963)
44. J. C. GIDDINGS, *J. Chromatog.*, **13,** 301 (1964)
45. J. H. KNOX, *Anal. Chem.*, **38,** 253 (1966)
46. J. C. GIDDINGS, S. L. SEAGER, L. R. STUCKI and G. H. STEWART, *Anal. Chem.*, **33,** 768 (1960)
47. D. D. DE FORD, R. J. LOYD and B. O. AYERS, *Anal. Chem.*, **35,** 426 (1963)
48. D. D. DE FORD, See Chapter 1, ref. 68, p. 23
49. J. C. STERNBERG and R. E. POULSON, *Anal. Chem.*, **36,** 58 (1964)
50. J. C. STERNBERG, *Anal. Chem.*, **36,** 921 (1964)
51. W. FELLER, "*Probability Theory and its Applications*", J. Wiley, New York (1950)
52. J. C. GIDDINGS, *Anal. Chem.*, **36,** 1170 (1964)
53. D. H. EVERETT, Chapter 1, ref. 70, p. 219
54. M. J. E. GOLAY, Chapter 1, ref. 59, p. 1
55. M. J. E. GOLAY, Chapter 1, ref. 60, p. 36
56. R. P. W. SCOTT and G. S. F. HAZELDEAN, see Chapter 1, ref. 63, p. 144
57. D. H. DESTY and A. GOLDUP, Chapter 1, ref. 63, p. 162
58. I. HALÁSZ and G. SCHREYER, *Z. Anal. Chem.*, **181,** 367 (1961)
59. D. H. DESTY, A. GOLDUP and W. T. SWANTON, Chapter 1, ref. 67, p. 105
60. J. C. GIDDINGS, *Anal. Chem.*, **34,** 458 (1962)
61. I. HALÁSZ, K. HARTMANN and E. HEINE, Chapter 1, ref. 70, p. 38
62. A. T. JAMES and A. J. P. MARTIN, *Biochem. J.*, **50,** 679 (1952)
63. R. P. W. SCOTT (Ed.), "*Gas Chromatography, 1960*", Third Symposium, Edinburgh, June 1960. Academic Press, New York, Butterworths, London (1960)
64. J. H. PURNELL, *Nature*, **184,** 2009 (1959)
65. J. H. PURNELL, *J. Chem. Soc.*, **54,** 1268 (1960)
66. L. ROHRSCHNEIDER, *Chromatographia*, **1,** 108 (1968)
67. H. A. SZYMANSKI (Ed.), "*Lectures on Gas Chromatography — 1962*", Plenum Press, New York (1963)
68. S. DAL NOGARE and R. S. JUVET, "*Gas-Liquid Chromatography. Theory and Practice*", Interscience, New York (1962)
69. B. L. KARGER, *J. Gas Chromatog.*, **5,** 161 (1967)
70. L. ROHRSCHNEIDER, *J. Chromatog.*, **22,** 6 (1966)
71. J. H. PURNELL and C. P. QUINN, Chapter 1, ref. 63, p. 154
72. I. HALÁSZ and H. O. GERLACH, *Anal. Chem.*, **38,** 281 (1966)
73. J. H. KNOX, *J. Chem. Soc.* **55,** 433 (1961)
74. B. O. AYERS, R. J. LOYD and D. DE FORD, *Anal. Chem.*, **33,** 986 (1961)
75. J. C. GIDDINGS, *Anal. Chem.*, **34,** 314 (1962)
76. B. L. KARGER and W. D. COOK. *Anal. Chem.*, **36,** 991 (1964)
77. G. GUIOCHON, *Anal. Chem.*, **38,** 1020 (1966)
78. R. KAISER, *Z. Anal. Chem.*, **189,** 11 (1962)

4. Gas Chromatography Apparatus

4.1 General aspects of apparatus construction

The first commercial gas chromatographs appeared on the market in 1955–56. Since that time gas chromatographic equipment has undergone considerable development and there are at present about 40 or 50 larger and many smaller firms engaged in the manufacture of gas chromatographs. From the point of view of purpose of application gas chromatographs may be classified into three groups, namely

analytical (laboratory) apparatus;
preparative apparatus;
process chromatographs.

There is no sharp difference between the first two groups, with appropriate supplementary devices the analytical apparatus may also be used for preparative purposes. In general the preparative equipment is specially designed for this end with extensive automation of sample introduction and fraction collection.

A new trend in the development of apparatus is represented by those special types which though working on the gas chromatographic principle are not aimed at separation and all of which may be classified into a fourth group. These instruments are devised for various purposes and their most important representatives are the analyzers for the micro-determination of C, H and N and the sorptometers for the rapid measurement of the specific surface areas of solids.

In this chapter we wish to deal only with the general description of analytical apparatus and trends in apparatus development. The construction and operation of preparative and process chromatographs are discussed in the appropriate chapters. The most important special applications of the gas chromatography are the subject of a special chapter which deals with the problem of the development of appropriate apparatus.

The development of analytical instruments has gained a considerable impetus in the 1960's when a number of new instrument manufacturers began marketing gas chromatographic and ancillary equipment. This last period witnessed significant progress in the performance and accuracy of the apparatus and in the versatility of the equipment due to new detectors and auxiliary equipments.

At the first symposium of ISA in 1957 A.J.P. Martin, one of the inventors of GLC, expressed his view on the urgent necessity for cheap and simple gas chromatographs which would help to spread the method. However, development in apparatus took another course and the new types are more versatile, complicated and expensive. This trend is primarily because along with the solution of complicated separation and qualitative analytical problems, there was an increasing demand for higher accuracy, reproducibility and sensitivity which raised new demands which the apparatus had to meet.

For quantitative analysis the apparatus must provide for separation and for the production of signals corresponding to the various components (the chromatogram) and also reliable operation of digital integrators and data processors which may be equally important for the calculation of the composition.

Today it is no longer enough to be acquainted with the fundamental correlations of gas chromatography if one wants to select the appropriate apparatus for various analytical tasks and to operate it optimally. The practising gas chromatographer has to acquire more and more knowledge about the mechanical, pneumatic, thermal and electronic systems which are incorporated in the complete gas chromatographic equipment to be able to form a comprehensive view on the possibilities and limits of the method.

In this chapter we shall attempt to survey briefly the most important components of the analytical gas chromatograph and the present state of its development. From the aspect of the general trend in apparatus development laboratory gas chromatographs can be classified into the following groups:

a) Simple, cheap isothermal apparatus with lower performance intended for the solution of the most frequently encountered routine analyses. Some of them can be used for gas analysis only (Hewlett Packard F. & M. Model 100; Precision Scientific Co. Chronofract, etc.), while with others the analysis of both gases and liquids can be performed up to 200–300 °C (F. & M. Model 700, Carlo Erba Model M, Beckman GC 2A, Becker Unigraph, etc.).

b) Versatile universal apparatus with isothermal and programmed heating and interchangeable parts (packed and open tubular columns, several types of detectors) and a great number of ancillary devices (e.g. automatic device for attenuation, microreactors, pyrolysis units, various injectors, etc.). These are devised mainly for research and are suitable for the solution of manifold analytical problems and for the study of the structure of chemical compounds (e.g. Perkin–Elmer types F 6 and F 11, Beckman GC 5, Pye Series 104, Varian Aerograph 1520, Philips PV 4000, Barber Colman Series 5000, etc.).

c) Special high precision and highly sensitive apparatus devised mainly for research (e.g. Perkin–Elmer 801 and 880, F. & M. 1720 and 400, Varian Aerograph 400).

d) Special types of apparatus have been evolved lately for biochemical and forensic analyses in which the decomposition sensitive sample is in contact only with glass components in the apparatus (e.g. Perkin–Elmer 811, Carlo Erba GB, Microtek MT 200).

With respect to their general construction and the arrangement of their parts the commercial gas chromatographs are extremely varied. From the point of view of general construction the apparatus can be classified into four groups:

1. All parts are built into a single unit or cabinet (e.g. Beckman GC 4, Microtek MT 200, F. & M. 7508).
2. The thermostat, controls and electrical parts are in one cabinet, while the recorder forms a separate unit (e.g. Varian Aerograph 1520, Beckman GC–M).
3. The thermostat, the control device and the electric and recording parts form three separate units (e.g. Perkin–Elmer 226 and 800).
4. Parts with different functions are on separate racks and the apparatus is built of modules which can be piled one upon the other or pushed like drawers into a common frame (e.g. Perkin–Elmer F 11 and F 6, Beckman GC 5, Philips PV 4000).

The latest trend in development is the apparatus built of modules, (paragraph 4) though constructions of the other three types are still to be found among new equipment. Construction from modules allows many different arrangements and the connection of many ancillary devices. The latest types can be operated in 20–30 variations. From the point of view of the customer this construction has the great advantage that, on the one hand, he can buy the basic unit and parts necessary for his particular problem, while on the other hand,

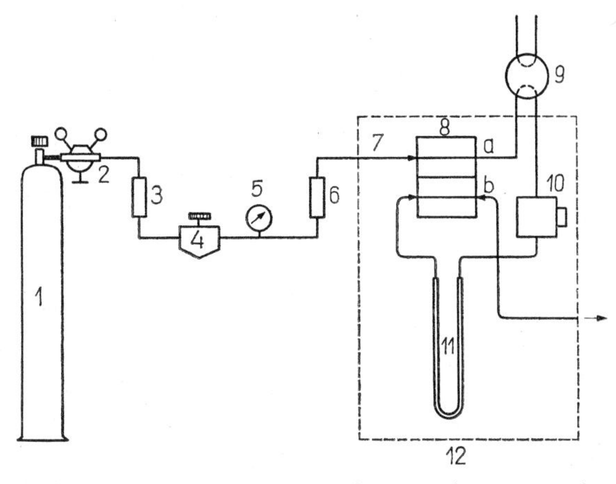

Fig. 4.1. Schematic diagram of a gas chromatography apparatus. 1. Carrier gas cylinder; 2. reducing valve; 3. gas purifier unit; 4. flow regulator; 5. manometer; 6. flowmeter; 7. carrier gas pre-heater; 8. thermal conductivity cell; a) reference branch, b) measuring branch; 9. gas sampling valve; 10. liquid sample injector head; 11. column; 12. thermostat

he can expand the performance of his apparatus to new tasks quite simply by buying the other necessary units.

Choice of analyzer unit depends largely on whether analysis will be performed under isothermal conditions or by programmed temperature. There are two basic types of analyzer units, namely single and dual channel arrangement. The dual channel type is essentially the arrangement of two independent parts of the analyzer in a common thermostat and can be combined in different ways for the solution of certain tasks.

Figure 4.1 shows the block diagram of a simple gas chromatography apparatus. The main parts of the laboratory analyzer were already mentioned in Chapter 1.

In the following sections we shall examine the demands which arise in connection with the apparatus and its operation conditions. Scott[1] has presented a general survey of the points to be considered in the selection of the apparatus and has listed the most important parameters of a great number of gas chromatographs of different types.[1, 2]

4.2 Carrier gas system

The carrier gas ensures the migration of the components of the sample to be separated. The high performance of gas chromatography is based on the fact that the moving phase is gaseous, thus enabling a rapid equilibration to take place between the moving and stationary phases. Because of the low flow resistance, long columns with high separating powers can be used and the components in the gas emerging from the column can be detected by simple and highly sensitive methods. The quality and purity of the carrier gas are important factors here, as are the control and measuring of the flow rate. To ensure constant and reproducible flow conditions of the carrier gas its system is composed of the following parts: gas cylinder or generator, purifier, pressure and/or flow control, pressure measuring device (manometer), flowmeter and pre-heater.

Nature of the carrier gas. In principle any gas which does not interact either with the stationary phase or the components of the sample would be a suitable carrier gas. However, the properties of the carrier gas affect separation in the column as well as detection of the emerging components. The effect of the carrier gas on separation will be discussed in Chapter 5 in considering column operation. In selecting a carrier gas detection is the primary factor, as there are other means available to improve separation (more effective packing, more selective stationary phase, or longer column).

In spite of the great variety of detectors which has been invented and marketed in recent years many gas chromatographs still operate with a katharometer or thermal conductivity cell. This type of detector measures the difference between the heat conductivity of the pure carrier gas and that containing the components of the sample. The greater the difference between the heat conductivity of the carrier gas and of the components the greater will be the observed signal. As known from kinetic gas theory the thermal conductivity of gases is inversely proportional to the square root of their molecular weight, and consequently hydrogen and helium are the most suitable carrier gases for thermal conductivity detection as their thermal conductivity is almost ten times higher than that of other gases and vapours. Because of the reactivity and inflammability of hydrogen, helium is the ideal carrier gas. Owing to its easy accessibility and low price in the USA, in that country helium is the most widely used carrier gas, but in Europe hydrogen and nitrogen are normally used, though recently the application of helium is spreading there too.

Nitrogen is easily available and handled, but its thermal conductivity differs only slightly from that of the components under investigation which is a disadvantage from the point of view of both sensitivity and quantitative analysis.

The use of argon as a carrier gas spread with the development of β-radiation ionization detectors. The other important application of this gas is in the gas density balance. As high purity argon can be relatively easily and cheaply purchased its use is on the increase. With radioactive detectors helium and neon may be used with advantage for certain tasks.

Carbon dioxide has been used as a carrier gas with detectors operating on the absorption principle (Janák's method) and more recently with gas density balance detectors.

The thermal conductivity detector (katharometer) and the flame ionization detector are the most widely used. The latter is based on the combustion of the components in a hydrogen flame of a microburner and the measurement of the electrical conductivity of the flame. The hydrogen and oxygen (air) necessary to maintain the flame are usually supplied by a separate gas system, while in the column argon or nitrogen is a convenient carrier gas.

In certain special cases oxygen or compressed air may also be used as carrier gas, but their use is limited by the possibility of reactions with the stationary phase or the components of the sample.

Gas cylinder or generator. In most cases the carrier gas comes from commercial 10–40 litre cylinders which can be joined through a reducing valve to the carrier gas system of the chromatograph. When hydrogen is used the cylinder can be replaced by recently evolved electrolytic hydrogen generators (Varian Aerograph, Milton Roy Co., etc.) which produce high purity hydrogen at a controlled rate. When carbon dioxide is chosen as carrier gas dry ice cylinders are preferred to the older *in situ* production methods.[3]

Purification of the gas. Technical grade gases contain various impurities of which oil vapours, oxygen and water are the most detrimental in gas chromatography. These impurities reduce the activity and life of the column, and may cause disturbances in the functioning of the detector whereby quantitative evaluation will be either difficult or impossible.

Various methods are applied to the purification of the carrier gas. Oxygen is most efficiently removed by a catalyst, e.g. in reactors packed with copper or nickel catalysts, oxygen is practically completely removed at 100–105 °C. At higher temperatures (600 °C) these catalysts are also suitable for the removal of carbon monoxide, methane and traces of other hydrocarbons. The oxygen impurity in hydrogen can be removed at room temperature by passing the gas over a palladium catalyst. By these catalytic methods the oxygen content is reduced to values below 1 ppm.

Carbon dioxide is removed by passing the gas through a tube filled with soda lime or soda asbestos.

For the removal of oil vapours and other heavy contaminants active charcoal adsorbers are most useful. Adsorption on a molecular sieve (4A, 5A, 13X) is the most efficient way to remove water vapours. When purity requirements are very high, the active charcoal and molecular sieve adsorbers are operated at low temperatures (−40 to −180 °C), though according to general practice their operation at room temperature is quite satisfactory. Active charcoal and molecular sieves should be incorporated into the same adsorber so that the entering gas comes in contact first with the charcoal bed. The adsorbers must be regenerated from time to time, this regeneration is in the case of active charcoal by heating at 300–350 °C for two to three hours *in vacuo*, and in the case of molecular sieves by treatment for four to five hours at 400–450 °C *in vacuo*.

Recently a special adsorption cartridge appeared on the market (Multiform Desiccant Products Inc.). This cartridge is filled with various adsorbers and is suitable for the removal of water, oil and other impurities down to 1 ppm values.

Some commercial gas chromatographs are not provided with a gas purifying unit, though generally the instructions attached to the apparatus give suggestions how to purify the carrier gas.

Pressure and flow control. The constant value of the flow rate of the carrier gas is of paramount importance in gas chromatographic analysis. Some detectors are sensitive to changes in flow rate which cause a drift of the base line and raise difficulties to both qualitative and quantitative evaluation. Changes in gas velocity alter retention data and affect the identification of the components, while on the other hand through changes in the areas below the peaks they greatly reduce the accuracy of quantitative evaluations. For an accuracy over

1% the flow rate has to be controlled at a constant value within ± 0.5–1%.[4] To achieve an accuracy of this degree of flow rate is no small task. The realization of a constant flow rate is further impeded by the fact that the flow resistance of the column changes with the analytical conditions and the physical state (particle size) of the packing. Thus, to eliminate the effect of changes in column resistance the volume rate of the carrier gas has to be adjusted to a constant value before the gas enters the column. The constancy of the flow rate can be ensured either by the adjustment of the pressure or of the flow rate.

Pressure control. The simplest way to keep the flow rate at a constant value is to produce such a large pressure drop outside the column that compared to it the resistance change within the column is negligible. This large pressure drop can be realized by the inclusion of a metal capillary or a precision needle valve in the apparatus. If accurate and sensitive control is required the needle valve has to meet a stringent specification of both material and machining. The poly-fluorohydrocarbon plastics have been found to be the most satisfactory for sealing the stuffing box of such flow regulators.

To keep flow rate at a constant value another pressure regulator must be inserted before the needle valve to ensure a high (at least 10 atm) outlet pressure. The regulators on the gas cylinders are quite adequate for this purpose. For more accurate measurements special pressure regulators with membranes and spherical or conical valves are used. To achieve constant flow rates membrane pressure regulators are usually embodied in commercial equipment. In this case the outside pressure drop is provided by the regulator on the gas cylinder.[5] Pressure control is applied in analyses under isothermal conditions, but is also necessary when a stream splitter is incorporated in the system. In the case of programmed temperature the pressure drop in the column changes to such a degree that pressure control is no longer sufficient for ensuring a constant flow rate. In this case flow regulators are used.

Flow control. The membrane flow regulator shown in Fig. 4.2 is suitable for keeping the flow rate of the carrier gas at a constant value.[6] Modern flow regulators work within wide limits (between 1 and 500 ml/min) and are capable of maintaining the flow rate at a constant value with ±1% accuracy even when the inlet pressure is doubled. In recent years the method of programmed carrier gas flow has been evolved which implies a gradual increase in flow rate during analysis.

Special flow programming units have been worked out for this and these can be obtained as ancillary equipment and directly incorporated in the carrier gas system. In isothermal analysis the flow regulator may be superfluous and cannot be used with stream splitting. Up-to-date equipment is provided with both pressure and flow regulators.

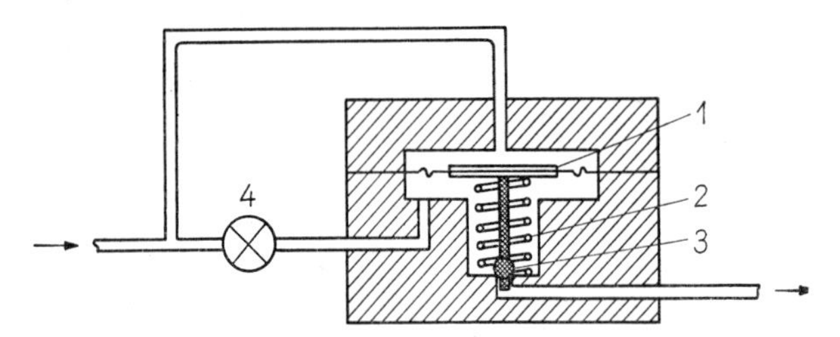

Fig. 4.2. Schematic diagram of a diaphragm flow regulator. 1. Diaphragm; 2. spring; 3. ball valve; 4. needle valve

When flame ionization detectors are used the hydrogen and air which are led to the detector are controlled by separate pressure regulators. In dual channel or dual detector apparatus the carrier gas as well as the hydrogen and air are controlled by dual pressure and flow regulators.

Measurement of the pressure. For the measurement of pressure any sufficiently accurate manometer can be used. The latest type of apparatus is exclusively provided with aneroid manometers; their range is generally between 0 and 3 atm, and more recently between 0 and 7 atm (100 lb/in^2). The pressure and flow regulating units of the carrier gas as well as those of the hydrogen and air are provided with separate manometers. The carrier gas manometer is inserted before the inlet of the column thus showing the inlet pressure. Outlet pressure can usually be taken as atmospheric. If a restriction is incorporated after the column then the outlet pressure must also be recorded for the correction of the retention data.

Measurement of the flow rate. Flowmeters operating by various principles can be used for the measurement of flow rate. Measurement can be either continuous or intermittent. The flowmeter can be fitted either before the column or at the carrier gas outlet. The usual flowmeters are: capillary differential flowmeters, flowmeters based on the measurement of thermal conductivity, ionization flowmeters, rotameters and soap film flowmeters. In commercial equipment the two last types are used almost exclusively. In some equipment the carrier gas system is fitted with a rotameter before the column, this

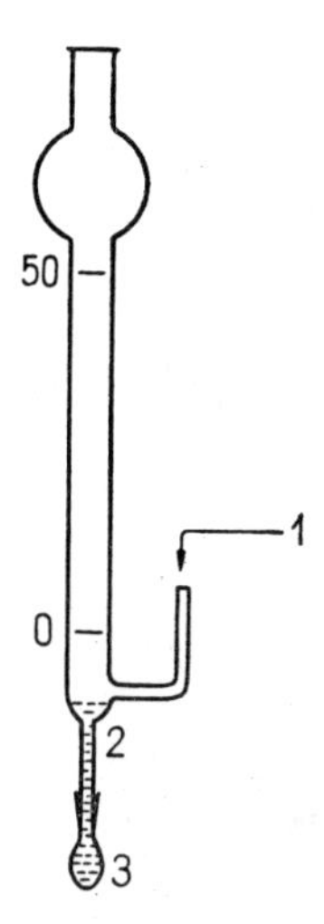

Fig. 4.3. Soap film flowmeter. 1. Gas inlet; 2. soap solution; 3. rubber bulb

however has the disadvantage that the indicated values will depend on the absolute pressure and also on the nature of the carrier gas.

The soap film flowmeter is a calibrated glass tube into which a soap film is blown by the gas stream. By measuring the progress of the membrane in the tube with a stopwatch the volume rate is obtained independently of the nature of the gas. The soap film flowmeter can be joined to the effluent carrier gas system, and so will function at atmospheric pressure. Figure 4.3 shows a very simple flowmeter. The soap film flowmeters can have different volumes, for packed columns their volume is usually 50–100 ml, for capillary columns 5–10 ml, glass tubes. The film can be formed from tensile solutions. The soap film flowmeter can be operated intermittently, is extremely simple and accurate and requires no calibration. This is the most common type of flowmeter in the latest equipment, but some kinds of apparatus are provided with both rotameters and soap film flowmeters.

Heating the carrier gas. In the case of katharometers, to ensure identical conditions in the reference and measuring branches the carrier gas must be heated to the temperature of the thermostat before it enters the reference branch of the detector. Heating the carrier gas before it enters the sample injector is advisable also in the case of other detection methods. The pre-heater is a U-shaped or helical tube hanging in the air space of the thermostat or wound around the detector. When liquid samples are analyzed it is advisable to provide the effluent gas line with a special heating system. This serves to prevent the condensation of high boiling components and blocking of the outlets and is further necessary when preparative fraction collectors are joined to the apparatus to ensure efficient removal of the separated components or fractions.

4.3 Thermostats

The migration rates of the components through the column depend on their partition coefficients between the moving and the stationary phase. These coefficients are temperature dependent. The migration rates of the components increase with increasing temperature, while at higher temperatures the separating power of the column decreases. Temperature should be chosen in such a way that the vapour pressure of the heaviest component shall be at least 5–10 mm Hg to ensure at the same time the highest speed of analysis and best separation efficiency. The effect of temperature on analysis will be discussed in the chapter on column operation. The constancy as well as the control of temperature is also of paramount importance for both qualitative identifications and accurate quantitative analysis. The temperature of the column must be kept constant within ± 0.1–0.5 °C, this tolerance depends on the temperature range.

Not only the column, but the detector, too, must be kept at a constant temperature to ensure adequate accuracy. In addition all parts of the gas system which are in contact with the sample must be kept at a constant temperature; the appropriate temperature of the sample injector is of special importance.

In the construction of the thermostat the following aspects must be borne in mind: a) ample effective space and accessibility for column assembly; b) satisfactory external heat insulation; c) the lowest possible heat capacity; d) maximum temperature at least 300 °C, often 500 °C; e) uniform temperature distribution; f) accurate temperature control.

In the case of programmed temperature rapid heating and cooling, including control of the heating rate and its adjustment according to the various programs, are added to the requirements a) to f). Programmed temperature requires that the detector should be kept at a constant temperature, consequently the thermostats of the column and of the detector must form separate units. The third heated unit is the sample injector.

In the first stage of the development of gas chromatography both liquid and vapour thermostats were used, mainly in "home-made" apparatus. In the latest apparatus metal block or air thermostats are employed; in the majority of commercial chromatographs, air thermostats.

The metal block thermostat has its advantages in isothermal operations. The column is placed in a hole drilled in the metal block, or between two blocks or is wound on a metal cylinder. Because of the considerable heat capacity of the metal block temperature fluctuations are slight and the high thermal conductivity of the metal ensures uniform temperature in the entire column. Usually aluminium blocks are employed. With metal thermostat it is possible to keep the temperature accurately and uniformly on the required level but with the disadvantage of a long heating period and an even longer cooling period. (The Pye Argon Chromatograph needs about twelve hours to cool down from 200 °C to room temperature.) For these reasons metal block thermostats cannot be used with programmed temperature. The apparatus of Glowall Corp. represents a novelty in this field as it applies a cylindrical metal block thermostat with low heat capacity with an air gap between the block and the column where the constant temperature is provided by compressed air. This solution is applicable to programmed temperature with rapid cooling through compressed air.

To thermostat heat conductivity cells smaller metal block thermostats with separate heating and thermometer devices are used. Quite often the head of the sample injector is also placed in a smaller electrically heated metal block, though for this end electrical heating wound directly on the head is quite satisfactory. If necessary the flame ionization or electron capture detectors are also provided with similar heating jackets.

In practice air thermostats were found to be the most satisfactory and the majority of commercial designs employ air thermostats.

They have the advantage of simple construction coupled with large air space and short heating and cooling periods. The simplest form of an air thermostat resembles the laboratory drying cabinet. A simple gas chromatograph mounted in a drying cabinet with sufficient air space and temperature control can quite easily be constructed in the laboratory.

For a uniform temperature distribution rapid air circulation has to be realized within the thermostat, because of the low specific heat of air, there may be large temperature differences within the thermostat. The older air thermostats all suffered from the drawback of uneven internal temperature. The new types provide for the rapid exchange of air so that the air space of the thermostat is exchanged six to eight times in one second. There is of course always a minor temperature gradient which however can be kept at a minimum by rapid air circulation and, the construction of the thermostat including the proper arrangement of the heaters and of the air circulating ventilator. A temperature gradient of 1–2 °C in the air space causes only a negligible fluctuation in column temperature.

The construction of the air thermostats is different in different models with respect to shape, arrangement of the heater and ventilator, and temperature control. The volume of the air space also varies and is in the case of spiral columns 3–5 litre, and 8–15 litre in the case of U-shaped columns. The shape of the thermostat depends on the shape of the column which will be inserted into it. For spiral columns to fit on a bench top, the thermostat has the shape of a cube (Aerograph 1520) or oblong box (Perkin–Elmer 800), in some chromatographs it is cylindrical (Kovo Chrom 3). In the case of U-shaped columns the thermostat may be an upright oblong, usually in a form to be placed directly on the floor (Becker 1452). The ventilator is at the bottom, top or side of the thermostat, the motor outside the insulation. Air circulates either upwards or downwards in an air channel or direct air circulation is used. The heaters are on the side in the air channels or directly in front of the ventilator. To achieve uniform heating several heaters fixed on the side of the thermostat are the most satisfactory. The incorporated thermal power is 800–1200 W in isothermal apparatus and 1500–2000 W in programmed temperature for rapid temperature increase and will obviously depend on the size of the thermostat. Figure 4.4 shows a diagram of the two most commonly used arrangements.

In earlier equipment the maximum temperature was 250–300 °C, but in later types the upper temperature limit reaches 400–500 °C, with special apparatus for higher temperatures. It should be noted here that such an increase of the upper temperature limit does not

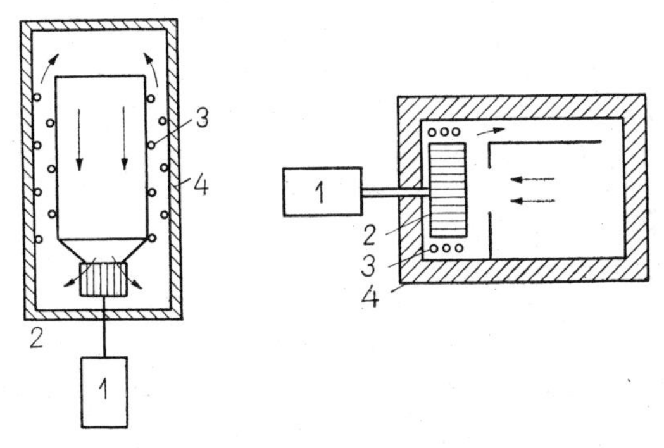

Fig. 4.4. Schematic arrangement of an air thermostat. 1. Motor; 2. ventilator; 3. heating elements; 4. heat insulation

seem to be justified from the practical aspect, partly because there are very few substances which can be analyzed without decomposition at 400–500 °C, and partly because there are but few packings which can be used at temperatures above 300–350 °C.

Efficient heat insulation is indispensable for the satisfactory operation of the thermostat. The most commonly used insulations are the air gap, 40–80 mm thick glass wool, Fuller's earth or glass foam. In the case of a temperature of 500 °C in the air space the temperature of the external wall of the apparatus must not exceed 40–50 °C. Heat insulation is of particular importance from the point of view of the protection of the pneumatic and electrical parts in contact with the thermostat.

The accuracy of temperature control is the most important parameter of thermostat operation. Control may be intermittent or continuous. For heat sensing contact thermometers, thermoelements or resistance thermometers can be employed. In air thermostats heating is usually in two stages. The temperature is raised with the maximum power and once the desired temperature is reached heating is controlled with 300–400 W. The heat senser is located in the vicinity of the heating elements. In the newest equipment continuous control proportional to the performance is employed almost exclusively which eliminates fluctuations due to switching on and off. In the case of continuous control the heat senser is connected to an electronic control unit. In commercial equipment the senser controls the temperature with an accuracy of 0.1–0.5 °C, while for some apparatus it is supposed to be even more accurate. The manufacturers quote an accuracy of ±0.05–0.1 °C for the constancy of column temperature measured on a metal column. The geometrical arrangement of the column or columns is also an important factor from the point of view of uniform and constant column temperature, as part of the column may shield some other part from the circulating air, while any accidental direct heat radiation may cause local overheating.

In the case of programmed temperature the apparatus contains a special heat programming unit which ensures the desired heating rate. In general the temperature program is performed at a 0.2–20 °C/min rate; in some apparatus the maximum rate may reach 40–50 °C/min. Programmed temperature stresses the importance of the rapid cooling of the thermostat after analysis to prepare the apparatus for its next job. Rapid cooling can be achieved by opening the door of the thermostat and circulating cold air through it; in the latest types the apparatus is cooled from 300–400 °C to room temperature within 5–10 min.

Some of the latest chromatographs are provided not only with a high temperature but also with a low temperature thermostat for analysis down to –100 or even –180 °C (e.g. Beckman GC 4). In some apparatus the air thermostat is constructed in such a way that it can be used down to –190 °C with the help of liquid air or nitrogen and is also suitable for temperature programming (Victoreen Series 4000).

4.4 Sample injector

The task of the sample injector system is to introduce the gaseous, liquid or solid sample rapidly and in a reproducible manner into the column. The efficiency of separation and the accuracy of the results depend on the manner in which the sample is introduced into the column and on the construction of the sample injector. The sample injector system should be constructed in such a way that a) the sample shall reach the column like a plug in the shortest possible time; b) there shall be no change in the flow and thermal conditions of the system during the injection of the sample and c) the quantity of the sample and the manner of injection, i.e. the process in time shall be perfectly reproducible. There is a considerable difference between the manner of sample injection and the injecting system itself for packed and open tubular columns. While in packed columns the quantity of e.g. a liquid sample is 1–10 μl, into open tubular columns only one hundredth to one thousandth part of this quantity is injected. The direct dosage of such very small quantities is extremely difficult, so that a larger quantity is originally injected which is then divided by a stream splitter in such a way that only one-hundredth to one-thousandth part of the total gas stream reaches the column. The stream splitter is built in after the injector head. We shall first deal with sample injectors designed for packed columns and separately with injectors suitable for open tubular column operations.

The sample injector systems can be classified according to the physical state of the sample. A great number of publications deal with the special injector heads and injection methods evolved in different laboratories. Many of these are quoted in Purnell's book.[7] In this chapter we shall describe only those injection systems which were found the most satisfactory in practice and which are generally used in commercial equipment.

Gas sampling. The quantity of the injected gas sample varies usually between 0.5 and 5 ml; in the case of highly sensitive detectors the sample may be 0.01 ml, while in trace analysis 10–20 ml samples may occur.

Gas sampling valve. This is the most generally used method for the injection of the gas sample. The first devices were glass tubes with a by-pass in which the measuring of the sample took place in the volume bordered by two stopcocks and the sample was swept into the column by switching into the carrier gas stream.[8] Commercial development has brought small metal injectors which are easy to operate. These injectors contain rotating or sliding valves; in the case of a sliding valve this is placed into a cylindrical space and when moved forward or backward it closes or opens the bores of the adjoining tubes. Efficient sealing is achieved by rubber sealing rings which move together with the shaft.[9] The reproducibility of the injection is $\pm$ 0.2%.

In spite of its relatively simple construction this type is not often used and in commercial equipment mainly rotating valves are found.

There are two main types of rotating valve. One type is made of metal only with both the rotating and the fixed part of tempered,

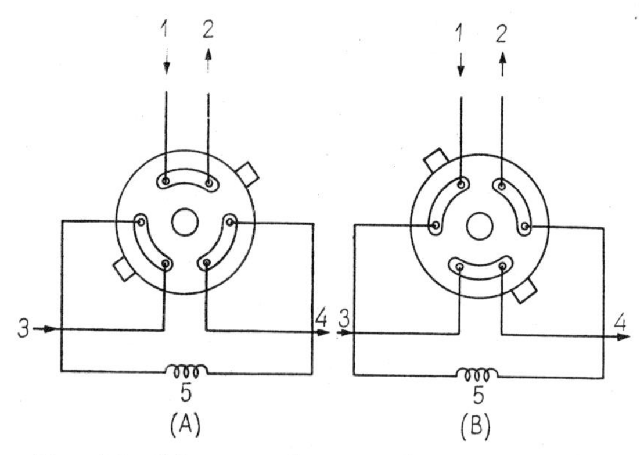

Fig. 4.5. Schematic diagram of a gas sampling valve. (A) Sampling position; (B) injection position; 1. carrier gas inlet; 2. carrier gas outlet; 3. sample inlet; 4. sample outlet; 5. sample volume

polished stainless steel. The parts are in contact over a large surface and lubrication is in principle unnecessary. Tubes with different sample volumes can be fitted to the rotating part. In the latest valves the rotating part is made of Teflon and the fixed part of metal. Teflon provides gas sealing without lubrication. Helices with different sample volumes can be joined to the valves. The sampling valves have 4, 6 or 8 ports. The schematic diagram of a sampling valve is shown in Fig. 4.5. The bore of 0.01–0.1 ml volume in the Teflon body is the sampling volume.

The gas sampling valve is mounted in the carrier gas system before the liquid sample injector and can be fixed to the front of the apparatus. With most gas chromatographs a gas sampling valve is included as a special ancillary device. When the valve is not mounted the carrier gas system can be short circuited with a capillary tube.

The gas sampling valve is easy to operate and ensures reproducible and readily altered sampling. By increasing the number of ports or by the simultaneous use of two valves it may be converted to special tasks, e.g. in combination with a pre-column or a microreactor. The further advantage of this valve is that it may be adapted relatively easily to automatic injection.

Gas sampling with a hypodermic syringe. The gas is sucked into the syringe whose needle is then pushed through the rubber septum which closes the top of the column, then the gas is injected into the carrier gas stream. Earlier this was a commonly used method, but it is not sufficiently accurate. The common hypodermic syringes are not perfectly gas tight and require some lubrication. Syringes coated with Teflon requiring no lubrication and provided with stainless steel pistons appeared recently on the market. With these syringes the reproducibility of sample injection may reach $\pm 1\%$.[10] The volumes of the syringes vary between 0.05 and 50 ml and are produced in a form which provides also for rinsing. The injection head is the same as that used for the introduction of liquids.

Liquid sampling. The introduction of liquid samples into the column is a far more difficult task, partly because of the very small sample size (usually between 1 and 10 μl) and partly because the sample must be evaporated rapidly and completely. The too early (prior to the introduction of the sample) partial evaporation of the lighter components or the adhesion of the heavy components to the walls of the sample injector or their too slow evaporation may cause an alteration in the composition and quantity of the sample. The sample injector head for the introduction of the liquid sample is mounted directly before the column inlet in such a way that there should be the smallest possible dead volume between the injector head and the column, and the carrier gas should sweep the vapour of the liquid directly and rapidly into the column.

Several methods and injector head constructions have been tried for the introduction of liquid samples; today sampling with precision syringes is accepted as the simplest and most reliable method which is now practically the only one used in commercial apparatus. Consequently we shall deal with this method in detail and touch only briefly upon some other solutions.

Construction of the sampling head and the method of sampling. In the older types of gas chromatographs and sampling methods, sample injection was performed by the rapid evaporation of the liquid sample and by sweeping the vapour into the column. Evidence collected in recent years indicates that this method is not always satisfactory.

In the case of samples with a wide boiling range or those containing heavy components the lack of uniformity in evaporation has a detrimental effect on both separation efficiency and the accuracy of analysis.[11] Evaporation at high temperature may induce the thermal degradation of the sample.[1] To eliminate these detrimental effects the liquid sample is fed directly on to the top of the column, 1–2 cm below the top of the packing. Thus there are two ways of sample injection, namely by evaporation or on column. The syringe is introduced in both cases through the rubber septum, but the construction of the sampling head is different. A further difficulty is encountered in the introduction of samples of certain substances (e.g. steroids, amino acids) by their decomposition at the temperature of sampling, even in the liquid state when in contact with the metal components of the sample injector. For this reason the column should not be made of metal but of glass. When analyzing any of these substances the sample injector head should be constructed in such a way that the sample should be in contact with the glass parts only. The outlet end of the column should be joined in a similar way to the detector, so that the glass column shall protrude into the inside of the detector. Figure 4.6 shows diagrams of the two types of sampling heads. The evaporation type sample injector (A) is placed in an aluminium block with controlled electric heating. The temperature of the sample injector is usually kept at a temperature 30–50 °C higher (in some cases 100 °C higher) than the temperature of analysis to ensure rapid and complete evaporation of the sample. The sample injector should be constructed in such a way that the rubber septum made of silicone elastomer shall suffer no thermal damage.

The latest chromatographs are provided with three exchangeable types of sample injectors (e.g. Philips PV 4000). The evaporation type sample injector may also be constructed in such a way that when heating is switched off the device is ready for on column injection. The inside of the injector provides for the rapid and efficient purging of the gas space.

Syringes. Syringes were first used by Ray,[12] but because of their imperfect sealing and large volumes the hypodermic medical syringes were used only in the early days of gas chromatography. In later years the method of sample introduction and the syringes have undergone considerable development.[13] Special gas chromatography microsyringes are now available which are provided with polished stainless steel or Teflon pistons and interchangeable needles of special shapes. The syringes manufactured by Hamilton Ltd. are the most widely used, they are marketed with 1–500 μl volumes.[10] Syringes of 1, 10 and 50 μl volumes are the most common in gas chromatography, With a

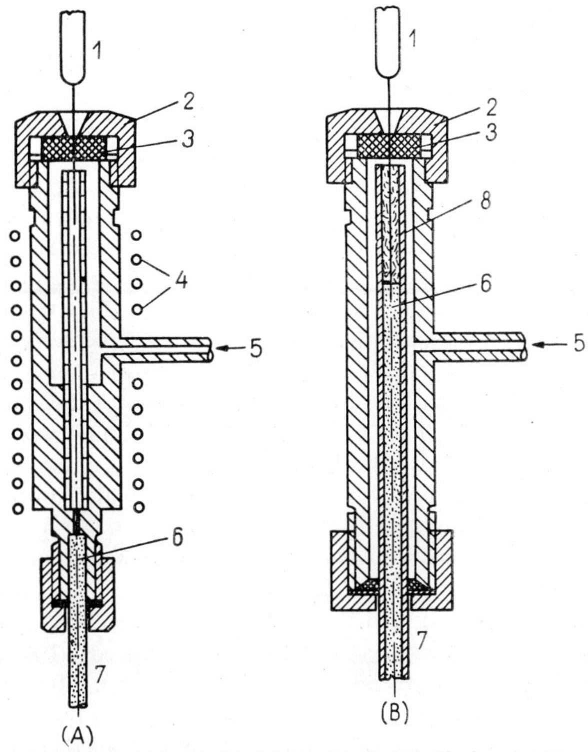

Fig. 4.6. Liquid sample injector head. (A) Flash evaporation; (B) on column injection; 1. syringe; 2. septum cover; 3. silicone rubber sheet; 4. heating; 5. carrier gas inlet; 6. packing; 7. column; 8. glass wool

1 μl syringe a sample of 0.01 μl can be directly measured. To ensure reproducibility the syringes may be fitted with a micrometer (Chaney) adaptor which provides for the preliminary adjustment of sample quantity. The accuracy and reproducibility of these precision syringes are below 1% and with care 0.5% reproducibility may be achieved.

Special syringes with cooling or heating jackets are available for various purposes.[10]

With the precision syringes liquid samples can be injected simply and in a reproducible manner even against high pressures (5–6 atm). The injector is usually sealed with a soft silicone rubber septum. The method has the disadvantage that after 20–30 injections the elastomer becomes gas permeable and has to be exchanged. The new Hamilton sealing rubber is a great advance in this field; this septum is built of three layers, one thick, soft silicone rubber layer sandwiched between two thin hard silicone rubber layers. This arrangement ensures good sealing and a long life.

Other liquid sample injectors. Glass micropipettes are also used for the introduction of liquid samples. When the sample is introduced with a pipette the ground glass joint in the carrier gas stream has to be disconnected at the head of the column and the contents of the pipette blown on the packing. This method is cumbersome and with many detectors impracticable. The microdipper method worked out by Tenney and Harris[14] ensures the introduction of the sample in a closed system in which the carrier gas blows the liquid with the help of the injector from a hollow metal rod which is provided with a capillary bore on to the column. The injector is sealed on the side in contact with the atmosphere.

Sample injection with sealed ampoules. This method is used in the analysis of liquids which are sensitive to moisture or when a highly accurate dosage is required.[15, 16] The sample is sucked into a glass capillary which is then sealed and accurately weighed. This capillary is introduced into a device at the top of the column. After equilibration of the flow and heat conditions the capillary is broken and the evaporating liquid reaches the column. A capillary made of low melting metal may also be used instead of the glass capillary. In the heated space the metal ampoule melts and the liquid evaporates.[17] Solid substances can also be introduced by this method.

Injection as a gas. The liquid is evaporated not after introduction into the sample injector but in an external injector system and the gaseous sample or part of it is introduced into the column.[18] This method is aimed at further improvement in the injection of very small vapour quantities, e.g. direct injection into open tubular columns,[19] and is suitable mainly for the injection of low boiling point (100–150 °C) liquids, but is less reliable for high boiling point or wide boiling range samples. This method is highly convenient for the analysis of volatile liquids, e.g. liquified gases or gasoline.[20]

Introduction of solid samples. Theoretically there are three possibilities for introducing solid samples into the gas chromatography column.

a) *Introduction in solution.* The solid substance is dissolved in an appropriate solvent and the solution is injected by one of the methods used for the injection of liquid samples.[21] The solvent should be chosen in such a way that it shall not interfere with the detection of the sample components. The method has the advantage that depending on the degree of dilution very small quantities of the solid sample can be injected.

b) *Introduction of melts.* Low melting point substances are fused prior to injection and introduced with a heated syringe or micropipette in liquid form.[22] Some special metal syringes have been developed for the injection of melts.[10]

c) *Direct injection.* For the direct injection of solid substances the capillary breaking device is quite satisfactory in the case of both glass capillaries[23] and low melting metal capillaries.[17] Injectors for solid samples working on a similar principle are commercially available; these can be fitted instead of the liquid sample injector (Hewlett Packard F. & M. Model SI–4). Evaporation in the injector tube is another solution for the introduction of volatile solid substances. The sample may be evaporated in a stationary gas or in the gas flow.[24] This injector system may also be used as a pyrolysing fixture.

The injection of viscous substances may be performed in a manner similar to the injection of solids.

Injection of the sample into the open tubular column. The quantity of sample which can be injected into the open tubular column is 10–100 μl of gas or 0.01–0.001 μl of liquid. Special sampling valves with very small bores are available for the injection of microquantities of gases. Some special systems were worked out for the direct injection of liquid samples, though the evaporation technique is the most frequently used. Several practical solutions have evolved since Desty *et al.*[25] first worked out the evaporation injection method. This method consists essentially in leading the carrier gas which contains the sample from the sample injector into a smaller expansion and mixing chamber from where it leaves by two outlets with different resistances. These resistances are adjusted in such a way that only

0.01–0.001th part of the carrier gas enters the column and the rest is blown off to waste. Figure 4.7 shows two different stream splitters. In the first stream splitter the proportion of the streams changes considerably depending on the conditions of column operation. For a

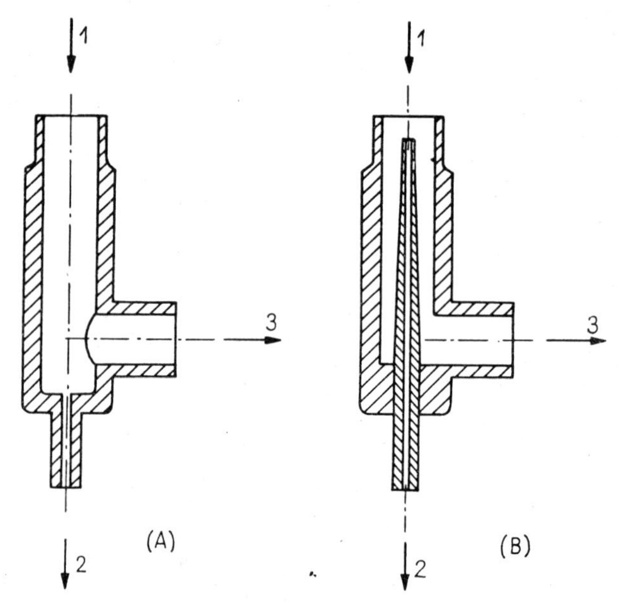

Fig. 4.7. Schematic diagram of a stream splitter. (A) Old type; (B) new annular type; 1. carrier gas + sample inlet; 2. to open tubular column; 3. blowing off (through adjustable choke)

linear distribution in a wide temperature and molecular weight range and for the "plug-like" feeding of the sample into the open tubular column the proper construction of the stream splitter is of paramount importance. In the latest apparatuses the stream splitters have variable resistances which makes possible highly accurate stream splitting.

Stream splitters of a similar type are also needed after the column, when two parallel detectors or a packed column and highly sensitive ionization detector are used. Stream splitters should be used with destructive (e.g. flame ionization) detectors when the separated components are subjected to further tests.

Special sample injection. Injection of a gas sample from a system under vacuum presents a special problem for which several methods are available.[26, 27]

Analysis of gases dissolved in liquids is a new field in gas analysis, and several devices have been constructed for the solution of the problems involved. These devices are fitted before the sampling valve. From the small quantity of liquid injected into the gas expeller the dissolved gases are desorbed either under vacuum or by passing the carrier gas through it and the gases thus liberated can be injected directly into the column. Some commercial gas chromatographs are provided with an ancillary device which is suitable for the injection of dissolved gases.[28]

Automatic injection. The latest achievement in sample injectors are systems for the automatic injection of gases, liquids and solids, so that the laboratory gas chromatograph can perform serial analyses without supervision or handling e.g. at night.[29, 30] Automatic sample injectors are also needed with preparative and process chromatographs and will be discussed in the relevant chapters.

4.5 Columns

In this section the column will be discussed only from the point of view of its construction, its arrangement in the thermostat and of the way it is joined to the other parts of the apparatus. A special chapter deals in detail with the various types of columns, the packing of columns and various column combinations.

With respect to their dimensions and arrangement the two basic types of columns, the packed and the open tubular (capillary) column, are rather different, though the latest type (packed capillary columns) represents a certain transition between the two.

Material of the column. The first packed columns were made of glass and later of metal, such as aluminium, copper and stainless steel. For handling and assembly the metal columns are more convenient, especially at higher temperatures and pressures. In the case of certain samples the aluminium and copper columns may be reactive or have a catalytic effect, so that the stainless steel column can be considered to be the most satisfactory. There are several publications discussing the merits of the various structural materials, but there are no clear-cut conclusions.[31] For the analysis of high molecular weight, easily decomposed substances (steroids, amino acids, etc.) glass columns are suggested. For corrosive gases nickel or Monel metal columns are best.

For capillary columns aluminium, copper, stainless steel, glass or plastics (polyamide) are used. At high temperatures here again stainless steel is the best material.[32] Glass and nylon capillaries give the most uniform liquid films. To increase the quantity of the liquid coating the internal surface area of the tube is increased by various methods. In the case of glass tubes the glass is etched,[33] while irrespective of its material any tube may be coated with a thin solid layer (support or adsorbent).[34]

Dimensions. In the 1950's the internal diameter of packed columns was 6–8 mm. To increase efficiency the diameter of the later columns was 3–4 mm, and today columns of 1–2 mm internal diameter are also used. The length of the column varies depending on the analysis it has to perform. The length of packed columns is usually 1–6 m. For most tasks 2–3 m columns usually suffice, while in certain cases a column length of 10–15 m may be necessary, though the trend in recent analytical methods is to apply shorter columns and to improve separation by some other way. In any case the thermostat provides for the acceptance of long columns and also for the operation of two or more columns in various arrangements. The larger thermostats built originally for U-shape columns are also suitable for preparative columns which have a greater diameter (10–20 mm).

Open tubular columns are made of capillaries with 0.2–1 mm internal diameter. The length of open tubular columns is usually 30–100 m, but occasionally capillaries of several 100 m length are used. Their assembly presents no problem, for they are wound into tight coils and occupy but little space.

Shape. Packed columns may be straight tubes, U, W, spiral or flat spiral shaped. The shape of the column has to meet two contradictory requirements: it should, on the one hand, occupy only a small space and in this respect the coiled columns are the most favourable, while, on the other hand, packing should be uniform and compact which is best realized with straight and U-shaped columns. In the case of long U-shaped columns the volume of the thermostat must be larger. Investigation of this problem has revealed that there will be no reduction in column efficiency when the ratio coil diameter:

external tube diameter is at least 10 : 1. Spiral columns demand very careful packing and winding, but when the necessary specifications are observed their efficiency will be practically the same as that of U-shaped columns. The latest types of gas chromatograph include

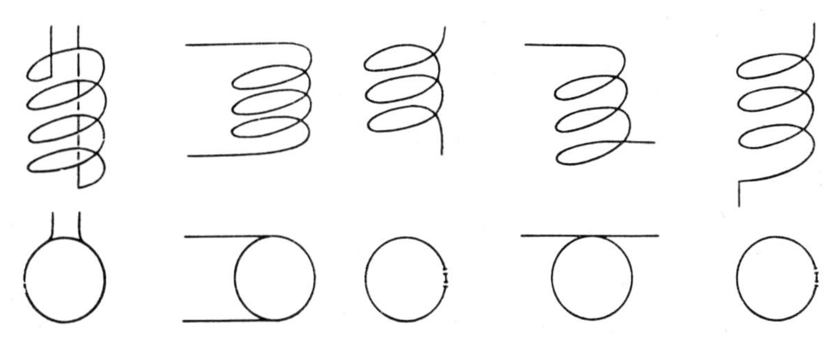

Fig. 4.8. Schematic diagram of various coiled columns

mostly spiral columns. Figure 4.8 shows the most frequently used arrangements and indicates the position of the ends.

The coil of capillary tubing is conveniently fixed on a spool or some other holder to protect it from external mechanical effects and to hold the ends in position.

Fixing and connecting the column. In setting up the column it is important to ensure high-temperature resistant, gas-tight seals; rapid dismantling for quick column change; minimum dead volume at the joints and connecting lines; possibility of assembling adequately long columns and of various column arrangements. The column inlet is joined to the sample injector, its outlet to the detector. The connections and joints should be arranged in such a way as to ensure the smallest possible dead volume. The dead volumes cause peak broadening and poorer separation. In this respect proper securing of the column packing with an end-plug is highly important and can best be ensured with a metal sieve disk. In the analysis of readily decomposed substances the sample should come into contact with glass only and the glass column should protrude directly into both the sample injector and the detector (see sample injectors).

There are two arrangements for mounting columns: in one the entire column section can be removed in one piece, in the other each column can be dismounted separately. In the latest equipment the second method is used when the columns are fixed to the lid of the thermostat or to the distributor block in the thermostat, or side-leads hold the column. There are also different ways to join the columns to each other and to the other parts of the apparatus: blocks with appropriate ports which can be either of one or several pieces, capillaries with ground metal ends, flare nut joints with soldered ends (Swagelok fittings) or flare nut joins with O-ring sealing. Open tubular columns are joined with flare nuts with soldered ends where higher pressures than in the packed columns (usually 3–6 atm) are used.

Dismantling columns, switches. The versatility of gas chromatography apparatus is considerably increased by the various column arrangements. In the case of samples which contain high boiling point components or components with widely different boiling points, provision for the back-flush of the column is highly important. Back-flush means the reversal of the direction of the carrier gas stream. With a valve operated from outside (provided with separate heating) the direction of flow of the carrier gas in the column can be reversed, whereby the heavy components can be swept out rapidly.

Several columns with different types of packings can be built into the same apparatus and depending on the analytical task on hand the columns can be operated in series or parallel or by-passing one or the other column. This may be realized by various switch blocks (e.g. Carlo Erba) which are provided with switch valves to be operated from outside. In many analyses various pre-columns are necessary which are either placed in the thermostat or joined to the sample injector system and fitted on the front panel. The pre-columns are responsible for the removal of contaminants which may interfere with the analysis or for the conversion of certain compounds into others more easily amenable to analysis.

4.6 Detectors

4.6.1 The purpose, classification and operation parameters of detectors

Detectors have the task to sense continually, rapidly and with high sensitivity the components which appear in the carrier gas as it emerges from the column. The detector senses the changes in a certain physical or chemical property of the effluent gas stream on the appearance of the components. With the help of an appropriate current source and signal transmitter the detector gives a direct electric signal or the response of the detector is converted into an electric signal by an appropriate electronic system. Usually this electric signal is a voltage which may be led directly, or through an amplifier, for recording or data processing.

One of the chief factors in the wide-spread application of gas chromatography is the availability of a great variety of highly efficient detectors. There is no ideal detector which may satisfy all requirements, but the detectors of various types supplement each other satisfactorily and greatly differing demands may be met with a few basic types. This is reflected in the construction of up-to-date gas chromatographs which can be operated with several exchangeable detectors, moreover the latest types provide for the simultaneous use of two or even three detectors. In recent years improvement of the existing detectors and the introduction of new detector types constitute a significant part in the development of gas chromatography. Next to the columns and packings most publications deal with the construction and operation of various detectors.

The requirements which the detectors have to meet and the properties of the detectors were discussed by several authors[4, 35, 36] and in recent years some excellent summaries on the comparative examination of the various detectors were published.[37–40] In general, the properties which a good detector should possess may be summed up in the following:

1. High sensitivity (in some cases selectivity) towards the various components.
2. The response should follow rapidly the actual change.
3. Should give in a wide concentration range signals proportional to the quantity of the component (linearity).
4. Stable operation with respect to both noise and drift.
5. Low sensitivity to fluctuations in operation conditions (carrier gas rate, pressure and temperature).
6. Should give signals which can be easily recorded.

In addition to the above basic operation aspects it is important that the detector should be easy to build cheap, of rugged construction, easily operated, safe from dangers of explosion or radiation and should if possible not require expensive and complicated auxiliary fixtures.

In the literature detectors are classified in diverse ways, of which only the most characteristic and most widely accepted will be discussed here. A uniform and detailed classification of the working principles of detectors remains a task to be solved in the future.

In the earlier literature from the point of view of operation, differential and integral detectors were distinguished. As the emerging signal can be differentiated or integrated electronically or by mathematical methods this distinction is of no special significance. Today the terms "differential" and "integral" are replaced by the terms "instantaneous" and "cumulative". The detector which works on the "instantaneous" principle measures the instantaneous response to the component when this passes through the detector. The "cumulative" type detector on the other hand measures the total quantity emerging during analysis and its response records at any given time the total quantity of substance which has entered the detector up to this time. In this case the chromatogram is made up of successive steps as shown in Fig. 1.5. Such "cumulative" detectors were used in the first gas chromatographic set-ups, of which the recording titration unit of James and Martin[41] and Janák's alkaline burette for the absorption of CO_2 should be mentioned.[42]

In commercial gas chromatographs detectors which sense instantaneous changes are used almost exclusively, so that in the following we shall restrict ourselves to the detailed discussion of this type. These detectors furnish a differential chromatogram as shown in Fig. 1.5 which consists of a series of successive peaks. In quantitative work the differential curves, i.e. the measured analogue signals are converted into integral signals corresponding to the area below the peak with different types of integrators or calculation methods.

Another classification of detectors is based on the fact that some detectors sense the change in a certain overall property of the carrier gas and the component (e.g. thermal conductivity, density, dielectric constant), while other detectors measure a specific property of the component which the carrier gas does not possess (e.g. flame ionization, photo-ionization, electron capture). Detectors with normal sensitivity belong to the first group, because the change in the overall property is relatively smaller. Detectors in the second group ensure very high sensitivity, as there is a difference of 3–7 orders of magnitude between the properties of the carrier gas and of the component. This classification has the drawback that detectors working by the same principle, e.g. ionization detectors, belong to both groups.

From the point of view of their effect on the component under investigation a distinction is made between destructive (e.g. flame ionization) and non-destructive detectors (e.g. heat conductivity cells).

Many of the detectors produce the signal which corresponds to the component directly. There are however detectors which first convert the component into a chemically different compound (e.g. catalytically or by pyrolysis) and detect the product obtained in this way.

Detectors may be classified and compared most conveniently on the basis of the work of Sternberg[38] and Halász.[39] Essentially the detectors may be classified into two groups, namely concentration sensitive and mass sensitive detectors. For concentration sensitive detectors the instantaneous response is proportional to the concentration of the component in the carrier gas

$$R = Sc \qquad 4.1$$

where R is the response in mV, S is the so-called response factor in mV × ml/mg, and c the concentration in mg/ml.

For some detectors the concentration refers to the carrier gas (e.g. heat conductivity, density measuring detectors) while for others c is the volume concentration e.g. in mole per litre (electron mobility, ultraviolet absorption measuring detectors). The response is independent of the mass velocity w of the sample and in general the sample undergoes no chemical change.

For mass sensitive detectors the response is proportional to the mass velocity of the sample which reaches the detector, i.e. to the mass of the sample reaching the detector in unit time, and is independent of its concentration in the carrier gas. This type of detectors usually involves also a chemical change (flame ionization, β-radiation, etc.).

$$R' = S'w \qquad 4.2$$

where R' is the response in mV, S' the response factor in units of mV × sec/mg and w the mass velocity of the sample in mg/sec.

It may be expedient to give the response of ionization detectors not in voltage but in current intensity when the unit of the response factor will be A × sec/mg or C/mg.

It appears from the characteristics of the two groups that the response of the detectors in the first group should refer to the concentration in the emerging carrier gas while the response of the detectors in the second group refers to the mass velocity of the component.

It is now obvious that there can be no direct comparison between the detectors in the two groups, that is to say, they can be compared only when both the concentration in the carrier gas and the mass velocity are fixed.

The choice of detector depends on the given analytical problem; the field of application of the various detectors will be dealt with later. For the choice of detectors and the characterization of their operation the following have to be considered:

1. sensitivity;
2. time constant;
3. linear dynamic range or linearity;
4. quantitative evaluation of the response.

Sensitivity. There are several formulae in the literature to characterize sensitivity, nevertheless there is no uniform and unambiguous equation which might be applicable to both types of detectors. The sensitivity of the detector, S, is in its general formulation characterized by the ratio of the change in the response and the associated change in quantity:

$$S = \frac{\Delta R}{\Delta Q} \qquad 4.3$$

Dimbat *et al.*[4] were the first to give an explicit correlation for the determination of sensitivity. This correlation has since found widespread acceptance. The sensitivity S can be calculated from the peak of a chosen component on the recorder:

$$S = \frac{A C_1 C_2 C_3}{Q} \qquad 4.4$$

where S is the sensitivity in mV × ml/mg; A is the area below the peak in cm^2, C_1 is the reciprocal value of the chart speed in min/cm; C_2 the sensitivity of the recorder in mV/cm; C_3 the flow rate of the carrier gas converted to the temperature of the detector and atmospheric pressure in ml/min and Q is the quantity of the component in mg.

The S value on the left-hand side of Eq. 4.4 is nothing but the response factor in Eq. 4.1 and depends not only on the geometrical parameters of the detector, but also on operating conditions and the nature of the component under investigation. The S value is not suitable for comparison of various detectors, since a highly sensitive detector is not necessarily suitable for sensing very small quantities. The use of the value S is particularly cumbersome in the case of highly sensitive ionization detectors whose S values are 4–5 orders of magnitude higher than the S values of other detectors.

The Gas Chromatography Discussion Group suggested as far back as 1957 that the detectable vapour concentration should be used to characterize sensitivity.[43] Johnson and Stross[44] characterize sensitivity by the detectability limit for the determination of which they have worked out an accurate statistical method which however has not gained acceptance because of the long time required.

Young[45] studied the correlation between sensitivity and the limit of detectability which includes the noise level of the detector system. (By noise level we understand the fluctuations of the base line due to fluctuations in operating conditions.) If twice the noise level is arbitrarily chosen as the detectability limit Q_0 then:

$$Q_0 = \frac{2R_n}{S} \quad \text{mg/ml} \qquad 4.5$$

where R_n is the noise level of the detector, and S is the sensitivity. Young gives the comparative S and Q_0 values of six types of detectors and points out considerable deviations in the differences between the S and Q_0 values. If for the calculation of S from Eq. 4.4 the quantity of the component is given not in mg but in millimoles, then the dimensions of Q_0 will be mmole/ml which is more suitable for comparison.

The detectability limit in the form given by Eq. 4.4 is valid for concentration sensitive detectors. The corresponding correlation for mass sensitive detectors is:

$$Q_0 = \frac{2R_n}{S'} \quad \text{mg/sec} \qquad 4.6$$

In the literature and in the specifications of commercial gas chromatographs sensitivity and detectability limit are given in all sorts of units: ppm, ppb, mmole, mg, mg/ml, mmole/ml, mg/sec, mmole/sec, etc. As the detector response depends on the quality of the component, the reference substance should be included in the definition. The first four units when standing alone are meaningless, for it follows from the aforesaid that depending on the type of the detector either the concentration or the mass velocity must be given.

The sensitivity of the detector can be determined by one of two frontal methods, namely the exponential retardation method worked out by Lovelock[46] or the diffusion dilution method described by Desty *et al.*[47]

Time constant. The time constant gives the speed of response and is usually characterized in practice by the time required for 90% of the total response. The time constant of the detector system is determined by the geometrical response time of the detector (the effective detector volume divided by flow rate), from the time of signal transfer (e.g. in the heat conductivity cell change in the temperature of the filament), the time constant of the electrical system and mainly from the time constant of the recorder. When detectors are compared the time constant of the recorder is not considered.

Linear dynamic range. Starting from the detectability limit this is the region of sample quantity in which the detector response is directly proportional to the quantity of sample which had been injected, that is, its concentration or mass velocity. Generally an accuracy of a maximum of ±3% deviation from linearity is required. Linear response is particularly important from the point of view of detector calibration and quantitative evaluation on the basis of the areas below the peaks. No accurate quantitative analysis is possible in the non-linear range.

Quantitative evaluation of the detector signal. A detector response corresponding to a given quantity of sample depends on the quality of the component and, in the case of most detectors, also on the quality of the carrier gas, the operating conditions of the detector (temperature, gas rate), the operating conditions of the signal transmitting system (voltage, bridge current, etc.) and the operating conditions of the recorder (sensitivity, chart speed). In quantitative evaluation the absolute or relative quantities of the components are determined from the magnitude of the response, that is from the areas under the peaks and from these quantities the composition of the injected sample is then calculated.

For concentration sensitive detectors the quantity of the component will be from Eq. 4.4:

$$Q = C_1 C_2 C_3 \frac{A}{S} \qquad 4.7$$

where C_1, C_2 and C_3 are constant parameters for a given analysis. Thus the relationship between quantity and area is determined by the value S which is also called the response factor.

For mass sensitive detectors the modified form of Eq. 4.7 may be used:

$$Q = C_1 C_2 \frac{A}{S'} \qquad 4.8$$

where S' is conveniently given in units of current density per mass per time (e.g. A × sec/mg).

As the value of S changes depending on various apparatus factors and operating conditions, the absolute detector response depends on a great number of factors. For practical quantitative work and for comparing different detectors it is more expedient to use the so-called response factors which give the responses to the various components in their relation to an identical quantity of a standard reference substance (e.g. with katharometers benzene is the most frequently used reference substance).

In the case of most detectors the response factor cannot be determined by preliminary calculation or is obtained only approximately, so calibration is necessary for the desired accuracy. Only the gas density balance and the latest coulometric detectors make possible an accurate determination of the response factors by calculation alone.

In the following we wish to discuss in detail those detectors which are most frequently applied and encountered in commercial apparatus, while the other types will only be described briefly.

4.6.2 The thermal conductivity detector (katharometer)

In the 1950's the bulk of gas chromatographs operated with detectors which measured thermal conductivity (katharometers) and which had been in use for years in certain gas analyses. The construction of the katharometer is very simple, it is relatively cheap, its operation is easy and it can be used for many versatile tasks. Though its sensitivity does not reach that of the highly sensitive ionization detectors, it is nevertheless suitable for most analytical work. In spite of the spectacular advance of ionization detectors about 70–80% of commercially available gas chromatographs are also provided with katharometers. The katharometer has a very extensive literature including several general reviews.[48–51]

Principle of operation. The heat conductivity cell is usually a metal block with holes bored through it, through which passes the carrier gas stream. A heated resistance element (resistance filament or thermistor) is placed inside the channel. The resistance of this element is highly temperature dependent. The gas flowing through the channel causes heat conductance from the heated resistance to the wall of the hole which always has a lower temperature. When a component appears in the carrier gas, this will alter the heat conductivity of the gas stream and consequently the conducted heat, resulting in a change of the temperature and resistance of the heated element. By suitable measuring circuit the resistance change is converted into a voltage change and recorded as such. Changes in heat conductivity are measured by the differential method. The cell has two or four such channels bored in it with identical resistances in each. The pure carrier gas flows through one or two of the channels, the carrier gas containing the component flows through the other, or other two. The resistance elements are connected to a Wheatstone bridge. If pure carrier gas flows through all the channels, the bridge is balanced at zero. When a foreign component appears in the measuring channel, the resistance of the element in this one will change and deflect the balance on the bridge. This deflection is plotted by the recorder connected to the bridge as a zero instrument.

Construction. The most important point in cell construction is to ensure a constant wall temperature for a low time constant and low current sensitivity.

Constant wall temperature is linked to the heat stability of the cell. Contrary to earlier practice, at present the cells are made almost exclusively from a metal block with several bored holes. The metal is usually brass or stainless steel, though other metals and even Teflon are occasionally used. The higher the heat capacity of the cell the more stable will be its operation. The symmetrical arrangement of the channels and the heated elements is highly important for uniform heat loss. The cell is placed in a block thermostat; in the case of programmed temperature the cell must operate in a separate isothermal thermostat. To achieve the desired accuracy the temperature of the

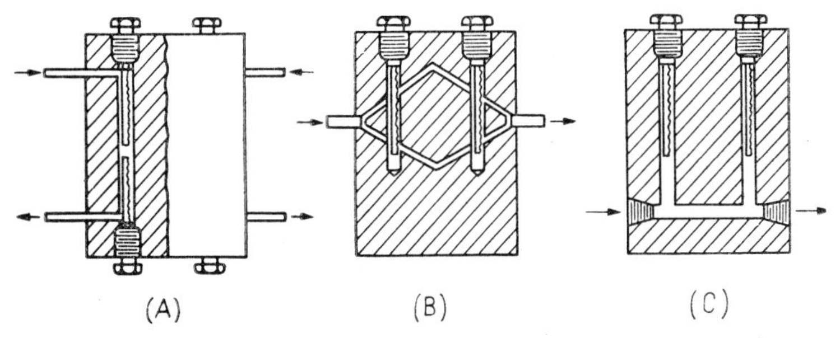

Fig. 4.9. Schematic diagram of thermal conductivity cells of different flow types.[51] *(A) Through flow; (B) partial flow or semi-diffusional; (C) diffusional*

block must be kept at a constant value with ±0.1 °C accuracy. The simplest construction contains two channels, one reference and one measuring branch. The two resistance elements with two constant resistances form the four branches of the Wheatstone bridge. The stability and sensitivity of the cell improves by the use of four channels when the heated resistance elements form all the four branches of the bridge. Most of the commercial cells have four channels, but the latest microcells are constructed with two channels. The time constant and sensitivity changes in the flow rate depend on the geometric arrangement of the holes and on the flow conditions within the cell. These two parameters raise different demands on the construction of the cell which can only be met by a compromise. This compromise will determine the construction of the channels. There are three different types of cells from the point of view of flow, their diagrams are shown in Fig. 4.9. Cell (A) provides for direct passage of the gas and has the lowest time constant (1/4–1 sec), but is sensitive to even minor changes in the flow rate. The diffusion cell type (C) is insensitive to changes in the flow rate, but has a high time constant (5–10 sec). Type (B) is a semi-diffusional type which embodies the advantages of the first two types, has a low time constant (about 1 sec) and is not sensitive to changes in the flow rate. The latest commercial gas chromatographs use mainly semi-diffusion cells.

The time constant and the sensitivity of the cell depend considerably on the internal volume of the cell. The gas volume of modern

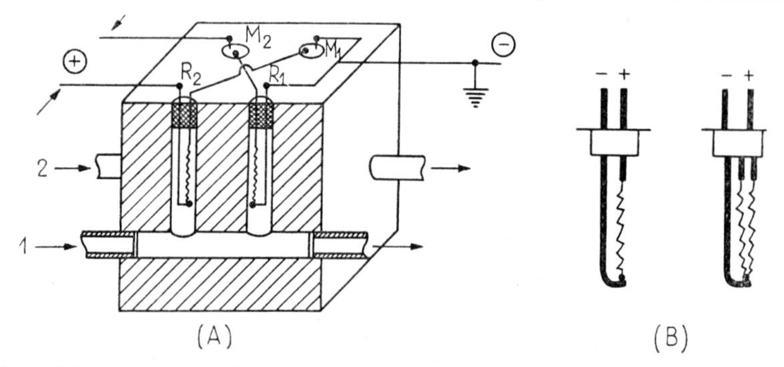

Fig. 4.10. Schematic diagram of four-element Gow Mac cells (A) and the arrangement of the resistance filaments (B). M_1, M_2—measuring branches; R_1, R_2—reference branches; 1. carrier gas; 2. carrier gas + sample

cells is in general 1–3 ml.[52] The gas volume of the recently evolved microcells for open tubular columns and highly sensitive analysis varies between 2.5 and 250 μl.[52–55]

Quality and arrangement of the resistance elements. The resistance element is either a heated resistance wire or thermistor, a semiconductive metal oxide.

The material of the resistance wire may be either Ni, Pt, Pt–Ir, W or W–Rh. For analyses at higher temperatures (> 250 °C) generally W, above 350–400 °C W–Rh filaments are used. The cold resistance of the wires varies between 10 and 60 Ω, often between 20 and 40 Ω. The resistance is built either by stretching the wire into the bore or by winding it into a coil and fixing it on a rod. In the latest cells mainly coiled resistance wires are used. Figure 4.10 shows the diagram of a Gow Mac cell with four channels and two common methods for fixing the resistance wires. The two coils on a V-shaped common holder ensure higher resistance and thereby also a higher sensitivity. The wires must be chosen in such a way that the difference between their resistances should not be over 0.1 Ω. The electric terminals of the wires are small copper rods of 0.2–0.5 mm diameter to which the resistance wires are hard soldered. The gas-tight sealing and electrical insulation of the metal block at high temperature from the terminals is no easy task. Teflon or Teflon–asbestos seals can be used up to 250–300 °C, at higher temperatures (400–500 °C) metal oxide or high melting glass seals are recommended.[56]

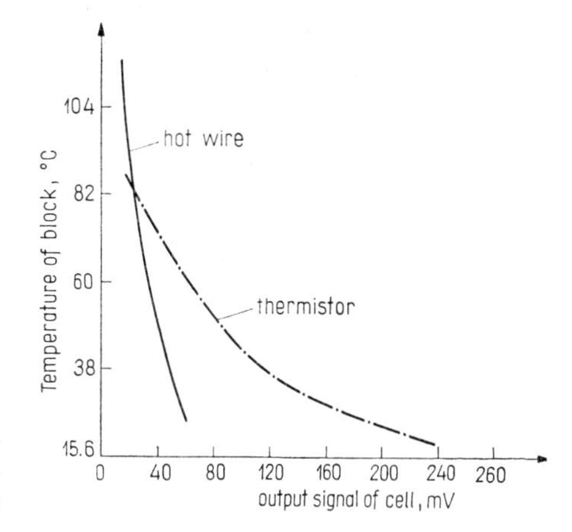

Fig. 4.11. Changes in the response of hot wire and thermistor cells with temperature

The other quite widely used resistance element is the thermistor, a semiconductor consisting of various metal oxides whose resistance is highly temperature dependent. The basic difference between the two resistance elements is the negative resistivity coefficient of the thermistor, that is, its resistance decreases with increasing temperature, contrary to the positive resistivity coefficient of the hot wire the resistance of which increases with increasing temperature. Consequently, thermistors are used mainly at low temperatures where their sensitivity may be several times higher than that of the hot wire. Figure 4.11 illustrates the temperature dependence of the response of the two types. Above 90–100 °C the sensitivity of the hot wire cell exceeds that of the thermistor. Thermistor cells can be used up to a maximum of 180 °C, as their characteristics change at higher temperatures. The cold resistance of the thermistors which are used in heat conductivity cells changes in general between 6000 and 10,000 Ω. Thermistors with identical resistances and characteristics are necessary for the Wheatstone bridge, though in practice it may be very difficult to select four thermistors with identical temperature *vs.* resistance characteristics over a wide temperature range. The characteristics of initially identical thermistors may alter in different ways during "ageing" which leads to the deterioration of cell operation. Several publications deal with the selection and optimum operation of thermistors.[57–60] The thermistor cell has the advantage that because of

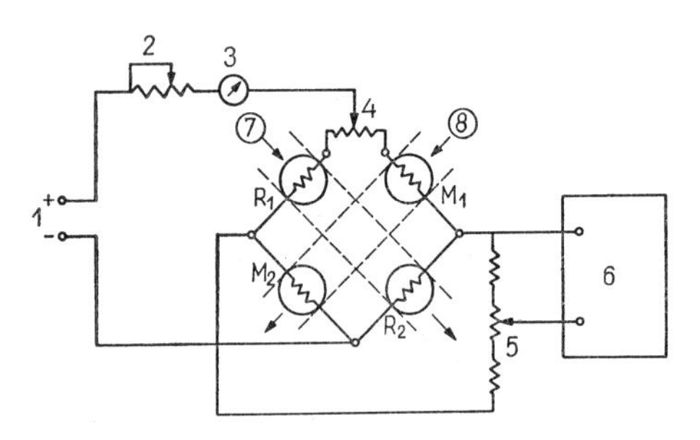

Fig. 4.12. Electrical circuit of the four element thermal conductivity cell. 1. Power supply; 2. potentiometer; 3. milliammeter; 4. zero control; 5. attenuator; 6. recorder; 7. carrier gas; 8. carrier gas + sample; R_1, R_2—reference cells; M_1, M_2—measuring cells

the small dimensions of the thermistor the cell will have a very small dead volume (0.025–1 ml). With thermistor cells the constancy of the temperature is of particular importance. Hydrogen carrier gas is not recommended for thermistor cells because of the partial reduction of the metal oxide, though most of the thermistors used in gas chromatography are glass coated. Thermistor cells are used mainly in analyses below 100 °C and especially below 0 °C. Microcells with small dead volumes are also produced with thermistors, though recently suitable microcells with hot wires have also appeared.

Electrical circuit. The resistance elements of the cell are connected to a Wheatstone bridge. The electric circuit of a cell with four channels is shown in Fig. 4.12. Cells with two channels are used less frequently and mainly in microcells; in this case the other two branches of the bridge are constant resistances. At first the current for heating the resistance was taken from a battery, today stable electronic power sources fed from the mains are used. The intensity of the current depends on the resistance and quality of the element and on the quality of the carrier gas. For Ni and Pt wires usually higher current intensities are used than for W and W–Rh wires. Current intensity depends strongly on the heat conductivity of the carrier gas, and is for W wires 200–500 mA, for Ni and Pt wires 300–800 mA in the case of H_2 and He carrier gas; when the carrier gas is argon, carbon dioxide or nitrogen the respective current intensities will be 100–200 and 100–350 mA. For a thermistor cell with average resistance (8000 Ω) the current is 10 mA in the case of helium, and 4–6 mA in the case of nitrogen, carbon dioxide or argon. For cold balance, the base line is set with a rough and a fine potentiometer. The sensitivity range of the cell can be adjusted with the help of an attenuator. For adequately stable operation the electrical circuit of the cell requires careful design, highly accurate control elements and precise construction. The optimum solution will depend on the geometry and resistance of the cell, on the temperature, current uptake and the parameters of the recorder. Electric noises contributing to the noise level of the detector have their origin in the unstable operation of the power source, in the imperfections of the insulation, but mainly in contact faults of the circuit, i.e. in dirt settled on the moving contacts.

Auxiliary devices. The cell requires no auxiliary devices beyond the power source and the bridge; the outgoing signal can be recorded directly with the generally used 1–10 mV recorders. To ensure the undisturbed operation of the cell, the heating of the thermostat, the heat regulator, the current supply and the recorder should be connected to a common earth terminal.

Choice of the carrier gas. For the operation of thermal conductivity cells the carrier gas plays an important role as the cell measures the difference between the heat conductivities of the pure carrier gas and of the carrier gas which contains the components of the sample. Table 4.1 shows the heat conductivities of several gases and vapours. It appears that the heat conductivities of H_2 and He are nearly ten times higher than the heat conductivities of other gases and vapours, among which the differences are smaller. Some earlier work indicated that carrier gases with lower heat conductivities ensure higher sensitivities,[61–63] but later detailed examinations proved[64–67] that the use of high thermal conductivity carrier gases, namely of H_2 and He is advantageous from the point of view of sensitivity and the dynamic linear range. In the thermal conductivity cell, changes

Table 4.1

Thermal conductivity of gases and vapours

Component	*Thermal conductivity 10^5 cal/cm sec °C*	
	at 0 °C	*at 100 °C*
Hydrogen	39.60	49.93
Helium	33.60	39.85
Nitrogen	5.68	7.18
Argon	3.88	5.09
Carbon dioxide	3.39	5.06
Ammonia	5.14	7.09
Carbon monoxide	5.42	—
Hydrogen sulphide	3.04	—
Nitric oxide (NO)	5.55	—
Methane	7.20	—
Ethane	4.31	7.67
Ethylene	4.02	6.36
Acetylene	4.40	—
i-Pentane	2.91	5.11
n-Pentane	3.27 (20 °C)	—
n-Hexane	2.85 (20 °C)	—
n-Heptane	—	4.14
Benzene	2.09	4.14
Methyl alcohol	3.36	5.16
Ethyl alcohol	3.58 (20 °C)	4.98
Diethyl ether	3.10	5.28
Ethyl acetate	—	3.86
Acetone	2.30	3.96
Butylamine	3.00 (6.5 °C)	—
Methyl chloride	2.22	3.84
Chloroform	1.52	2.33
Carbon tetrachloride	1.67 (46 °C)	2.05
Water vapour	4.58 (46 °C)	5.51

in resistance depend on the thermal conductivity of a binary gas mixture, namely of the carrier gas plus the component. The heat conductivity of such binary mixtures changes differently depending on the composition and the temperature, and this is a considerable obstacle in quantitative evaluation. Recent extensive tests with mixed carrier gases offer a possibility of more sensitive detection and more accurate quantitative analysis by the tailor-made adjustment of the thermal conductivity of the mixture.[68–70]

Operating conditions. The temperature of the resistance wire is another important factor in the operation of the cell. Schmauch and Dinerstein[71] have subjected to detailed investigation the factors which may have an effect on the response and found a correlation by which these changes may be described. The sensitivity of the cell increases with the 3/2 power of the wire temperature. Generally the wire temperature is 100–150 °C higher than the temperature of the detector block. The practical upper limit of the wire temperature is 600–700 °C, though at this high temperature the life of the wire is rather short. The resistance of the wire changes practically linearly with its temperature according to the following relationship:

$$R_T = R_0(1 + \alpha T) \qquad 4.9$$

where R_0 is the resistance at 0 °C, R_T the resistance at T °C and α the temperature coefficient resistance of the wire. Since in practice the temperature of the wire cannot be measured current intensity is measured instead. The correlation between current intensity and wire temperature can be calculated by accounting for the thermal conductivity of the carrier gas.[62]

The operation of the cell changes sensitively with any temporary fluctuation in the flow rate or pressure of the carrier gas and this will be manifest in a higher noise level or in drift of the base line. Changes in the temperature of the detector block have a similar influence so that this temperature has to be kept at a constant value with ±0.5 °C accuracy. Here are some useful directives for the operation of the cell: 1. It should be repeatedly checked whether the system is still gas tight as the slightest leak will cause a fluctuation of the base line. 2. Carrier gas rate has to be controlled with high precision. 3. To prevent condensation the temperature of the detector should be at least as high as that of the column. 4. The cell should operate with the lowest possible heating current intensity to prevent wear of the resistance wires. 5. The current must not be switched on unless the carrier gas flows through the cell or the wire will burn out.

Quantitative evaluation and calibration. There are many contradictory statements in the earlier literature on the evaluation of the areas below the peaks, i.e. of the corresponding percentage areas.

Some authors maintain that the areas are proportional to the mole concentration,[72, 73] while according to others to the weight concentration[57, 74, 75] of the sample. More detailed studies have demonstrated that for accurate evaluation a correction factor has to be introduced for each component or calibration for each component will be necessary.[76–78] Today mainly the empirical relative molar response factors, RMR, are used[77, 78] which refer the response factors to the response factor of a standard substance, namely benzene (RMR = 100). Within a homologous series there is a linear correlation between the molecular weight of the compounds and their RMR values, thus for members of a homologous series quantitative analysis can be performed directly from the percentage areas. When the components belong to different homologous series the relative error without RMR correction may amount to 50%. If we wish to obtain the results not in molar but weight per cents, another correction factor has to be applied which is called the substance specific correction factor, f, and when the RMR value is known, it can be calculated as follows:

$$f = \frac{100 \times M}{M_S \times \text{RMR}} \qquad 4.10$$

where M is the molecular weight of the component, and M_S the molecular weight of the standard.

For quantitative analysis instead of these correction factors, direct calibration for each component or the so-called internal standard method may also be used. These methods will be discussed in detail in connection with quantitative evaluation, it need only be pointed out here that the application of correction factors furnishes the most reliable results.

Sensitivity. The sensitivity of the cell increases with the 3/2 power of the wire temperature or the 3rd power of current intensity. For most substances the maximum sensitivity is a few μg. The value of S, as defined by Eq. 4.4, varies between wide limits, in the range of 10 to 10^4, depending on the components and on the construction and operation conditions, mainly on the current intensity of the cell. The lowest detectable concentration depends on the noise level which is controlled by mechanical, thermal and electrical influences. Mechanical influences may be reduced by the stable, shock-free fixing of the wire, while the appropriate arrangement of the wire and the thermal stability of the block reduce thermal influences and carefully planned and assemblied electrical circuits diminish the electrical effects. With careful work the noise level can be kept at 1–10 μV which at the customary flow rates and with highly sensitive detectors corresponds to 1–10 ppm lowest detectable concentration in the carrier gas.

Dynamic linear range. With He or H_2 as carrier gas, response changes linearly up to about 10 volume% concentration in the carrier gas. With other carrier gases the dynamic linear range is considerably more limited.

Time constant. Depending on the construction the time constant of the heat conductivity cells varies usually between 0.25 and 10 sec.[79]

4.6.3 Gas density balance

Martin and James[80] used the gas density balance as far back as 1956, but due to difficulties of construction it has only acquired wide-spread use in the 1960's. Compared to other detectors the gas density balance has the great advantage of measuring an additive physical property of the gas which depends only on the kind and number of atoms in the molecule and is suitable for the detection of any organic or inorganic vapour or gas. The other great advantage of the method is that with it accurate quantitative analysis can be performed without preliminary calibration.

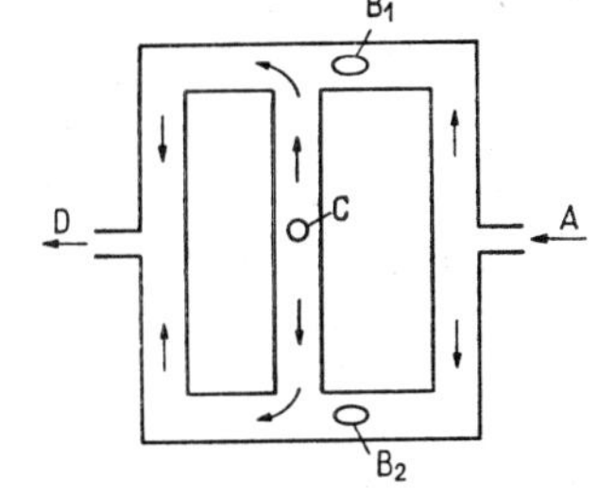

Fig. 4.13. Schematic diagram of the gas density balance. A—pure carrier gas inlet; B_1, B_2—flowmeters; C—carrier gas + sample inlet; D—gas outlet

Principle of operation. Figure 4.13 shows a schematic diagram of the gas density balance. The reference gas, the pure carrier gas, enters the detector at A where it is split into two streams, and unites at D where it leaves the detector. The gas emerging from the column enters the detector at C, is also split into two streams and leaves united at D. As shown by Munday and Primavesi[81] the flow diagram of the gas density balance is the same as that of an electric bridge. The system is in equilibrium when the gas flow in the reference and measuring branches is the same. When a foreign component appears in the effluent carrier gas the molecules which are heavier than those of the carrier gas will flow downwards in the vertical branch, consequently less gas will flow through the AB_2D and more through the AB_1D branch. When the molecular weight of the component is lower than that of the carrier gas the shift in flow is reversed. Changes in flow rates are sensed by the sensitive flowmeters B_1 and B_2 in the two branches. Flow rates change within a wide range linearly with changes in gas density. The response of the flowmeters B_1 and B_2 is proportional to the gas density, that is, to the molecular weight of the component.

Construction. The first gas density balance of Martin and James[80] was made of a brass block with an intricate system of bores. Munday and Primavesi[81] constructed a gas density balance from copper tubes, but the sensitivity of this balance was below that of the bored block. In later years construction was greatly simplified (Fig. 4.13, a schematic diagram of the simplified type). In Great Britain the firm Griffin and George manufactures gas density balances of the bored block type, while in the USA the Gow Mac Company markets tube type balances as patented by Standard Oil Co.[82, 83] The material of the detector is brass or stainless steel, but if necessary other metals (e.g. nickel or Monel metal) may also be used. There are two methods for measuring the flow rate, namely sensing by thermocouple or by the resistance wire.

For sensing with thermocouples (Griffin and George) B_1 and B_2 are sensitive thermocouples made of nichrome – constantan, and the detector is surrounded by a thin heating filament. This filament is heated with a low voltage to 600–800 °C. Changes in the flow rate of the carrier gas cause changes in the heat given off by the two heating wires resulting in a change in the temperature of the thermocouples.

When changes in resistance are measured (Gow Mac) the detector is a resistance wire or a thermistor. As in the heat conductivity cell the two resistance elements are connected to a Wheatstone bridge. Changes in the flow rate alter the resistance of the sensing elements in an opposite way and thus deflect the balance of the bridge, this deflection can be recorded directly. The resistance wires and thermistors described in the section on heat conductivity cells are used here as resistance elements. To increase sensitivity, cells with four resistance wires were designed. With the resistance wire the detector can be used up to 300 °C. This construction has the great advantage that the components of the sample are not in contact with the high temperature elements only the pure carrier gas is in contact with it, which prevents decomposition, coke formation or other undesirable reactions leading to contamination.

Electrical circuit. The electrical circuit, including the resistance of the flowmeters and current intensity, are the same as those used in the heat conductivity cell.

Auxiliary devices. Compared to the heat conductivity cell a new demand arises here because of the necessity of a reference carrier gas. The apparatus contains two separately controlled carrier gas systems. The flow rate of the reference gas is in general 50–100 ml/min, while the rate of the carrier gas which travels through the column is 10–15 ml/min lower to prevent back-diffusion of the components.

Choice of the carrier gas. Guillemin and Auricourt[84, 85] have dealt in detail with the influence of the nature of the carrier gas. In the choice of the carrier gas, or more correctly, of the reference gas, it must be borne in mind that the density of this gas must differ markedly from the densities of the components. Essentially the most intensive response is obtained with a high molecular weight, but low viscosity, gas. Usually N_2, Ar or CO_2 is used as reference gas, but the highest sensitivity can be achieved with CO_2 because of the high density and low viscosity of this gas. SF_6 has been suggested as carrier gas in the analysis of permanent gases to improve sensitivity.[85] If the sample contains components of both higher and lower molecular weights than the molecular weight of the carrier gas, peaks of opposite directions will appear on the chromatogram. He and H_2 carrier gases have also been used for the solution of certain problems.

Operation conditions. Guillemin *et al.*[86] have studied in detail the operation conditions of the gas density balance. The sensitivity of the detector depends on the nature of the carrier gas and also on flow conditions in the reference and measuring branches, on the temperature and the parameters (quality) of the flowmeter. These authors claim[85, 86] that a maximum response is obtained with a Reynolds number of 20 in the reference branch. Now in the case of a constant reference gas rate V_R and for a given quantity of sample the detector response decreases with increasing gas rate in the measuring branch, V_M. For satisfactory operation the ratio V_R/V_M should be greater than one and depending on the properties of the carrier gas, maximum response will be obtained at $V_R/V_M = 1.3–3$. When examining the flow rate in the measuring branch it must be borne in mind that the rate of the carrier gas has to correspond to the optimum range for separation of components.

The sensitivity of the detector decreases with increasing temperature. This is due partly to the sensing method based on the measurement of heat conductivity and is partly the result of the higher gas viscosities at higher temperatures. Thus for instance under otherwise identical conditions the sensitivity of the detector is five times higher at 50 °C than at 250 °C. As flow rate is measured on the basis of heat conductivity, as with the heat conductivity cell, sensitivity here too greatly depends on the number of sensing elements (two or four) and on their type (resistance wire or thermistor).

Quantitative evaluation. The most important characteristic of the gas density balance is the fact that provided the component is known the intensity of the response can be accurately calculated and no calibration will be necessary. In the linear range of the detector the quantity of the component is proportional to the area below the peak:

$$Q = fA \qquad 4.11$$

From the molecular weights of the carrier gas and of the component the proportionality factor f can be calculated:

$$f = \frac{M_C}{M_C - M_R} \qquad 4.12$$

where M_C is the molecular weight of the component and M_R that of the carrier gas.

From the corrected areas under the various peaks the weight per cent composition of the sample can be calculated directly.

The gas density balance may also be applied to the determination of the molecular weight[87, 88] of an unknown component with the help of Eq. 4.12. By performing the measurement with two carrier gases with different molecular weights (if possible the molecular weight of one of the gases should be higher, that of the other lower than the molecular weight of the component), the molecular weight of the unknown component can be calculated.

Sensitivity. The sensitivity of the gas density balance is the same as that of the heat conductivity cell employing the same resistance elements. The noise level of the detector and consequently the limit of detectability are similar or lower than those of the heat conductivity cell. The factors affecting sensitivity were described there. The noise of the detector is due partly to internal stream splitting and partly to pressure fluctuations.

Dynamic linear range. The dynamic linear range of operation also agrees with that of the heat conductivity cell and ranges from the limit of detectability to 10^3–10^4 concentration range. Maximum permissible concentration can be calculated by the method of Tóth *et al.*[89]

Time constant. The time constant of the gas density balance is the same as that of a normal heat conductivity cell.

Applications. The gas density balance is not selective, it may be used for any gaseous substance. It has no detrimental effect on the sample as the sensing elements are not in contact with the components. It is highly versatile and permits accurate quantitative analysis without calibration. It may also be used for the determination of the molecular weight, i.e. identification of unknown components.

4.6.4 General characteristics of ionization detectors

Ionization detectors are next in importance after heat conductivity cells as sensing devices. The sensitivity of ionization detectors is several powers of ten higher than the sensitivity of the other detectors, their dead volume and time constants are extremely low and in general their dynamic linear range is very wide (10^4–10^7). They are only moderately sensitive to changes in the flow rate, pressure and temperature. Owing to these favourable properties the ionization detectors have manifold applications of which the most important are those with open tubular columns and for trace analysis. Some detector types (e.g. electron capture) are outstandingly sensitive to certain types of components and are thus suitable selective detectors for qualitative identifications. Before dealing with the most important ionization detectors we shall sum up briefly the common physical fundamentals of the ionization methods and the general characteristics of these detectors.[90]

Operation of the ionization detectors is based on the electrical conductance of gases. At normal temperatures and pressures the gases

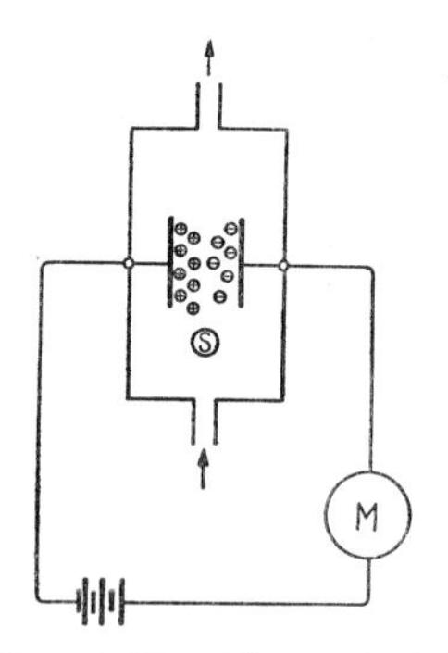

Fig. 4.14. The principle of ionization detectors.[90] *S–ion source; M–ammeter*

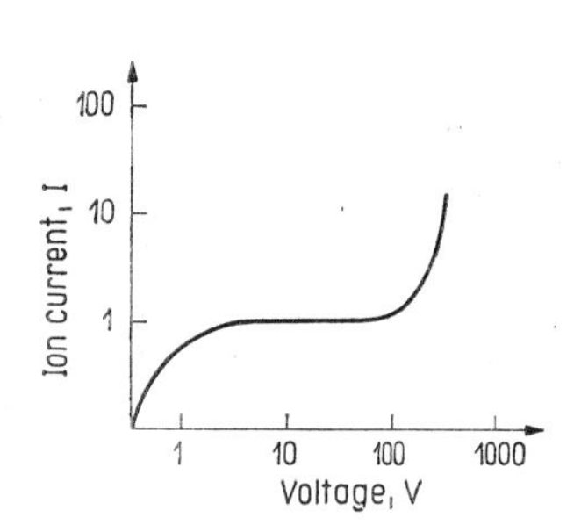

Fig. 4.15. Ion current vs. electrode voltage for ionization detectors

are perfect insulators, but when electrically charged atoms, molecules or free electrons are in some way produced in the gas, the charged particle will be displaced under the action of an external electric field and the gas will become conductive. This effect appears even at a very low quantity of the ionizable component, often not more than a few ions, which explains the high sensitivity of ionization detectors. Figure 4.14 illustrates their principle of operation. The gas flowing in the space between two electrodes is ionized by the energy sources (e.g. a flame or radioactive emitter). The ion current which is produced by the voltage on the electrodes is sensed by the instrument M, after adequate amplification, in the electrical circuit.

There are various ways of causing ionization according to which the detectors are classified into the following groups:

a) Flame ionization detectors; burning in the hydrogen flame.

b) Radioactive radiation detectors; application of α-ray or β-ray radiating preparations. The most important types are: ionization cross-section detector, argon detector, electron capture detector, electron mobility detector and helium detector.

c) Thermal ionization detector; electrons emerging from a heated cathode.

d) Discharge ionization detectors; effect of the electrical discharge between two electrodes.

e) Photo-ionization detectors; ionization with photons of adequate energy.

All these methods of ionization have their limits, as well as their specific advantages, and often complement each other. Detectors of groups a) and b) have the greatest practical importance and will therefore be discussed in greater detail.

Figure 4.15 shows the general relationship between the ion current in the ionization detector and the voltage applied on the electrodes. The horizontal section represents the range in which all the ions reach the electrode (saturation current). Owing to secondary effects at higher potentials the ion current will suddenly rise. In the horizontal section the ion current depends only on the number of ions present and is proportional to the quantity of the component in the detector. Depending on the type of detector the voltage applied on the electrodes varies between 10 and 2000 V. Table 4.2 sums up the general characteristics of the most frequently used ionization detectors.

Electrical circuit. The ion current generated in the ionization detector is very small, in general 10^{-6}–10^{-14} A, which necessitates for its recording amplifiers with high input impedance.

Beside their numerous advantages the ionization detectors require a considerably more complicated and expensive electronic system than the other detectors. The d.c. amplifiers used for the amplification of very low intensity ion currents have to be highly sensitive, possess a low time constant, wide dynamic linear range and good stability. Amplifiers with oscillating capacitors[91] or with electrometer tubes[92–94] are used for this purpose. The electrometer tube amplifiers are the most widely used; a simple circuit of this type is shown diagrammatically in Fig. 4.16. Amplification is controlled by two range switches. The input impedance of the amplifier can be adjusted in steps of powers of ten between 1 and 1000. The output voltage divider reduces the output voltage of the amplifier in binary steps (usually ten steps). The output voltage can be connected directly to the recorder.

Usually the amplifier is placed in a separate module. This unit contains the variable d.c. power supply, the electronic amplifier with

Table 4.2

General characteristics of ionization detectors

Parameter	*Flame ionization detector*	*Cross-section detector*	*Argon detector*	*Electron capture detector*	*Electron mobility detector*
Carrier gas	H_2	H_2	Ar	Ar	Ar
Dynamic linear range	10^6	10^4	10^5	10^3–10^5	10^3
Noise level (A)	10^{-14}	10^{-12}	10^{-12}	10^{-12}	10^{-13}
Background current (A)	10^{-12}	10^{-9}	10^{-8}	10^{-9}	10^{-10}
Limit of detectability (g/sec)	10^{-13}	10^{-7}	10^{-14}	10^{-13}–10^{-14}	10^{-12}
Detectable substances	All organic compounds (except HCOOH)	All compounds	Most organic and inorganic compounds	Halogen, metal, sulphur etc. organic compounds	Permanent gases, organic compounds

an attenuator, zero point set and polarity change. In the latest apparatus the amplifier unit can be equally well used for different types of ionization detectors. Dual channel amplifier units provide for the simultaneous operation of two ionization detectors. The trend is in

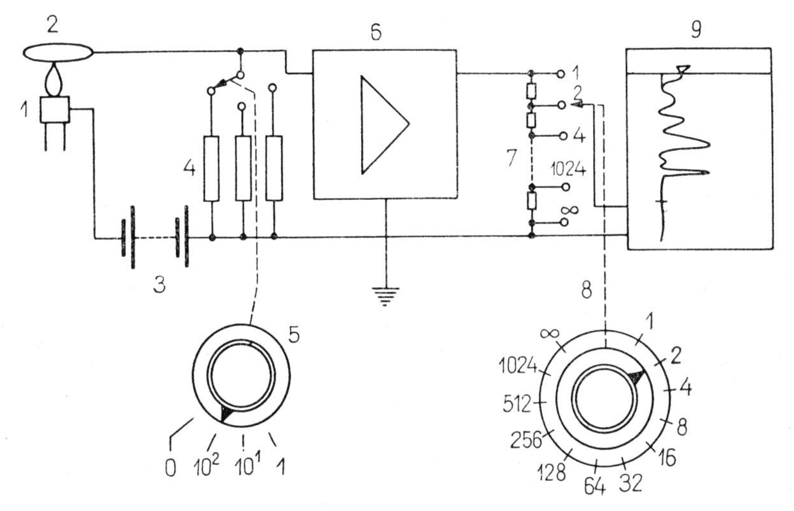

Fig. 4.16. Schematic diagram of the amplification of the ion current. 1. Jet; 2. collector electrode; 3. power supply; 4. input resistance; 5. range switch; 6. electrometer amplifier; 7. output attenuator; 8. range switch; 9. recorder

the direction of electrometer amplifiers producing linear, logarithmic and integral (digital) responses. Essentially the logarithmic response is the result of a continuous attenuation. Beside the generally used linear electrometer amplifiers in recent years a so-called digital log electrometer has been marketed which eliminates the range switch and the attenuator. The logarithmic response can be recorded simply and the simultaneous digital signal can also be recorded with a less expensive counting device instead of an integrator.[95]

4.6.5 Flame ionization detectors (FID)

Principle of operation. Figure 4.17 shows a schematic diagram of the principles of flame ionization detectors. The detector consists of a diffusion type hydrogen burner so that the flame is burning between two electrodes the potential difference between which is 100–300 V. The effluent gas from the column is mixed with an accurately controlled hydrogen stream. Air or oxygen which is required to burn the hydrogen is led through a dust filter at the bottom of the cell into the detector and the combustion products leave the cell through its perforated top. At the temperature of the hydrogen flame (2000–2200 °C) hydrogen too is somewhat ionized producing a constant basic current. When the column effluent contains organic substances these will burn in the hydrogen flame of the detector and produce ions, changing the conductivity of the flame and consequently the intensity of the ion current. The ion current between the two electrodes, one of which is usually the jet itself, can be recorded after appropriate amplification.

Construction. The first flame ionization detectors were constructed almost at the same time by McWilliam and Dewar[96] and Harley, Nel and Pretorius.[97] Since then a great number of researchers have been engaged in the study and development of flame ionization detectors,

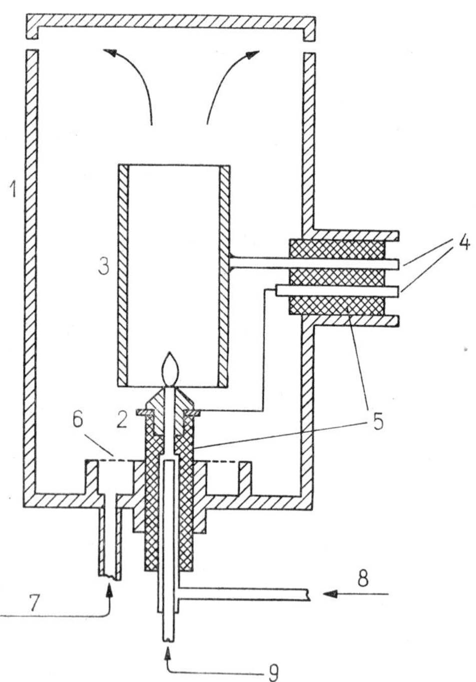

Fig. 4.17. Schematic diagram of a flame ionization detector. 1. Detector body; 2. jet; 3. collector electrode; 4. electrode leads; 5. Teflon insulation; 6. air diffuser; 7. air or oxygen inlet; 8. hydrogen; 9. effluent carrier gas

mainly in connection with the operation of open tubular columns.[47, 98–103] In the 1960's the flame ionization detector rose to the rank of a standard part in gas chromatographic equipment and almost all gas chromatograph manufacturers now have flame ionization detectors of their own construction. The various designs differ

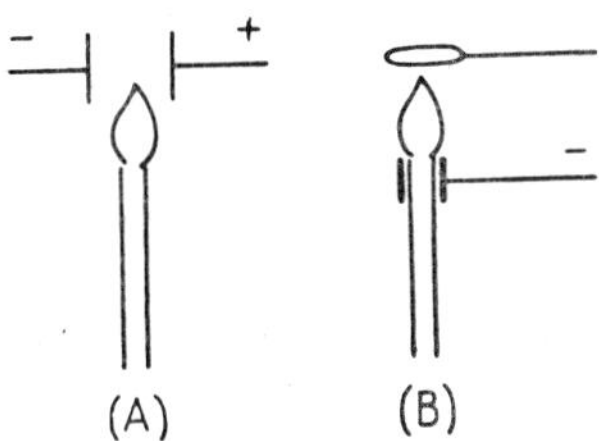

Fig. 4.18. Electrode circuits of flame ionization detectors. (A) Lateral electrodes; (B) the jet as one of the electrodes

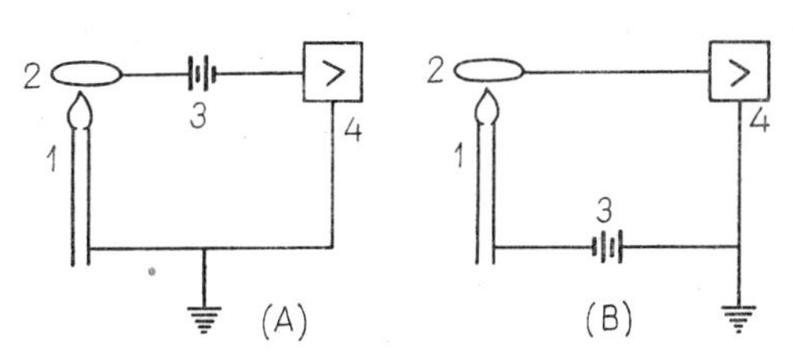

Fig. 4.19. Electrical circuit of flame ionization detectors. (A) Jet and amplifier grounded; (B) jet under voltage; 1. jet; 2. collector electrode; 3. power supply; 4. electrometer amplifier

in the construction of the jet, the shape and arrangement of the electrodes and gas inlets and of the insulation of the electrodes and in the electrical circuit. Bruderreck *et al.*[104] subjected to detailed study the general aspects of the construction of flame ionization detectors and the influence of construction on their operation. The construction of the jet is highly important from the point of view of detector noise. The jet can be made of metal, quartz or ceramics. When the jet is one of the electrodes it must be made of metal, preferably of stainless steel. For satisfactory operation it is important that the outlet end of the tube should not be glowing, that is to say heat conductivity should be adequate. In addition the outlet opening must be of a shape allowing the flame to "sit" as little as possible on the jet, while turbulence at the outlet should be minimum. Precipitates (e.g. oxides) and internal unevenness of the tube may also disturb operation.

The shape and arrangement of the electrodes are important factors from the point of view of detector linearity and quantitative evaluation of the response. Two basic types of electrode arrangements are shown in Fig. 4.18. In type (A) the flame burns between two plate or ring-shaped electrodes, while in type (B) one of the electrodes is the jet and the other electrode is placed above it. In this construction variously shaped (rod, plate, cylindrical, net) counter-electrodes of different dimensions are used, the distance between the two electrodes is 5–15 mm where the optimum value depends on the construction. The electrodes are usually insulated with Teflon and for higher temperatures (> 250 °C) with ceramics or special glass. Perfect insulation is of fundamental importance. The jet is ignited after the removal of the external cylinder or with the help of an incorporated platinum wire which can be made to glow. For programmed temperature analyses a dual channel apparatus has been constructed which is provided with two flame ionization detectors. Usually the two detectors are placed in a common detector block. For the satisfactory operation of the detector the gas inlets have to be adequately constructed to ensure the proper mixing of the carrier gas with the hydrogen. Here adequate feeding of the gas which supports burning (air or oxygen) is of paramount importance to ensure that the flame receives oxygen uniformly from all sides.

Electrical circuit. For the operation of the flame ionization detector a 100–300 V external power supply is required, and for the amplification of the ion current a d.c. amplifier with high input impedance (see above) is needed. Several solutions have been proposed for the connection of the detector to the amplifier, diagrams of the two most frequently used circuits are shown in Fig. 4.19.[105] In type (A) both the jet and the amplifier are grounded, while the polarizing voltage is connected to the grid circuit of the electrometer tube. In type (B) the amplifier and the polarizing power supply are grounded and the jet is at a potential. Circuit (B) is the more frequently used because in circuit (A) the noise level is higher due to the connection between the power supply and the grid circuit. In circuit (A) the voltage source must be a battery, while in (B) any electronically stabilized current source can be used.

Auxiliary devices. The power supply for operating the flame ionization detector and the electronic amplifier for recording were described in the preceding paragraph. The jet must be fed with accurately controlled quantities of hydrogen and air (or oxygen); these are ensured by a system of needle valves, flow regulators and manometers which are built into a separate unit.

Choice of the carrier gas. In principle all carrier gases already mentioned can also be used here. When the carrier gas is hydrogen no other hydrogen stream is needed. Usually nitrogen or argon are used to improve separation and the stable operation of the column and the hydrogen to the jet is admixed before the latter. The perfect removal of all impurities from the carrier gas and from the gases used for combustion is extremely important in the case of the flame ionization detector, just as with all other highly sensitive ionization detectors, as the presence of impurities causes higher noise levels and consequently diminished sensitivity.

Operation conditions. The dynamic linear range and the sensitivity of the flame ionization detector depend on several factors. As ionization efficiency in the flame is low, it is of fundamental importance that the mass velocity of the combustible substance which reaches the detector should also be low. In general the dynamic linear range lies at 10^{-7}–10^{-6} g of carbon per sec mass velocity,[104] and the quantity of the sample, i.e. the mass velocity of the substance in the detector, must be chosen accordingly. The constancy of the ionization current depends also on the polarizing voltage in accordance with the general relationship illustrated in Fig. 4.15. With flame ionization detectors the horizontal range extends to about 300 V. The flow rates of the carrier gas and of hydrogen, and the carrier gas: hydrogen ratio and the nature and quantity of the combustible gas have a significant influence on the noise level and sensitivity of the detector. With increasing hydrogen rates the detector response increases, but so do the basic current and the noise level of the detector. Depending on the construction, 15–50 ml/min of hydrogen rate are considered adequate values. McWilliam[106] has studied the effect of the carrier gas : hydrogen ratio. If this ratio has a value of approximately 1, as for instance in the case of packed columns, response will increase and so will sensitivity fluctuations in the flow rate. With open tubular columns carrier gas rate is of the order of 1–2 ml/min, the effect of which is negligible. For some detectors the optimum N_2 : H_2 ratio is also given. When the quantity of the combustible gas is increased the ion current will first increase steeply and will reach a constant level after a certain value, usually at 400–500 ml/min flow rate.[104] The curves for oxygen and air have a similar profile, but in the case of oxygen the saturation interval ensures about six times higher response than with air. In spite of this fact, mainly air is used as combustion gas today because with oxygen the basic current increases steeply even at 15–20 ml/min hydrogen flow rates, which on the one hand reduces the response: noise ratio and on the other hand raises difficulties in the way of quantitative evaluations. It should be mentioned that maximum sensitivity and quantitative responses make different demands on optimum operating conditions and these can be established only after detailed studies.

Another extremely important factor from the point of view of detector operation is the influence of the liquid which evaporates from the liquid coated columns, since even very low concentrations of this liquid will cause a considerable rise in the basic current. To eliminate this factor the liquid phase and the temperature must be chosen in such a way that evaporation should be negligible. Another recently adopted solution which is especially important in the case of programmed temperature involves the parallel operation of two columns and two detectors when only the carrier gas passes through one of the branches. By differential operation of the two detectors it is possible to suppress the noise due to bleeding.

Quantitative evaluation and calibration. Provided the operating parameters described above are suitably adjusted, the detector response is proportional to the number of ions, that is, to the mass of the component, which reaches the detector. A great number of scientists have been engaged in the study of problems involved in quantitative evaluations with flame ionization detectors, the results are often contradictory. The response to hydrocarbons is strongest. The intensity of the response depends on the number of carbon atoms and on the structure of the compound, intensity decreases with increasing numbers of heteroatoms. The following conclusions can be drawn from the works on the problems of quantitative evaluation and on the determination of response factors.[103, 104, 107–113] For hydrocarbons with carbon numbers over five the response calculated for the unit weight of the component is practically constant and the experimental data show deviations of 2–5%. Thus the weight per cent composition of the sample can be calculated directly from the areas below the peaks. Only in the case of benzene, toluene and of compounds containing triple carbon bonds are the deviations higher. A correction factor has to be applied, should a higher accuracy be required, or should the hydrocarbons differ considerably in number of carbon atoms and structure. The use of relative response factors is in practice the most convenient expedient. The reference component is usually n-heptane. The relative response gives the ratio of the signal corresponding to the unit mass of the component and the response corresponding to the unit mass of n-heptane (n-heptane = 1.00). The weight per cent composition of the sample can be calculated by dividing the areas under the peaks by the relative response values of the components and totalling the areas which had been corrected in this way. The normalization or area factor used by Halász *et al.*[103, 104] is the reciprocal of the relative response factor.

Another method for calibrating detector responses is the use of molar response factors. It was demonstrated that in the case of hydrocarbons the relative molar response (with n-heptane = 700 as reference component) is directly proportional to the number of carbon atoms. This is however strictly valid only within the same homologous series and for different homologous series different linear correlations are obtained. For organic compounds which contain hetero-atoms the response is in general less intense than for hydrocarbons with the same number of carbon atoms. Within the same homologous series the relative molar response is again linearly proportional to the number of carbon atoms.

Responses for different types of compounds can be best compared on the basis of their effective carbon number.[108] This represents the number of aliphatic carbon atoms which gives the same response as the compound under investigation. The effective carbon number can be approximately calculated from the data shown in Table 4.3.

Sensitivity. The sensitivity of the detector depends on the construction and on the operating conditions. Under optimum operating conditions at about 10^{-4} A noise level the limit of detectability is about 8×10^{-4} g atom carbon per sec, or about 10^{-14} mole heptane per sec.

Dynamic linear range. The dynamic linear range is very wide, usually of the order of 10^6–10^7. Its upper limit is a vapour concentration corresponding to 2–3% in the gas stream.

Time constant. The detector has a very low time constant of about 1–10 msec. The time constant of the electrometer is between 0.01 and 1 sec. Thus the time constant of detection is primarily determined by the time constant of the recorder.

Applications. The flame ionization detector is suitable for sensing every organic compound with the exception of formic acid. The detector is insensitive to permanent gases and water vapour, a very important characteristic in the analysis of air contaminations or of very low concentration components in aqueous solutions. Because

Table 4.3

Contribution of bonding types and hetero-atoms to the effective carbon number[103]

Atom	*Bond type*	*Effective carbon number*
C	Aliphatic	1.0
C	Aromatic	1.0
C	Olefinic	0.95
C	Acetylenic	1.30
C	Carbonyl	0.0
C	Nitrile	0.3
O	Ether	—1.0
O	Primary alcohol	—0.6
O	Secondary alcohol	—0.75
O	Tertiary alcohol, esters	—0.25
Cl	Two or more on single aliphatic C	—0.12 each
Cl	On olefinic C	+0.05
N	In amines	similar to 0 in the corresponding alcohol

of its simple construction, high sensitivity and stable operation the flame ionization detector is next to the thermal conductivity cell the most generally used detector.

4.6.6 Modified flame ionization detectors

The "singing flame" ionization detector. The sensitivity of the flame ionization detector can be doubled or trebled by using a so-called singing flame. By an appropriate modification in construction and adjustment of flow rates the diffusion flame begins to vibrate resulting in a much higher ionization efficiency and detector sensitivity.[109] This type of detector is not yet used in commercial equipment.

Sodium thermal ionization or phosphorus detector. This thermionic detector is the most important modified form of the flame ionization detector which is increasingly used in commercial chromatographs.

Principle of operation. A sodium salt is placed above the flame of the flame ionization detector and this salt is thermally ionized, producing thereby a constant ion current between the two electrodes.

This basic current is about 100 times higher than the basic current of the flame ionization detector. If a component containing phosphorus is introduced into the detector, ionization efficiency and the intensity of the ion current both rise steeply. The cause of this phenomenon is yet unknown. Under appropriate operation conditions the detector is fully specific and senses only compounds containing phosphorus. Under certain operation conditions it may however be used for the detection of halogen compounds. The phosphorus detector has a very high sensitivity, the limit of detectability is 10^{-13} g/sec of phosphorus compound, or, in other words, components in ppb concentrations can be detected in 1 μl samples. Because of its selectivity and sensitivity it is an indispensable tool in the analysis of pesticides. In earlier models the electrode above the flame was coated with a sodium salt,[115, 116] but this resulted in a short lifetime and continually diminishing sensitivity. In more recent designs[117] a ceramic cylinder filled with a sodium salt (usually NaBr) or a compressed salt cylinder is placed above the outlet end of the jet. Figure 4.20 shows schematic diagrams of the various types. The latest detectors possess a long lifetime and constant sensitivity.

Dual detectors are built by placing two flame ionization detectors, one on the top of the other, when the lower detector senses the organic and the upper the phosphorus and halogen compounds.[118] Janák and Svojanovsky have investigated the details of the construction and operation problems of dual detectors.[119]

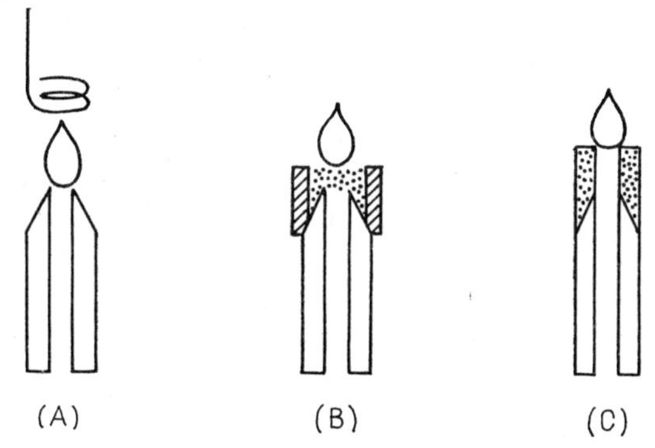

Fig. 4.20. Jet designs for the sodium ionization detector.[117] *(A) Coiled wire coated with salt; (B) ceramic cylinder filled with salt crystals; (C) compressed salt cylinder*

Detection of metal organic compounds. The analysis of alkyl silanes and tin alkyls with the help of the flame ionization detector has already been investigated.[120] Garzó and Fritz[121] described in detail the construction and operation conditions of a detector for the qualitative and quantitative analysis of metal organic compounds. When the flame ionization detector is operated with a hydrogen plus methane flame, a negative peak is obtained in the presence of metal organic compounds and sensitivities in the trace detection range can be achieved.

4.6.7 Ionization cross-section detector

Of the detectors which operate by radioactive radiation, the ionization cross-section detector or by its older name, the β-radiation ionization detector, was the first used in gas chromatography,[36, 122, 123] but later radiation detectors have largely replaced it.

Principle of operation. A β-ray radiating preparation is placed in the detector cell. The gas entering the cell is ionized by this radiation and after appropriate amplification the ion current which is produced by the potential difference between the two electrodes is recorded. The intensity of the ion current is proportional to the number of collisions, that is to the molar concentration and to the ionization cross-section of the molecule. Ionization cross-section depends primarily on the electron shell configuration of the atoms, while the ionization cross-section of the molecule is the sum of the ionization cross-sections of the atoms in the molecule and can be calculated from literature data.[122] This is the only type of ionization detector which is suitable for the analysis of all components including permanent gases. Its sensitivity is lower than that of the other ionization detectors, but may be applied to the detection of high concentrations up to 100%.

Construction. The detector cell should be constructed and the radiation source chosen in such a way that ionization intensity shall be adequate at the lowest possible gas volume of the detector. Perfect insulation of the electrodes is highly important for the measurement of low ionization currents. Figure 4.21 shows a diagram of a typical cell. The frame is built of stainless steel or brass and itself acts as the cathode. The anode is made of the same metal and is separated from the body by a thick Teflon insulation. A thick anode should be used to avoid recombination ion losses, the ratio cell diameter : anode diameter should be a maximum of 3 : 1. Various designs differ in the dimensions of the cell and the anode, in the arrangement of the radiation source and of the gas inlets and outlets. Usually the ionizing

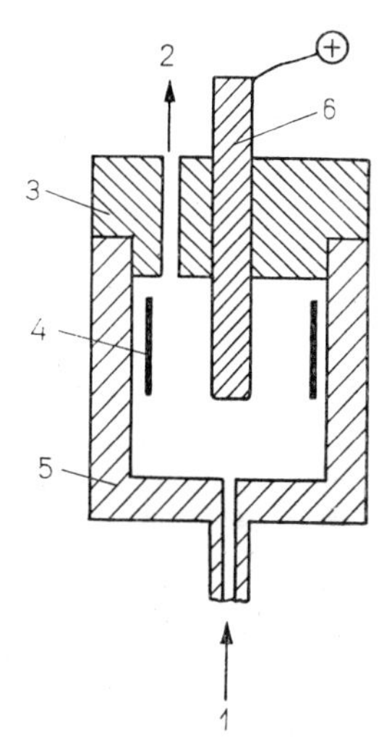

Fig. 4.21. Schematic diagram of an ionization cross-section detector.[90] *1. Gas inlet; 2. gas outlet; 3. Teflon insulation; 4. ion source; 5. brass or steel body (cathode); 6. anode*

source is precipitated on to a silver or gold plate and placed on two sides of the detector (see Fig. 4.21). Intensive β-emitters (^{90}Sr, ^{90}Y, ^{147}Pm, etc.) are the most frequently used radiation sources. In other types the source is placed over the bottom of the cell with lateral gas inlets and outlets or the preparation is precipitated on to the anode. The radiation intensities of the isotopes are 10–25 mC; when handled according to specification, a 10 mC ^{90}Sr preparation represents no radiation danger. The volume of the cell varies between 0.5 and 5 ml, the voltage on the electrodes between 300 and 1000 V. Because of their short free paths, isotopes with weak β-ray or α-ray radiations were found unsatisfactory for classical cross-section detectors. Lovelock *et al.*[124] developed a so-called micro-cross-section detector whose volume is about 10 μl and was able to use tritium or RaD as current sources (100–200 mC) because of the smaller distances. The micro-cross-section detector is today a standard part for the latest types of chromatographs.

Electrical circuit. A stable d.c. power supply is necessary to ensure the 300–1000 V potential difference on the electrodes. High impedance input d.c. amplifiers described earlier are used for the measurement of the very low intensity ion current.

Auxiliary devices. This type of detector requires no special auxiliary devices.

Choice of the carrier gas. The detector can be operated with any of the carrier gases, but H_2 or He provide for the most intense responses, since the ionization cross-section of H_2 and He is considerably smaller than that of other gases and so the first two gases absorb only a very small portion of the radiation and produce a very low basic current. With higher molecular weight carrier gases radiation absorption and basic current are higher, resulting in less sensitive detection of the components under investigation. Earlier mainly nitrogen was used as carrier gas.

Operation conditions. The construction and operation parameters which influence ionization have been described above. Unlike other detectors, the cross-section detector is only slightly sensitive to temperature and flow rate fluctuations. With more stable preparations of ^{90}Sr the highest permissible detector temperature is usually 250 °C, with tritium as radiation source this is about 200 °C.

Quantitative evaluation and calibration. The intensity of the ion current is proportional to the concentration of ions in the detector which is again proportional to the ionization cross-section of the molecules in the detector, and the cross-section is proportional to the molecular weight. As the carrier gas alone produces an ionization current, the concentration of the components can be calculated from the response with the help of calibration factors. The calibration factor is proportional to the difference between the ionization cross-sections of the carrier gas and of the component. If the difference between the molecular ionization cross-sections of the component and the carrier gas is given the symbol ΔQ and the areas under the peaks are multiplied by a factor ΔQ/molecular weight, the weight per cent composition of the sample can be calculated directly.

The ionization cross-section of the molecule can be predicted from the ionization cross-sections of the atoms in the molecule so that the detector can be calibrated without measurement.[36,123]

There are however no experimental data available which would confirm this for compounds of different types.

Sensitivity. The sensitivity of the detector depends on the intensity of the radiation source, the construction of the detector and the nature of the carrier gas. Since the intensity of the ion current depends on the ionization cross-section of the molecules, the detector has a relatively low sensitivity which is in general the same as that of the thermal conductivity cell. The sensitivity of the microdetectors is about ten times higher than that of the thermal conductivity cell. The limit of detectability is about 10^{-7}–10^{-8} g/sec which under the conditions of packed columns corresponds to about 10^{-4} volume % concentration in the carrier gas.

Dynamic linear range. From the limit of detectability up to 100% sample concentration the response is perfectly linear. This is the only ionization detector which is suitable for the sensing of high concentrations.

Time constant. The geometric time constant of the normal cross-section detector is a few seconds depending on the cell volume and flow rate. The time constant of the latest microdetectors is with packed columns 40 msec, with open tubular columns 1–2 seconds, depending on the flow rate of the carrier gas.

Applications. The most important feature of the cross-section detector is its applicability to the analysis of all organic and inorganic substances. This detector complements the other ionization detectors since its dynamic linear range begins at concentrations which are the upper limits of linearity of the other detectors. Because of this property the cross-section ionization detector is especially appropriate in preparative gas chromatography where higher sample concentrations are found.

4.6.8 Argon ionization detector

By operating the ionization cross-section detector with argon carrier gas at higher polarization voltages Lovelock[125] evolved a highly sensitive detector which he called the argon detector.

Principle of operation. The principle of the method is based on the work of Jesse and Sadauskis[126] who showed that before reaching their ionization potentials the rare gases form metastable atoms with long lifetimes and the energy of these atoms is transferred by way of collisions to organic molecules which are present in the gas. The ionization potential of argon is 15.7 eV and the excitation potential of the metastable argon atom is 11.7 eV. The ionization potentials of organic and of most inorganic vapours is less than 11.7 eV, so these vapours are ionized by their collision with excited argon atoms. Because of their higher ionization potentials H_2, N_2, O_2, CO, CO_2, H_2O and the fluorine compounds are not detected with the argon detector and CH_4, C_2H_6, acetonitrile and propionitrile give very low intensity responses.

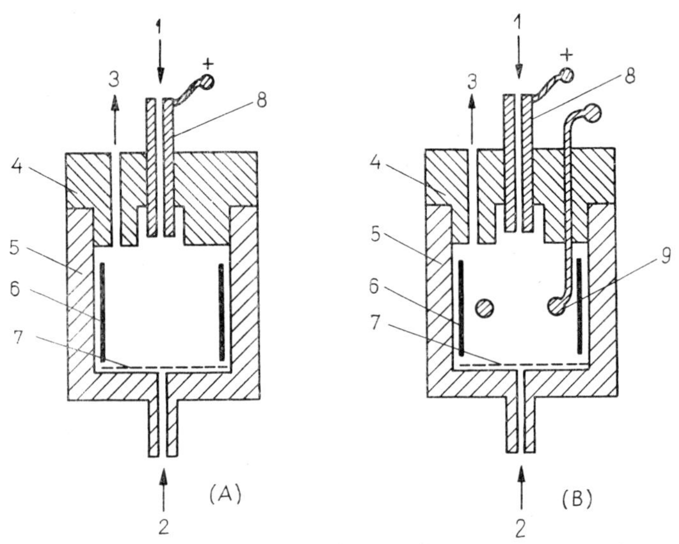

Fig. 4.22. Construction of argon detectors.[129] *(A) Micro-argon detector; (B) triode detector. 1. Carrier gas inlet; 2. sweeping gas inlet; 3. gas outlet; 4. Teflon insulation; 5. metal body (cathode); 6. radiation source; 7. gas diffuser screen; 8. hollow anode; 9. ring electrode*

The schematic diagram of the argon detector is the same as that of the cross-section detector (Fig. 4.21), but the presence of the argon carrier gas raises ionization efficiency and thus the sensitivity of the detector by several orders. Lovelock *et al.* have described in several papers the operation and advantages of the argon detector.[127,128]

Construction. Lovelock[129] studied in detail the construction problems of the argon detector and described several detector types. In the construction of the argon detector the principles already described for the cross-section detector are applicable. The most important point is to ensure as far as possible a uniform electric field in the cell. The argon detector similar to that in Fig. 4.21 was first produced primarily for packed columns, here the cell volume and the time constant are rather high. Depending on the carrier gas the detector may operate either as a cross-section or as an argon detector. Figure 4.22 shows a recent development of this type, the trend is towards highly sensitive microdetectors with small cell volumes mainly for open tubular columns. Diagram (A) in Fig. 4.22 represents a so-called micro-argon detector. The effluent gas from the column is led through a hole in the anode at a maximum rate of 20 ml/min. Through a stream splitter at the bottom of the cell pure argon gas is led in at 50–100 ml/min, this gas rapidly removes the ionized particles. If the flow rate of the carrier gas is higher than 20 ml/min in packed columns, the separate argon stream may be omitted and the carrier gas led in through the joint at the bottom of the cell. The triode detector in the same figure is a further development of the microdetector. With a third ring-shaped electrode between the cathode and the anode the basic ionization current can be suppressed, resulting in a lower noise level and higher detector sensitivity.

The β-ray preparations mentioned in the preceding paragraph, or in certain cases α-radiating preparations (Ra, RaD) can be used as radiation sources, but with the latter lower sensitivities are obtained. The latest detectors mainly use tritium (100–200 mC) sources, as these are the most reliable from both safety and sensitivity aspects.

Electrical circuit. The electrical circuit and the amplification of the response of the argon detector are the same as those of the cross-section detector with the difference of a higher polarization voltage (800–2000 V) in the first case.

Auxiliary devices. The high purity, primarily the dryness of the argon gas is of paramount importance.

Choice of the carrier gas. The detector operates in general with argon carrier gas, but other rare gases, mainly helium and neon are also used. These latter play an important role in the detection of permanent gases which cannot be detected with argon carrier gas (see below).

Operating conditions. The sensitivity of the detector depends on the flow rate of the carrier gas, as at low flow rates ionization efficiency is also low. The optimum range is in general between 20 and 50 ml/min flow rate depending on the volume of the detector. This explains the necessity for a separate argon purging stream in the case of open tubular columns. The detector is only slightly sensitive to fluctuations in flow rate, temperature or pressure. Small quantities of contaminants, primarily water vapour (0.005%), but also CO_2 (0.01%), O_2 (0.05%)

and N_2 (0.1 %) etc. also reduce the sensitivity of the detector. Contaminants liberated from the connecting tubes or moisture adsorbed on the walls of connecting tubes and the detector may also interfere with the operation of the detector. For similar reasons it is highly important that evaporation of the liquid phase should be kept at a minimum, so that e.g. with programmed temperature argon detectors are not used. The dynamic linear range of the detector extends to relatively low concentrations, with 10^{-2}–10^{-3} volume % component concentration the response is no longer proportional to the quantity of the component, while at higher concentrations the peak is split into two or three separate peaks. Consequently this detector can operate only with very small samples.

Quantitative evaluation and calibration. In general the intensity of detector response increases with the molecular weight of the component, but the degree of ionization depends on the probability of collision with the excited argon atoms which is greater in the case of smaller molecules. In addition the degree of ionization depends also on the ionization potential of the molecule, so that the relative response factors cannot be determined by calculation alone and for quantitative evaluation a separate calibration for each component is necessary. The response factors in the literature are uncertain for hydrocarbons and have little practical value.[111] Another obstacle in the way of quantitative evaluation is the fact that the various components and contaminants interfere with each other's ionization. Changes in polarization voltage also have a significant influence on the response factors.

Sensitivity. The argon detector has the highest sensitivity of general detectors with a detectability limit of 10^{-13}–10^{-14} g/sec. Because of this high sensitivity the noise level of the detector is strongly dependent on electrical noise, contaminants in the carrier gas or from the apparatus.

Dynamic linear range. There are different views in the literature on the dynamic linear range of the detector; in general the range 10^3–10^5 is accepted as dynamically linear.

Time constant. The time constant of a well-constructed detector is very low (10^{-3} sec), so that the time constant of the detector system depends on the amplifier and is usually about 1 sec.

Applications. Because of its high sensitivity the detector is primarily suited for trace analyses. Quantitative analyses require appropriate preliminary calibration. With argon carrier gas the detector can be used for the detection of all organic and inorganic compounds, except those listed above. With some other rare gases (helium or neon) as carrier gas its application can be extended to components which are not detectable with argon. Helium[130] and neon[131] ensure highly sensitive trace analysis, as the excitation potentials of these gases (20.8 eV and 21.6 eV, respectively) are considerably higher than that of argon. High purity helium and neon are needed for this purpose and the polarization voltages are considerably lower, in general 200–400 V. A maximum sensitivity was achieved with neon carrier gas containing 0.3% of argon.[132] A highly sensitive micro-helium detector with a wide dynamic linear range has been described for the trace analysis of permanent gases in ppb concentrations.[133]

This detector can be adapted to sensing permanent gases with argon carrier gas by mixing the carrier gas with a small quantity (1–100 ppm) of an organic component (acetylene, ethylene, etc.) which ionize and produce a high intensity ion current. On the appearance of the permanent gas the ion current drops and this change can be detected.[134, 135] This method is also known in the literature by the name of indirect electron mobility detection.[90]

4.6.9 Electron capture detector (ECD)

Principle of operation. The electron capture detector is a special application of the cross-section detector for the detection of compounds with high electron affinity. The method was first applied by Lovelock and Lipsky.[136] Under the influence of radiation free electrons and positive ions are produced in the carrier gas, usually nitrogen, and the charged particles migrate to the appropriate electrodes. The velocity of the electrons is about ten thousand times greater than the velocity of the positive ions so recombination is negligible and a constant ion current appears. When an electronegative compound reaches the detector this will capture the electrons with the production of negative ions. These negative ions combine with the positive ions which had been formed in the carrier gas and the ion current will diminish according to a correlation which resembles the Lambert–Beer law:

$$N = N_0 e^{-kxc} \qquad 4.13$$

where N is the number of electrons which reach the anode, N_0 is the basic ionization current, c the concentration of the electron acceptor component, x the electron capture cross-section of the component and k is a proportionality constant. Thus the detector system responds to the decrease in the constant ion current. The electron capture detector is highly sensitive and selective to certain molecules including halogen

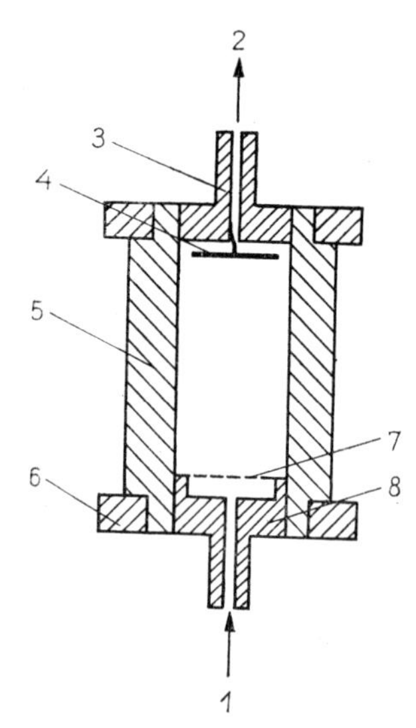

Fig. 4.23. Scehmatic diagram of an electron capture detector.[90] *1. Gas inlet; 2. gas outlet; 3. cathode; 4. radiation source; 5. Teflon insulation; 6. metal jacket; 7. stream splitter; 8. anode*

compounds, conjugated carbonyls, nitriles, nitrates, metal organic compounds, etc., while to other compound types, e.g. hydrocarbons, alcohols, etc. it is practically insensitive. The other important feature of the electron capture detector is the possibility of compensating the negative response by changing the voltage on the electrodes. For different types of compounds this voltage is different, so that by an appropriate adjustment of the voltage functional groups can be detected and unknown compounds identified.

Construction. The first tests were carried out with the standard argon detector whose construction has been described above. Instead of the rod-shaped anode, parallel electrodes provide a higher sensitivity. Figure 4.23 shows a detector with parallel electrodes. The gas enters through the anode and leaves the detector through the cathode. The distance between the electrodes has a significant influence on the intensity of the response.[137] In the latest types the radiation source is precipitated onto the cathode and the anode is made of a wire mesh through which the gas enters.

The preparations described as radiation sources in connection with the foregoing detectors are also used here: tritium was found to be the most suitable from the point of view of both safety and radiation efficiency. Most commercial detectors are equipped with a 100–200

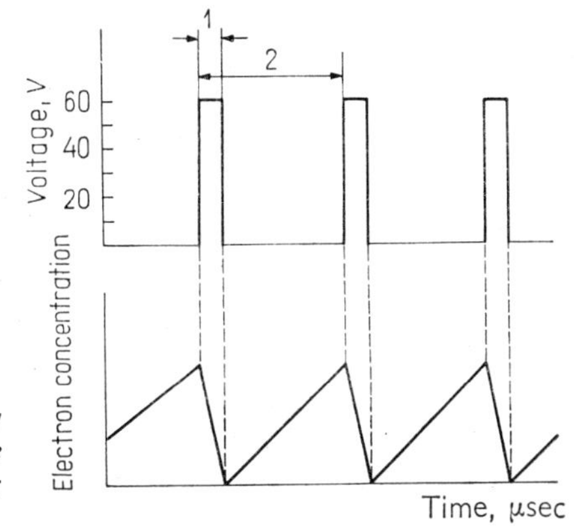

Fig. 4.24. The effect of pulsating voltage on the electron capture detector. 1. Pulse duration: 1 μsec; 2. Pulse interval: 5–150 μsec

mC tritium radiation source which has the single drawback that it cannot be used over 220 °C. Newer models have the temperature range extended to about 350 °C by the use of a ^{63}Ni radiation source extending its application to the analysis of high boiling components.

Electrical circuit. The electron capture detector can be operated in two ways, namely at constant voltage or with pulsating voltage. The power supply of the latest detector types is constructed in such a way that both methods can be applied by means of a switch. Pulsating voltage has the advantage of raising the efficiency of electron capture and consequently the sensitivity of the detector and the dynamic linear range.[138] With increasing polarization voltage the intensity of the response goes through a maximum which depends on the construction and on the operating conditions. Polarization voltages between 2 and 100 V, most frequently between 10 and 50 V are used. The pulse generator produces pulses of constant amplitude and width but of variable periods. The period of the pulses is about 1 μsec, the interval between the pulses generally varies between 5 and 150 μsec. Electron capture and the recombination of ions take place in the intervals and the short pulse periods provide only for the collection of the high velocity electrons. The effect of the other ions in the ion current is negligible. Figure 4.24 shows the effect of the pulsating voltage.

Auxiliary devices. Under some conditions a separate purging gas, the same as the carrier gas, is needed for the optimum operation of the electron capture detector requiring a separate flow control system. The polarizing power supply is either a d.c. power supply or a pulse generator. The voltage and the pulse period can be adjusted by steps. For the amplification of the ion current the usual electrometer is employed. The amplification unit usually contains also the power supply.

Choice of the carrier gas. Various aspects have to be considered in the choice of the carrier gas, the views in the earlier literature are not uniform on this point. Because of the effect of the excited ions the use of rare gases is not advisable. Owing to the low basic ion current intensity and the frequency of collisions, H_2 and He are unfavourable. The requirements which the carrier gas has to meet are different depending whether the constant voltage or the pulsating voltage method is used. For the system operating at constant voltage high purity nitrogen is the most suitable carrier gas, while in the pulse technique argon with 5–10% of CH_4 or 1–2% of H_2 or CO_2 content seems to be the most satisfactory, as the admixture of these components inhibits the formation of excited argon atoms. When using mixtures the constant concentration of the admixed components is of paramount importance.

Operation conditions. The construction of the cell and the effect of the electrical circuit have already been discussed. For the satisfactory operation of the detector the gas flow in it must be appropriate. At low flow rates, e.g. in the case of open tubular columns the detector may easily become overloaded with the sample. To eliminate this danger a separate pure carrier gas stream is added to the column effluent. With the application of sweep gas larger samples can also be analyzed and the dynamic linear range extended. A high purity carrier gas is absolutely necessary for the operation of the detector. Small quantities of the liquid phase in the effluent may also interfere with detector operation.

Quantitative evaluation and calibration. The response of the electron capture detector is proportional to the electron affinity of the component, which involves a very high sensitivity towards certain components and thus selective detection. The detector is suitable chiefly for the quantitative analysis of small amounts of organic and inorganic halogen compounds, certain types of oxygen compounds, and metal organic compounds which cannot be detected by other methods. Detector response depends on the construction of the cell and on the voltage applied, so preliminary calibration is necessary for quantitative evaluations.

Sensitivity. Sensitivity of the detector depends decisively on the nature of the component. Responses differing by a factor of 10^6 may be obtained for compounds with different electron affinities. The detector has the highest sensitivity towards halogen compounds, especially towards compounds containing several halogen atoms. For such substances the detectability limit is 10^{-12}–10^{-13} g/sec.

Dynamic linear range. The dynamic linear range of the detector operated with constant voltage is a maximum of 10^3. This range can be extended up to about 10^5 by applying the pulsating voltage technique and will depend on the magnitude of the pulse intervals.

Time constant. Theoretically the method involves a retardation of the order of milliseconds. In practice the time constant is about 0.5–3 sec, but lower carrier gas flow rates may result in higher values.

Applications. This highly sensitive detector may be applied to the detection of organic halides and other halogen compounds, metal organic compounds, conjugated carbonyls, nitriles, nitrates and sulphur compounds. As a qualitative detector it is suitable for the selective detection of the above compounds and quantitative trace analysis. Owing to these properties the detector is indispensable in chlorinated pesticide analysis, as well as in the analysis of biochemical substances, aromas, etc. In recent years the electron capture detector has become a standard accessory of commercial gas chromatographs. For testing mixtures which contain components of different types it is mostly used in dual channel arrangement with the flame ionization detector. In pesticide analysis a parallel operation of the phosphorus and of the electron capture detector was found to be the most useful.

Dual detector methods have been primarily evolved in connection with the above-mentioned ionization detectors and shall therefore be briefly discussed here. Figure 4.25 shows a diagram of the most frequently employed dual detector methods. Scheme (A) illustrates the differential arrangement used in programmed temperature methods. The two columns contain the same packing and the basic responses of the measuring and reference branches are balanced so that the response due to the evaporation of the liquid phase can be eliminated and a stable base line obtained. (B) is the schematic diagram of a complete dual channel apparatus where two different columns operate in the same apparatus. The two detectors might be of the same or of different types. (C) is the schematic diagram of a dual channel detection where the effluent carrier gas is split into two streams

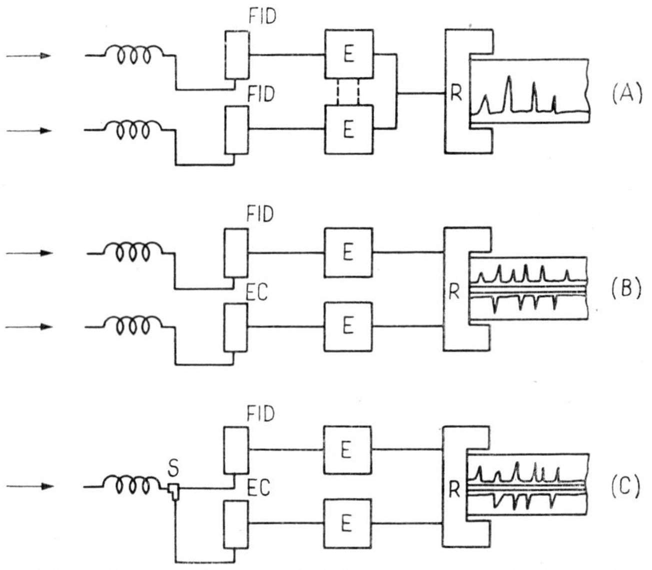

Fig. 4.25. Schematic diagram of dual detector methods. (A) Differential operation; (B) two different columns; (C) dual channel detection. E–Electrometer amplifier; R–recorder; S–splitter

which pass through two detectors of different types to ensure a selective detection of the components. The most frequently used detector combinations are the following: FID–E.C.; FID–phosphorus detector, E.C.–phosphorus detector, FID–cross-section detector. Sometimes three parallel detectors of different types are used for the solution of certain problems. For methods (B) and (C) two separate recorders or one dual channel recorder is required.

4.6.10 Other ionization detectors

The above-described ionization detectors are the most frequently used and are almost the only ones found in commercial equipment.

A great variety of ionization detectors operating on other principles have been made however; in the following we wish to deal with the best known among them.

Electron mobility detector. This detector is also known under the name of pulse or drift velocity detector.[139] Its schematic diagram is like that of the electron capture detector shown in Fig. 4.23, but the direction of gas flow is reversed and the detector operates with argon carrier gas. The electrodes are charged by short, repeated pulses whose period is chosen in such a way that in the case of pure argon the electrons shall not reach the anode. On the appearance of the foreign component the motion of the electrons slows down so that they reach the anode during the period of the pulse. The current which is generated in this way is proportional to the quantity of the component. This detector is specially suited to the trace analysis of permanent gases, CO, CO_2 and water vapour and to the detection of ppb concentration impurities. It is occasionally used also in commercial equipment (e.g. Varian Aerograph).

Glow discharge detector. In normal gas discharge between two electrodes the potential difference depends on the nature of the gas. Potential changes are measured in a Wheatstone bridge circuit.[140] A modified neon tube may also be used as a highly sensitive detector.[141]

Radio-frequency corona discharge detector. The presence of foreign gases changes the radio-frequency corona discharge in a helium atmosphere. Corona discharge is produced by a centrally situated helical wire heated with 40 Mc radio frequency. A potential difference of 50–60 V should be maintained between the wire and the metal wall of the cell.[142]

Spark discharge ionization detector. The electrodes are charged with a voltage to produce discharge only in the presence of some foreign vapour. The capacitor discharge is measured.[143]

Electron collision ionization detector. Its operation resembles that of the mass spectrometer; the gas molecules are ionized by collision with high velocity electrons which emerge from the heated cathode. In the prototype helium was the carrier gas and potentials were applied to ionize only the entering components.[144] Stable electron emission improves the operation of the detector and renders it generally applicable.[145]

Photo-ionization detector. When bombarded with photons having adequate energy the sample components in an inert carrier gas are completely ionized. A discharge tube is used as photon source when the generated ultraviolet light irradiates the carrier gas between the two measuring electrodes[141] and the generated ion current is measured after amplification.

4.6.11 Other methods of detection

a) Flame detectors

Scott's flame detector[147, 148] is based on the combustion of the organic compound in a micro-jet and the measurement of the flame temperature with a sensitive thermocouple. There is a linear relationship between the heat of combustion of the components and the sensitivity of the detector and the response also changes linearly with the concentration of the component. The detector has a simple construction and its sensitivity is the same as that of the thermal conductivity cell.

Light emission flame detectors are modifications of Scott's flame detector. Grant and Vaughan[149,150] measure with a selenium photocell the intensity of the light which accompanies the combustion of the organic compound instead of the temperature of the flame. Organic halogen compounds,[151] phosphorus and sulphur compounds[152] metal halides and chelates[153] can be analyzed in this way. More recently these detectors have been called flame photometric detectors and owing to their high sensitivity and selectivity their application is spreading.

b) Detectors operating by combustion or catalytic conversion

The common feature of these detectors is the chemical conversion of the effluent components prior to actual detection which may be performed by various methods.

This method was first applied in the analysis of organic compounds. When the effluent is led through a microreactor packed with cupric oxide at 750–800 °C the organic compounds are oxidized to carbon dioxide and water. The carbon dioxide is detected after the removal of the water vapour by infrared spectrometry,[154] by titration[155] or by the thermal conductivity cell.[156, 157] Because of its simplicity the thermal conductivity cell is preferred. Instead of combustion over cupric oxide combustion in oxygen at high temperature may also be employed.[158] The conversion of the combustible components into carbon dioxide has several advantages. The molecules of CO_2 generated correspond to the number of carbon atoms, so detection sensitivity rises. No calibration for each component is required for quantitative evaluation, from the area under the CO_2 peak the weight per cent composition can be calculated directly. As the heavy components fail to reach the detector, this can be operated at lower temperatures and with higher sensitivity, while it is also excellently suited to high temperature gas chromatography. The method can be applied to elementary analysis, giving qualitative identification of hydrocarbons if a second column separates the CO_2 from the water vapour. CO_2 determination by titration eliminates expensive integrators and offers a possibility of direct accurate quantitative analysis.[158] It is quite surprising that the practical application of the method is in no way proportional to its merits. Other catalytic methods apart from the combustion method have also been used without having gained general acceptance. Zlatkis *et al.* for instance have converted the hydrocarbons in a microreactor packed with nickel catalyst to methane and water and after the removal of the latter detected the methane with a thermal conductivity cell.[159] This method later became more useful when coupled with a flame ionization detector. CO and CO_2 in gas samples are not sensed by the flame ionization detector, but when reduced to methane they are detected with high sensitivity.[160]

Detection by *electrolytic conductivity* is one of the developments of the combustion method, whereby the carbon dioxide is absorbed in distilled water[161] or aqueous sodium hydroxide[162] and the conductivity of the solution measured. This method is sufficiently sensitive for use with open tubular columns. In the past few years the method has been improved and organic compounds containing S and Cl are now combusted at 800 °C in an oxygen stream, the CO_2, SO_2, SO_3 and HCl formed are absorbed in water and the conductivity of the water is measured in a flow-through cell.[163] The sensitivity of the method for S and Cl compounds is very high, about 10^4 times higher than for C, offering a possibility of selective detection and trace analysis. A highly sensitive and fully selective method has been worked out for the analysis of nitrogen compounds by replacing oxidation with hydrogenation over a catalyst of Ni and absorbing the NH_3 formed.[164] Electrolytic conductivity measurement is insensitive to changes in flow rate, temperature or pressure, and provides a selective detection method for S, Cl and N compounds. Its use is mainly in the analysis of pesticides and similar compounds.

Detection by *coulometric titration* is suitable to the selective detection of S, Cl and P compounds. The oxidation or reduction products are led into a cell which is filled with a suitable titrating solution. This solution is continuously replaced with the help of the servomechanism of the generator electrode. The current necessary for this process is proportional to the quantity of the sample under investigation.[165] Chloride and sulphide ions can be titrated with silver acetate, SO_2 iodometrically and phosphorus in the form of phosphine also argentimetrically. Dohrman's microcoulometer which is provided with a pyrolysis reactor for both oxidation and reduction and special exchangeable titration cells for determining halogen, sulphur and nitrogen compounds is excellently suited for this purpose. Because of its high sensitivity the method is suited mainly to the analysis of pesticides, pharmaceuticals, biochemical substances, etc.

Recently a so-called *reaction coulometer* has been described for the analysis of organic compounds.[166] The column effluent is mixed with a constant stream of electrolytically generated oxygen and the organic components are combusted in a reactor. Oxygen consumption is sensed by a sensitive electrochemical cell and the oxygen is replaced by means of a feed-back from an oxygen generator. The method is highly sensitive and can be used in a wide concentration range. The coulometric method has the considerable advantage over the other methods that it provides means for highly accurate quantitative analysis and for the calculation of the response factors.

c) Measurement of physical properties

The measurements of many parameters of the carrier gas have been used so far for detection of which the more interesting are: the surface tension detector,[167] measurement of flow resistance, measurement of sound velocity,[168] application of the gas interferometer, and measurement of absorption in the far ultraviolet.[169] The ultrasonic detector is perhaps the most important[170, 171] in which ultrasonic frequency changes on the appearance of the component and the frequency difference can be measured with high accuracy. The construction of the detector is very simple and it is already used in commercial equipment.

d) Other methods

Various titration methods.[172] Absorption of carbon dioxide carrier gas in a burette filled with potash solution and measurement of the change in the gas volume[173] or in the pressure at constant volume[174] (Janák's method) has been used. There are on the market some simple gas analytical apparatuses using carbon dioxide as carrier gas. Gravimetric detection involves the adsorption of the effluent components and determination with a sensitive microbalance.[175] Biological detectors have also been described.[176]

4.6.12 Application of spectrometers

Gas chromatography is primarily a separation method and often the identification of the separated components requires specific physical and chemical methods. The most reliable and detailed information is obtained by spectrometric methods, and mass spectrometers, infrared spectrometers, ultraviolet and nuclear magnetic resonance spectrometers are equally used for the identification of the components and for the examination of unresolved peaks. The spectrometer is a secondary detector and is connected in series with the detector of the gas chromatograph, or operates parallel to the destructive detector. These methods require costly spectrometers and with the exception of the mass spectrometer large samples, while their analysis time is also rather long. The mass spectrometer has found the widest practical application and many publications deal with the combination of gas chromatography and mass spectrometry.

In the majority of these methods the effluent gas from the chromatograph is continuously fed into the ion source of the mass spectrometer.

Mass spectrometry can be performed by two methods:

a) The mass spectrometer is focused on an appropriately chosen fixed mass number and the analysis is repeated for different mass numbers.[177, 178]

b) A selected range of the spectrum or the entire spectrum is oscillographically[178] displayed or recorded on a photographic plate.[179, 180]

Recording on the photo-plate is most satisfactory with the Mattauch–Herzog type mass spectrometer, because it records simultaneously the entire spectrum. This eliminates any effect which changes in concentration with time may have and also the necessity of rapid scanning as the peak emerges.

With the high power, double focus adjustment mass spectrometers a very high resolution of 1 : 10,000 can be achieved; this is adequate for the detailed study of the structure of chemical compounds and is applicable up to mass numbers of at least 1000. However, considerably less expensive smaller mass spectrometers are quite adequate for identification purposes, for example, the Atlas GD 150 with a resolving power of 80 may be applied up to a mass number of 300.[181]

The sensitivity of the mass spectrometer is the same as that of the flame ionization detector, it also ensures a wide dynamic linear range and can be used with open tubular columns too. With its help compounds containing e.g. Si, B or metal atoms and compounds which interfere with the operation of other detectors can also be identified.

Recently several excellent reviews have been published on the joint use of the gas chromatograph and the mass spectrometer, including a detailed survey of their various fields of application.[182–184]

The infrared spectrometer is quite satisfactory for the identification of certain types of compounds. Earlier an intermittent method was used, the separated components (fractions) were collected in receivers and examined successively.[185] More recently fast scanning apparatus has been constructed for the continuous analysis of the gas stream.[186]

Quite recently the intermittent operation of the gas chromatograph has also been solved,[187] involving the automatic interruption of the carrier gas stream after each component and its parallel feeding into the mass spectrometer and infrared spectrometer. The method has the advantage that though it employs simple and cheap spectrometers it furnishes highly detailed data.

The possibilities and implementation of the joint application of gas chromatography and infrared spectrometry are presented in Littlewood's excellent review.[188]

4.6.13 Detection of radioactive compounds

Because of its high separation power and great sensitivity gas chromatography has acquired an important role in the analysis of compounds with radioactive components, i.e. labelled atoms (^{3}H, ^{14}C, ^{32}P, etc.).[189, 190] The effluent gas from the column is led through a second detector which is connected in series or parallel to the gas chromatographic detector. In this way a gas chromatogram and a radiochromatogram are obtained of the sample under investigation and the nature and quantity of the compounds with the labelled atom can be determined. The radioactive components may be detected with a Geiger–Müller counter, a ratemeter or scintillation counter. The detector is coupled to a pulse counter scaler, usually a ratemeter whose output signal is recorded.[191] The most up-to-date ratemeters produce directly an integral signal. In another method for the detection of compounds with ^{3}H and ^{14}C atoms the effluent components are led through a combustion chamber over CuO where they are oxidized to CO_2 and H_2O, the H_2O is led over an iron catalyst where it is reduced to H_2 and this mixture of CO_2 and H_2 is led to the radioactive detector.

An isotope exchange (^{3}H) on the prepared gas chromatographic column itself and the detection of the labelled compound by the radioactive detector is a new possibility of qualitative identification.[192]

4.7 Recorders and integrators

4.7.1 Recorders

The response of the detector is plotted by the recorder which furnishes the chromatogram, the result of the analysis. As recorders, electronic chart potentiometers are commonly used which are servo-operated potential balances. They operate on the principle of balancing the input signal by a feed-back signal of the same intensity but of the opposite sign. The difference between the input and feed-back signals reaches through an amplifier a servo-motor which moves the pen. There are many different chart potentiometers on the market and the parameters of about 100 types of recorders were described.[193] The recorder in gas chromatography has to meet the following requirements:

1. High sensitivity; the full-scale deflection should vary in general between 0.1 and 10 mV, the span of the most frequently used instruments is 1 mV.

2. High pen-speed, the time for scanning 90% of the scale should be 1–2 sec, but there are instruments which need only 0.25–0.5 sec.

3. High accuracy: 0.2–0.5% of the entire scale.

4. Accurate zero-line adjustment, continuously with a Hg cell or Zener diode.

5. The chart recorder should have a low inertia (small dead range) and should ensure undisturbed recording.

6. Simple adjustment of the chart speed between wide limits (usually between 100 and 1500 mm/hour).

7. Automatic reduction of sensitivity after 90% of the span has been reached.

8. Linear operation over the entire scale and in the different sensitivity ranges.

The recorders are of the box or bench top type. The width of the paper is generally 150 or 250 mm. More recently dual channel recorders with two pens have been produced which are particularly suitable for methods with two detectors. The useful paper width on one side is about 100 mm. The latest types produce both linear and logarithmic responses.

The most frequently used recorders are the Honeywell–Brown, Philips, Leeds and Northrup Speedomax, Texas–Servo–Ritter, Kent Mark 3, etc.

The practising gas chromatographer may find many valuable directives in Hay's report[194] describing the general operation of compensographs, their accurate setting and control, including the detection and elimination of the most frequently occurring errors.

Because of their low pen-speed these recorders are not suitable to the high speed analyses which have evolved in recent years and the chromatogram is then recorded with a cathode ray oscilloscope.

4.7.2 Integrators

The differential curve, the chromatogram, produced by the detector is quantitatively evaluated by measuring the area of the Gaussian curve. An essentially higher accuracy is achieved when, instead of approximate calculations and manual planimetry, automatic integrators are used. To improve the speed and accuracy of the analyses the use of integrators has been accepted in recent years as being indispensable for accurate quantitative analysis.

Mechanical, electromechanical and electronic integrators are equally used in gas chromatography.

Of the mechanical integrators, the disc integrator measures and records the number of rotations of a cylinder rotated by a friction gear of two balls on a rotating disc driven by a synchronous motor. The instantaneous rotation number of the disc is proportional to the height of the peak scanned at this instant, so that the total of the rotation numbers is proportional to the area under the peak.[193] The dis-integrator can be fitted into commercial recorders, the accuracy with which it measures the area is $\pm 0.1\%$, it produces a reproducible analogous signal which changes linearly in the entire range and plots these signals on the chart of the recorder under the differential curve.

The analogous output signal of the disc integrator can be converted with the help of an appropriate transformer into a digital signal which will give the numerical value of the area under the peak.

In the electromechanical integrator the potential signal which is proportional to the deflection of the recorder is converted by an electromechanical unit into a frequency signal proportional to the voltage and this signal operates a mechanical counter thus producing a digital signal. Frequently a low inertia d.c. motor is used. The rotation number of the motor is directly proportional to the voltage, the axis of the motor operates the counter. In another design the light pulses obtained through the opening of a rotating disc driven by the motor are sensed by a photocell.

The electronic integrators operate on a similar principle, namely on voltage-frequency transformation, but contain no moving parts, have no inertia and work at a great speed. There are certain differences in the integrating circuits of the various electronic integrators. In one arrangement the voltage is totalled up, in another, more commonly used, the pulse frequency changes. The voltage totalling circuit produces a signal[195] which can be recorded e.g. with a compensograph or converted to a digital signal by a separate signal transformer. The design based on the change of pulse number furnishes digital signals directly. Because of its speed and accuracy the latter is now generally used and is known by the name of digital electronic integrators.

The trend in the development of integrators is to separate the integrator system from the recording potentiometer, to improve the accuracy and speed of the integrator system and to ensure digital signals which can be readily processed. The range of integrators on the market has greatly increased and many gas chromatograph manufacturers have produced their own types.

The most modern high power electronic digital integrators possess a wide dynamic linear range and high counting speeds of 10^3 to 10^6 signals/sec.

These integrators are provided with automatic zero correction and base line drift correction. A memory unit is inserted between the integrator and printer to increase operation speed. The integrator also has a built-in or connectable printing unit which operates automatically when the signal reaches the base line or the lowest point of the valley between two peaks. Some types are capable of adding up the areas under the peaks and to record the retention time of the peaks, i.e. the position of the peak maximum. With these integrators analysis is fully automatic after the injection of the sample and requires no further supervision. The reproducibility of the areas and area ratios is $\pm$0.2–0.4%. These integrators may operate also without a separate recorder.

Several reports have been published recently on the operation and advantages of up-to-date digital integrators.[196, 197]

There is a considerable difference in price between the various integrators; a disc integrator may cost £ 250 or about 600 U.S. dollars, while the price of electronic digital integrators varies between £ 500 and £ 3000 or 1200 and 7000 U.S. dollars.

In the past few years new types of integrators have also been evolved, of which the electronic integrator operating on the basis of light impulses and the application of an optical follow-up system should be mentioned.[198]

The application of digital computers is a further development in the evaluation of gas chromatographic data.[199,200] The digital computers in general use may be employed for the simultaneous integration of signals from several gas chromatographs and the simultaneous calculation of percentage compositions. The signals can be fed either directly or from a magnetic tape store. Magnetic tape signal storage is also used in combination with high power integrators which are capable of servicing several analyzers simultaneously. Computers are used in the USA in refineries and chemical plants which operate a great number of gas chromatographs.

The most recent trend of development aims at the elaboration of special systems for the automatic processing and evaluation of the gas chromatographic signal.[201–204] Achievements in the field of digital computers have led to the elaboration of relatively cheap small gas chromatographic computers[205, 206] which may be accessible even for medium size laboratories. These systems are capable of serving simultaneously several gas chromatographs. By using appropriate correction factors the output data refer directly to the percentage composition of the sample and to other characteristics of the chromatogram (peak number, retention times, relative retention, component quality, etc.). A computer system and program have been worked out for the automatic identification of the components of complex mixtures from the retention data measured on two columns at the same time.[207] Computer processing of chromatogram data offers a possibility of calculating with high accuracy the H values and other column parameters for various components.

4.8 Auxiliary and ancillary devices

Because of the many variations and the great number of auxiliary and ancillary devices ordering a gas chromatograph is no easy task. By the basic unit we understand the analyzer without recorder. Many firms do not include the column in the basic unit. From the more recent equipment built of modules several combinations can be set up: isothermal and programmed temperature thermostat, with different types of single or dual detectors and relevant electronic units, with various sample injector and flow regulator systems. The most important alternatives and ancillaries are the following:

Carrier gas system. Flow regulator unit with manometer, pressure or flow control in single or dual channel construction: separate control unit for supplying H_2 and air to the flame ionization detector, in single or dual channel construction. Flowmeters.

Thermostat. Column thermostat for isothermal or for isothermal and programmed temperature. Separate detector thermostat for one or two detectors. Separate thermostat for the heated sample injector head.

Sample injectors. Heated or on column injector heads. Gas sampling valve with separate fitting. Injector for solid substances. Stream splitter for injection into open tubular columns. Stream splitter to by-pass the detector after the column or to operate two detectors.

Column. Metal and glass columns with various diameters and lengths. Back-flush valve to reverse carrier gas flow. Switch block or valve for the serial, parallel or alternative operation of two or more columns. Open tubular columns with appropriate fixtures.

Detectors. Various types of detectors for the alternative or parallel operation of two detectors. The most frequently used detectors: heat conductivity cell, flame ionization, electron capture, micro-ionization cross-section (micro-argon) detectors, gas density balance.

Electrical units. Temperature control unit to the thermostat with special regulators for thermostating the detector and the sample injector. Electric power supply for the heat conductivity cell. D.C. amplifier and power supply for the ionization detector.

The recorder and integrator require separate specifications.

Ancillary devices:

Temperature programmer.
Carrier gas flow rate programmer.
Automatic sample injectors.
Outer column or pre-column.
Switch valves for back-flush and change in column arrangement.
Pyrolysis unit for pyrolysis gas chromatography.
Microreactor for catalytic reactions.
Preparative column, with separate sample injector.
Sample collector system for preparative columns.
Cryostat for low temperature operations.
Electrolytic hydrogen generator for the flame ionization detector.
Carrier gas filters.
Reserve columns of various diameters and lengths.
Solid supports and adsorbents of various types and different mesh sizes.
Liquid phases of different types.
Packings coated with different liquid phases.
Vibrator for packing the column.

Spare parts:

Syringes of various volumes.
Silicone septums for the sample injector.
O-ring seals and joints for assembling.
Resistance thermometer.
Spare parts of the temperature control unit.
Flowmeter.
Thermal conductivity cell or resistance wires fixed on a holder.
Spare parts for the d.c. amplifier.

Spare parts for the recorder: chart paper, pen, ink, dry battery, electron valves, vibrator, etc.
Spares for the integrator.

References

1. R. P. W. SCOTT, *Laboratory Practice*, **15**, 882 (1966)
2. Anon. *Laboratory Practice*, **15**, 1017 (1966)
3. L. SZEPESY, S. KESZTHELYI and I. GULYÁS, Publications of MÁFKI (Hungarian Oil and Gas Research Institute) **2**, p. 302, Műszaki Könyvkiadó, Budapest (1961)
4. M. DIMBAT, P. E. PORTER and F. H. STROSS, *Anal. Chem.*, **28**, 290 (1956)
5. Beckman Gas Chromatography Applications Manual Bulletin 756-A, (1960)
6. L. GUILD, S. BINGHAM and F. AUL, Chapter 1, ref. 60, p. 226
7. J. H. PURNELL, "*Gas Chromatography*", J. Wiley, New York (1962)
8. F. VAN DE CRAATS, *Anal. Chim. Acta*, **14**, 136 (1956)
9. Wilkens Instr. Co., Aerograph Research Notes, Fall Issue (1961)
10. Hamilton Co. Bulletin H-65 (1965)
11. G. R. UMBREIT, *Facts and Methods* (Hewlett-Packard F. & M. Sci. Div.), **7**, No. 5 (1966)
12. N. H. RAY, *J. Appl. Chem.*, **4**, 21 (1954)
13. F. R. CROPPER and A. HEYWOOD, Chapter 1, ref. 58, p. 316
14. H. M. TENNEY and R. J. HARRIS, *Anal. Chem.*, **29**, 317 (1957)
15. S. W. MCCREADIE and A. F. WILLIAMS, *J. Appl. Chem.*, **7**, 47 (1957)
16. J. JODLIK and V. BAZANT, *Chem. Listy*, **53**, 277 (1959)
17. H. E. DUBSKY and J. JANÁK, *J. Chromatog.*, **4**, 11 (1960)
18. E. P. SAMSEL and J. C. ALDRICH, *Anal. Chem.*, **31**, 1288 (1959)
19. P. FEJES, J. ENGELHARDT and G. SCHAY, *J. Chromatog.*, **11**, 151 (1963)
20. R. A. MEYER, Chapter 1, ref. 59, p. 93
21. L. D. QUIN and M. E. HOBBS, *Anal. Chem.*, **30**, 1400 (1958)
22. J. L. OGILVIE, M. C. SIMMONS and G. P. HINDS, *Anal. Chem.*, **30**, 25 (1958)
23. R. L. BOWMAN and A. KARMAN, *Nature*, **182**, 1233 (1958)
24. Hamilton Co. Bulletin. H-65 SS (1965)
25. D. H. DESTY, A. GOLDUP and B. F. WHYMAN, *J. Inst. Petroleum*, **45**, 429 (1959)
26. S. P. VANGO, *Chemist-Analyst*, **52**, 53 (1963)
27. M. F. BEDNAS and D. S. RUSSEL, *Can. J. Chem.*, **42**, 1249 (1964)
28. Carlo Erba Short Notes 1–67 (1967)
29. Carlo Erba Short Notes 2–66 (1966)
30. M. W. RUCHELMAN, *J. Gas Chromatog.*, **4**, 265 (1966)
31. J. ARNOLD and H. M. FALES, *J. Gas Chromatog.*, **3**, 131 (1965)
32. R. D. CONDON, *Anal. Chem.*, **31**, 1717 (1959)
33. F. A. BRUNER and G. P. CARTONI, *Anal. Chem.*, **36**, 1522 (1964)
34. L. S. ETTRE *et al.*, *J. Gas Chromatog.*, **3**, 181 (1965)
35. A. T. JAMES, Chapter 1, ref. 58, p. 217

36. H. Boer, Chapter 1, ref. 58, p. 169
37. G. Schomburg, *Z. anal. Chem.*, **189,** 14 (1962)
38. J. C. Sternberg, Chapter 1, ref. 68, p. 161
39. J. Halász, *Anal. Chem.*, **36,** 1428 (1964)
40. K. Derge, *Chemiker Ztg.*, **89,** 247 (1965)
41. A. T. James and A. J. Martin, *Biochem. J.*, **50,** 679 (1952)
42. J. Janák, *Chem. Listy*, **47,** 464, 700, 817 (1953)
43. D. H. Desty, *Nature*, **179,** 242 (1957)
44. H. W. Johnson and F. H. Stross, *Anal. Chem.*, **31,** 1206 (1959)
45. J. G. Young, Chapter 1, ref. 31, p. 75
46. J. E. Lovelock, *Anal. Chem.*, **33,** 162 (1961)
47. D. H. Desty, C. J. Geach and A. Goldup, Chapter 1, ref. 63, p. 46
48. A. B. Littlewood, "*Gas Chromatography: Principles, techniques, and applications*", Academic Press, London (1962)
49. R. Kaiser, "*Gas Phase Chromatography*" (English translation of Vols I–III of "Chromatographie in der Gasphase"), Butterworths, London (1963)
50. R. E. Johnson, Chapter 1, ref. 25, p. 163
51. A. E. Lawson and J. M. Miller, *J. Gas Chromatog.*, **4,** 273 (1966)
52. Gow Mac prospectus.
53. J. A. Petrocelli, *Anal. Chem.*, **35,** 2220 (1963)
54. D. L. Camin, R. W. King and S. D. Shawhan, *Anal. Chem.*, **36,** 1175 (1964)
55. G. Préau and G. Guiochon, *J. Gas Chromatog.*, **4,** 343 (1966)
56. R. L. Faley and J. F. Long, *Anal. Chem.*, **32,** 302 (1960)
57. J. C. Browning and J. O. Watts, *Anal. Chem.*, **29,** 24 (1957)
58. C. E. Bennett, S. Dal Nogare, L. W. Safranski and C. D. Lewis, *Anal. Chem.*, **30,** 898 (1958)
59. S. Dal Nogare and C. E. Bennett, *Anal. Chem.*, **30,** 1157 (1958)
60. A. D. Davis and G. A. Howard, *J. Appl. Chem.*, **8,** 183 (1958)
61. G. Dijkstra, Chapter 1, ref. 58, p. 74
62. N. H. Ray, *Nature*, **182,** 1663 (1958)
63. N. H. Ray, Chapter 1, ref. 64, p. 127
64. L. J. Schmauch and R. A. Dinerstein, *Nature*, **183,** 673 (1959)
65. W. A. Wiseman, *Nature*, **183,** 1321 (1959)
66. E. M. Fredericks, M. Dimbat and F. H. Stross, *Nature*, **184,** 54 (1959)
67. A. B. Littlewood, *Nature*, **184,** 1631 (1959)
68. J. Jordan and B. B. Kebbekus, *Anal. Chem.*, **37,** 1348 (1965)
69. J. M. Miller and A. I. Lawson, *Anal. Chem.*, **37,** 1572 (1965)
70. B. B. Kebbekus *et al.*, *J. Amer. Chem. Soc.*, **88,** 2398 (1966)
71. L. J. Schmauch and R. A. Dinerstein, *Anal. Chem.*, **32,** 343 (1960)
72. H. H. Hausdorff, Chapter 1, ref. 58, p. 377
73. R. M. Soemantri and H. J. Waterman, *J. Inst. Petroleum*, **43,** 94 (1957)
74. J. S. Lewis and H. V. Patton, Chapter 1, ref. 59, p. 145
75. L. J. Nunez, W. H. Armstrong and H. W. Cogswell, *Anal. Chem.*, **29,** 1165 (1957)
76. A. I. M. Keulemans, A. Kwantes and G. W. Rynders, *Anal. Chim. Acta*, **16,** 29 (1957)
77. D. M. Rosie and R. L. Grob, *Anal. Chem.*, **29,** 1263 (1957)
78. A. F. Messner *et al.*, *Anal. Chem.*, **31,** 230 (1959)

79. L. J. SCHMAUCH, *Anal. Chem.*, **31,** 225 (1959)
80. A. J. P. MARTIN and A. T. JAMES, *Biochem. J.*, **63,** 138 (1956)
81. C. W. MUNDAY and G. R. PRIMAVESI, Chapter 1, ref. 58, p. 146
82. Gow Mac Bulletin SB 1065
83. A. G. NERHEIM, *Anal. Chem.*, **35,** 1640 (1963)
84. C. L. GUILLEMIN and M. F. AURICOURT, *J. Gas Chromatog.*, **1,** 24 (1963)
85. C. L. GUILLEMIN and M. F. AURICOURT, *J. Gas Chromatog.*, **2,** 156 (1964)
86. C. L. GUILLEMIN, M. P. AURICOURT and P. BLAISE, *J. Gas Chromatog.*, **4,** 338 (1966)
87. A. LIBERTI, L. CONTI and V. CRESCENZI, *Nature*, **178,** 1067 (1956)
88. C. S. G. PHILLIPS and P. L. TIMMS, *J. Chromatog.*, **5,** 131 (1961)
89. P. TÓTH, E. KUGLER and E. KOVÁTS, *Helv. Chim. Acta*, **42,** 2519 (1959)
90. J. E. LOVELOCK, *Anal. Chem.*, **33,** 162 (1961)
91. F. SCHUBERT, Chapter 1, ref. 65, p. 226
92. L. ONGKIEHONG, Chapter 1, ref. 63, p. 7
93. R. D. CONDON, P. R. SCHOLLY and W. AVERILL, Chapter 1, ref. 63, p. 30
94. M. KUHL, Chapter 1, ref. 65, p. 163
95. Victoreen Bulletin GC 4000, Victoreen Instr. Co.
96. J. G. McWILLIAM and R. A. DEWAR, Chapter 1, ref. 60, p. 142
97. J. HARLEY, W. NEL and V. PRETORIUS, *Nature*, **181,** 177 (1958)
98. L. ONGKIEHONG, "*The Hydrogen Flame Ionisation Detector*", Dissertation. Eindhoven (1960)
99. R. D. CONDON, *Anal. Chem.*, **31,** 1717 (1959)
100. I. HALÁSZ and G. SCHREYER, *Chemie Ing. Techn.*, **32,** 675 (1960)
101. J. O. WALKER, *Oil Gas J.*, **61,** (14) 78 (1963)
102. I. HALÁSZ and W. SCHNEIDER, *Anal. Chem.*, **33,** 978 (1961)
103. I. HALÁSZ and W. SCHNEIDER, Chapter 1, ref. 67, p. 287
104. H. BRUDERRECK, W. SCHNEIDER and I. HALÁSZ, *Anal. Chem.*, **36,** 461 (1964)
105. J. M. GILL, F. BAUMANN and C. H. HARTMANN, Aerograph Research Notes, Fall 66, Varian Aerograph (1966)
106. J. G. McWILLIAM, *J. Chromatog.*, **6,** 110 (1961)
107. R. A. DEWER, *J. Chromatog.*, **6,** 312 (1961)
108. J. C. STERNBERG, W. S. GALLOWAY and T. L. JONES, Chapter 1, ref. 67, p. 231
109. L. S. ETTRE and H. N. CLAUDY, *Chem. in Canada*, **12,** (9) 32 (1960)
110. L. S. ETTRE and W. AVERILL, *Anal. Chem.*, **33,** 680 (1961)
111. L. S. ETTRE, Chapter 1, ref. 67, p. 307
112. L. S. ETTRE, *J. Chromatog.*, **8,** 525 (1962)
113. A. J. ANDREATCH and R. FEINLAND, *Anal. Chem.*, **32,** 1021 (1960)
114. L. B. GREIFF, *J. Gas Chromatog.*, **3,** 155 (1965)
115. A. KARMEN and L. GIUFFRIDA, *Nature*, **201,** 1204 (1964)
116. A. KARMEN, *J. Gas Chromatog.*, **3,** 336 (1965)
117. C. H. HARTMANN, Aerograph Research Notes, Summer, 66. Varian Aerograph (1966)
118. A. KARMEN, *Anal. Chem.*, **36,** 1416 (1964)
119. J. JANÁK and V. SVOJANOVSKY, Chapter 1, ref. 73, Paper 13
120. F. H. POLLARD, G. NICKLESS and P. C. UDEN, *J. Chromatog.*, **14,** 1 (1964)
121. G. GARZÓ and D. FRITZ, Chapter 1, ref. 73, Paper 12

122. J. W. ÖTVÖS and J. P. STEVENSON, *J. Amer. Chem. Soc.*, **78,** 549 (1956)
123. C. H. DEAL, J. W. ÖTVÖS, V. N. SMITH and P. S. ZUCCO, *Anal. Chem.*, **28,** 1958 (1956)
124. J. E. LOVELOCK, G. R. SHOEMAKER and A. ZLATKIS, *Anal. Chem.*, **35,** 460 (1963)
125. J. E. LOVELOCK, *J. Chromatog.*, **1,** 35 (1958)
126. W. P. JESSE and J. SADAUSKIS, *Phys. Rev.*, **100,** 1755 (1955)
127. J. E. LOVELOCK, *Nature*, **182,** 1663 (1958)
128. J. E. LOVELOCK, A. T. JAMES and E. A. PIPER, *Ann. N. Y. Acad. Sci.*, **72,** 720 (1959)
129. J. E. LOVELOCK, Chapter 1, ref. 63, p. 16
130. R. BERRY, *Nature*, **188,** 578 (1960)
131. D. GNAUCK, *Z. analyt. Chem.*, **189,** 124 (1962)
132. G. GNAUCK and H. SCHON, Chapter 1, ref. 71, p. 143
133. C. H. HARTMANN and K. P. DIMICK, *J. Gas Chromatog.*, **4,** 163 (1966)
134. V. WILLIS, *Nature*, **184,** 894 (1959)
135. R. LESSER, *Angew. Chem.*, **72,** 775 (1960)
136. J. E. LOVELOCK and S. R. LIPSKY, *J. Amer. Chem. Soc.*, **82,** 431 (1960)
137. P. F. WASHBROOK, *Chemiker Ztg.*, *Chem. Apparatur*, **86,** 377 (1962)
138. J. E. LOVELOCK, *Anal. Chem.*, **35,** 474 (1963)
139. J. E. LOVELOCK, *Nature*, **185,** 49 (1960)
140. J. HARLY and V. PRETORIUS, *Nature*, **178,** 1244, (1956)
141. R. C. PITKETHLY, *Anal. Chem.*, **30,** 1309 (1958)
142. A. KARMEN and N. L. BOWMAN, *Ann. N. Y. Acad. Sci.*, **72,** 714 (1959)
143. J. E. LOVELOCK, *Nature*, **181,** 1460 (1958)
144. S. A. RYCE and K. A. BRYCE, *Can. J. Chem.*, **35,** 1293 (1957)
145. L. V. GUILD, M. J. LLOYD and F. AUL, Chapter 1, ref. 64, p. 91
146. J. E. LOVELOCK, *Nature*, **188,** 401 (1960)
147. R. P. W. SCOTT, Chapter 1, ref. 58, p. 131
148. M. M. WIRTH, Chapter 1, ref. 58, p. 154
149. D. W. GRANT and F. A. VAUGHAN, Chapter 1, ref. 58, p. 413
150. D. W. GRANT, Chapter 1, ref. 60, p. 153
151. F. A. GUNTHER, R. C. BLINN and D. E. OTT, *Anal. Chem.*, **34,** 302 (1962)
152. S. S. BRODY and J. E. CHANEY *J. Gas Chromatog.*, **4,** 42 (1966)
153. R. S. JUVET and R. P. DURBIN, *Anal. Chem.*, **38,** 565 (1966)
154. A. E. MARTIN and J. SMART, *Nature*, **175,** 422 (1955)
155. L. BLOM and L. EDELHAUSEN, *Anal. Chim. Acta*, **15,** 559 (1956)
156. S. D. NOREM, Chapter 1, ref. 59, p. 191
157. V. STUVE, Chapter 1, ref. 60, p. 178
158. L. BLOM, L. EDELHAUSEN and T. SMEETS, *Z. analyt. Chem.*, **189,** 91 (1962)
159. A. ZLATKIS and J. F. ORÓ, *Anal. Chem.*, **30,** 1156 (1958)
160. K. PORSER and D. H. VOLMAN, *Anal. Chem.*, **34,** 748 (1962)
161. O. PIRINGER, E. TATARU and M. PASCALAU, *J. Gas Chromatog.*, **2,** 104 (1964)
162. A. DIJKSTRA *et al.*, *J. Gas Chromatog.*, **2,** 180 (1964)
163. D. M. COULSON, *J. Gas Chromatog.*, **3,** 134 (1965)
164. D. M. COULSON, *J. Gas Chromatog.*, **4,** 285 (1966)
165. H. P. BURCHFIELD *et al.*, *J. Gas Chromatog.*, **3,** 28 (1965)
166. G. BURTON *et al.*, Chapter 1, ref. 73, p. 193

167. J. Griffiths, D. James and C. G. S. Phillips, *Analyst*, **77**, 897 (1952)
168. A. E. Martin, *Nature*, **178**, 407 (1956)
169. H. K. Testerman and P. C. McLeod, Chapter 1, ref. 67, p. 183
170. F. W. Noble, K. Abel and P. W. Cook, *Anal. Chem.*, **36**, 1421 (1964)
171. W. J. Kaye, *Anal. Chem.*, **34**, 287 (1962)
172. A. G. McInnes, Chapter 1, ref. 58, p. 304
173. J. Janák, Chapter 1, ref. 58, p. 247
174. F. van de Craats, *Anal. Chim. Acta*, **14**, 136 (1956)
175. S. C. Bevan and S. Thorburn, *Chem. in Britain*, **1**, 206 (1965)
176. E. Bayer, *Angew. Chem.*, **69**, 732 (1957)
177. D. Henneberg, *Z. anal. Chem.*, **183**, 12 (1961)
178. D. Henneberg and G. Schomburg, Chapter 1, ref. 66, p. 191
179. R. Ryhage, *Anal. Chem.*, **36**, 759 (1964)
180. C. Brunnee, L. Jenckel and K. Kronenberger, *Z. anal. Chem.*, **189**, 50 (1962)
181. J. T. Watson and K. Biemann, *Anal. Chem.*, **37**, 844 (1965)
182. J. A. Vollin, W. Simon and R. Kaiser, *Z. anal. Chem.*, **229**, 1 (1967)
183. K. Van Cauwenberghe, M. Vandewalle and M. Verzele, *J. Gas Chromatog.*, **6**, 72 (1968)
184. A. B. Littlewood, *Chromatographia*, **1**, 37 (1968)
185. J. C. Holmes and F. A. Morrel, *Appl. Spectr.*, **11**, 86 (1967)
186. A. M. Bartz and H. D. Ruhl, *Anal. Chem.*, **36**, 1892 (1964)
187. R. P. W. Scott *et al.*, Chapter 1, ref. 73, p. 20
188. A. B. Littlewood, *Chromatographia*, **1**, 223 (1968)
189. J. B. Ewans and J. E. Willard, *J. Amer. Chem. Soc.*, **78**, 2908 (1956)
190. R. Wolfgang and F. S. Rowland, *Anal. Chem.*, **30**, 903 (1958)
191. R. Pearson, W. C. Fink and A. A. Gordus, *J. Gas Chromatog.*, **3**, 381 (1965)
192. T. Bálint and L. Szepesy, *J. Chromatog.*, **30**, 433 (1967)
193. Disc Instr. Inc. Bulletin 200 (1965)
194. P. Hay, *Chromatographia*, **1**, 265 (1968) and **1**, 343 (1968)
195. Pye Bulletin, Pye Integrating Amplifier
196. F. Baumann and F. Tao, *J. Gas Chromatog.*, **5**, 621 (1967)
197. G. Vallet, *Chromatographia*, **1**, 336 (1968)
198. O. Weber, *Z. anal. Chem.*, **194**, 334 (1963)
199. R. D. McCullough, *J. Gas Chromatog.*, **5**, 635 (1967)
200. R. N. Sauer, Chapter 1, ref. 73a, p. 20
201. J. G. Karohl, *J. Gas Chromatog.*, **5**, 627 (1967)
202. J. T. Shank and H. E. Persinger, *J. Gas Chromatog.*, **5**, 631 (1967)
203. Y. Vollers, A. Schuringa and C. C. Horns, Chapter 1, ref. 73a, p. 19
204. E. J. Levy *et al.*, Chapter 1, ref. 73a, p. 24
205. F. Baumann *et al.*, Chapter 1, ref. 73a, p. 22
206. J. G. Karohl, Chapter 1, ref. 73a, p. 23
207. C. Merritt, J. T. Walsh and D. H. Robertson, Chapter 1, ref. 73a, p. 21

5. Choice of Columns and Stationary Phases

5.1 Column types

Gas chromatography is a separation method and the "heart" of the gas chromatograph is the column in which this separation is effected. The other parts of the equipment, the carrier gas system, sample injector, and thermostat serve to ensure the appropriate conditions for separation in the column to detect the separated components and to evaluate the response obtained by means of detector, recorder and integrator.

The fundamental condition required of the gas chromatograph is that it should operate smoothly at optimum separating conditions and under a constant work load. For a given analytical task the possible variations of the operating conditions are rather limited and for the gas chromatographer the choice of the column and of the stationary phase in the column are the most important factors. It should be pointed out here that even the most expensive apparatus will fail to give the desired results with an inadequate stationary phase or imperfect column preparation. The fact that a multitude of different solids (GSC) and liquids (GLC) may be used as the stationary phase has greatly contributed to the rapid acceptance and wide-spread application of gas chromatographic methods, so that a suitable column may be found even for the most highly complex separation problems.

Up to 1957 only packed columns were used in gas chromatography with small particle size adsorbents or liquid coated solid supports as packings. After 1957 the open tubular, or by its older name the capillary, column invented by Golay came into use; in this column the internal surface of the tube is coated with the liquid stationary phase. A few years later new, more efficient types were evolved within these two main types.

The most useful way to classify and compare the different types of columns is the method of Halász *et al.*[1] as illustrated by Fig. 5.1. The three basic types are the packed column, the packed capillary and the open capillary which differ considerably in their operating conditions, firstly in their flow parameters (so-called permeabilities), phase ratios (V_G/V_F) and mass transfer resistances. From the point of view of their characteristics and operation the packed capillary columns occupy a place between the packed and the open tubular

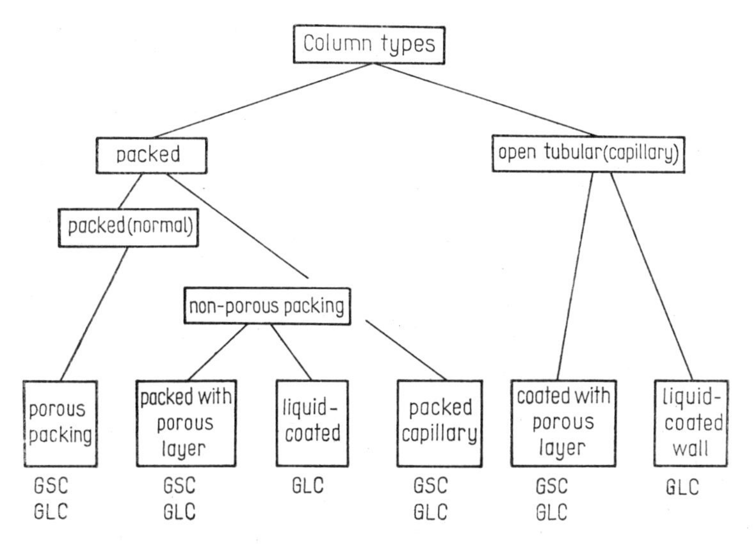

Fig. 5.1. Classification of column types

columns. Loose packing improves the internal contact surface and liquid distribution, while capillary columns lack the influences due to flow unevenness in the packed columns. The permeability of packed columns depends primarily on the particle size of the packing, while that of open tubular columns is a function of the internal diameter with a hundred to a thousand times higher permeabilities than in the first type. The essential difference between the packed columns and the packed capillary columns is the difference in the ratio of internal diameter to packing particle size. In the case of packed columns this ratio is 8–10, with an internal diameter of 2–4 mm in general. The packed capillary columns[2] are packed loosely and irregularly, the diameter to particle size ratio is less than 3–5 with internal diameters of 1 mm or less. Here packing serves primarily to increase the internal surface and the permeability is between that of the packed and the open tubular column. From the point of view of capacity ratio and plate number for separation (see Chapter 3) this latter type represents a transition between the other two with the important feature of requiring a shorter time for the same separation than the packed column.

Depending on the structure of the packing the packed columns may be classified into two groups. To the first group belong the most frequently used columns packed with a porous solid adsorbent (GSC) or a liquid coated solid support (GLC). In recent years non-porous packings (e.g. glass beads) have been developed which may be either coated directly with liquid or first with a thin layer of porous solid acting as an adsorbent or as a support for the liquid phase.

Capillary columns may also be used with adsorbent packings or with liquid coated packings.

In the open tubular columns with diameters between 0.2 and 0.5 mm the internal surface of the tube is coated with the stationary phase. As with non-porous packings two types may be distinguished, when the internal surface of the tube is coated with a thin layer of a porous solid (adsorbent or solid support) it may be used either for gas–solid or gas–liquid chromatography. In the "classic" type of open tubular column the internal wall of the tube is coated with a thin liquid film.

Of these column types the packed columns are in practice most important and are mainly used with porous packings. Many different types of packings are now on the market, but packings may easily be prepared in the laboratory and the filling of the column is not a difficult task. Only a few years ago the applicability of open tubular or capillary columns was the subject of much lively discussion, but today they are widely accepted. Their preparation is a considerably more complicated procedure, mainly from the aspect of reproducibility. The packed and open tubular columns more or less complement each other; very long, sometimes several hundred meters long, highly efficient open tubular columns may be constructed which are therefore suitable for the separation of complex mixtures, containing closely related components, while shorter open tubular columns with 0.5–1 mm internal diameter offer possibilities of very rapid analyses.

The more recent types, namely the packed column with non-porous packing, the packed capillary column and the open tubular column coated with a solid layer are used today mainly in the laboratory for the solution of special tasks. They were however found more effective for certain analyses than the older column types and it may be expected that they will soon find wider practical application.

In the following we shall scrutinize the general principles to be observed in the choice of the stationary phase of the most frequently used column types and in column preparation.

5.2 Packed columns

5.2.1 Adsorbents (GSC)

In gas–solid chromatography various adsorbents are used as column packings. Though in the first applications of gas chromatography adsorbent packings were used, with the growth of gas–liquid chromatography their application became restricted to a narrow field mainly in gas analysis. Recent years have however seen a considerable progress in the field of GSC with respect to both new adsorbents and applications.

Industrial adsorbents. Earlier exclusively commercial adsorbents with high specific surface were used as packings, of which the most important are: active charcoal (specific surface area 1000 m^2/g), silica gel (200–800 m^2/g), alumina (300–600 m^2/g), molecular sieves 4A, 5A, 10X and 13X type (artificial zeolites).

Because they have high specific surface the adsorbents possess a very high adsorption capacity. Energetically the surface of commercial adsorbents is inhomogeneous, which means that the most highly active areas are saturated first, followed by the gradual saturation of the other areas in the order of their decreasing activity. It has been demonstrated in the investigation of adsorption equilibrium (Chapter 2) that the adsorption isotherm may have different shapes, but is usually a concave curve, that is to say, the relationship between the concentrations in the adsorbed and gas phases is non-linear. As a result of the non-linear equilibrium correlation the gas chromatographic peaks are asymmetric with a sharp front and diffuse tail (tailing, see Fig. 2.6). Another consequence of the non-linear isotherm is the dependence of the retention volume, which characterizes the nature of the component, on the quantity of the sample. It should be pointed out here that all isotherms begin with a linear section at low concentrations, but in case of non-homogeneous surfaces this is restricted to a very narrow range. For any given adsorbate the linear range of the isotherm increases with increasing temperature and the influence of surface inhomogeneity diminishes. Accordingly adsorbent packings are mainly used at higher temperatures, usually 50–200 °C above the boiling points of the components to be separated. Thus, for instance, permanent gases are generally analyzed at room temperature, while with C_4–C_5 hydrocarbons it is best to work at 50–100 °C. These adsorbents are not suitable for the analysis of liquids, especially of polar liquids because of strong interactions, and are therefore primarily applied to the analyses of various gas mixtures. According to the selectivity of the adsorption process, polar and non-polar, or by the new terminology, specific and non-specific adsorbents are distinguished. Active charcoal is one of the non-polar adsorbents, though it is somewhat polar in practice because of the presence of the activating agent. On it the strength of adsorption, i.e. the order in which the components emerge, is determined by the boiling point or molecular weight of the components. Silica gel, alumina, etc. may be considered polar adsorbents on which the order of separation depends on the molecular structure. These adsorbents retain for instance unsaturated hydrocarbons in preference to the saturated ones with the same number of carbon atoms.

The molecular sieves (4A, 5A, 10X, 13X) are a special group of adsorbents with a well-defined pore structure containing 4, 5 and 11 Å diameter channels and effect separation according to the size of the molecule. Their most important application is the analysis of permanent and rare gases.

Adsorbent packings are important in the analysis of liquid samples when they are used for preliminary separation. In adsorbent packed short pre-columns, the heavy components of the sample may be retained. Another possibility is the extraction of certain classes of compounds from the sample, e.g. n-paraffins may be separated from branched-chain hydrocarbons with the help of a 5A molecular sieve pre-column, or the olefins may be retained on sulphuric acid treated silica gel. In recent years gas–solid chromatography, i.e. the application of adsorbents, has again gained in importance.[3] This is partly associated with the development in high temperature gas chromatography and programmed temperature methods, as at temperatures above 300 °C there are only few liquids which can be used as stationary phases and virtually only solid packings will stand temperatures above 400 °C. To extend the application possibilities of gas–solid chromatography adsorbents have been treated by various methods to reduce and homogenize their surface, new adsorbents have been developed and many solids tried as adsorbents.

Modification of the adsorbents. In the first trials the adsorbents were coated with small quantities of high boiling liquids to cover the most highly active areas.[4] A far more effective way of surface modification is sintering at high temperature, treatment with water vapour or by chemical transformation. Kiselev's research team[5, 6, 7] has carried out detailed work on the modification of silica gels. They have thoroughly studied the effect of surface and pore structure on the adsorption of compounds of various types and have prepared special adsorbents for the analysis of high boiling components.[8, 9] Scott has modified the surface of alumina by heating with 20–40% NaOH or KOH at 400 °C.[10] The surface of alumina may be modified in a similar manner by treatment with inorganic salts (NaCl, NaI).[3] These modified adsorbents produce symmetrical peaks at lower temperatures, and adsorbent packings with different selectivities as alternatives to liquid coated packings can be prepared.

Another trend in the modification of adsorbents is to coat them with some finely powdered solid or polymer, e.g. graphitized carbon black or Teflon.

Development of new adsorbent packings. Recently a magnesia–silica gel marketed by the trade name "Florosil" has found wide-spread application among commercial adsorbents, mainly in the analysis of saturated and unsaturated hydrocarbons. Many other new adsorbent packings have also been developed in the past few years. By the heat treatment of carbon black at 3000 °C a completely non-polar substance with highly homogeneous surface may be obtained which is also applicable to the analysis of strongly polar compounds (amines, alcohols, etc.).[7, 11] Kirkland prepared from fibrous boehmite ("Baymal" colloidal alumina) a selective packing[12] which he also used for coating glass beads and open tubular columns.[13] Porous glass beads were also used as adsorbent.[14] Non-porous glass beads coated with graphitized carbon black or metal oxide layer provide satisfactory packings, mainly for programmed temperature analysis.[15]

A new trend in the application of adsorbent packings is the use of inorganic substances, mainly of heavy metal salts.[16] Because of their small surface areas their capacity is low, but they furnish symmetrical peaks and for different components their selectivity may be different. Selective separation was observed also on the surface of organic crystals.[17]

Of the new adsorbent packings the polymer gels first introduced by Hollis are no doubt the most important.[18, 19] These polymer gels are firm, compact small particles with a regular microporous structure and are prepared from polyaromatic resins, obtained by the copolymerization of styrene and divinylbenzene or ethylvinylbenzene. Their specific surface varies between 50 and 700 m^2/g and their apparent density between 0.2 and 0.4 g/ml. The gel itself acts as the stationary phase without additional liquid coating, but the mechanism of its action is not yet clear. The polymer gel packings may be employed for many tasks when other packings raise great difficulties. At lower temperatures they are excellently suited to the analysis of various types of light hydrocarbons, at higher temperatures to various sulphur compounds, halogenated hydrocarbons, oxygen compounds, etc. It is one of their important features that with their help strongly polar compounds, e.g. alcohols, amines, water etc. may be eluted with symmetric peaks. They play an important role in programmed temperature analysis where they help to avoid bleeding of the liquid phase. Various trade names appeared in quick succession on the market (Porapak, Polypak, Phasepak, Chromosorb 102 and 101). Gels with different properties and selectivities are now sold, thus Porapak can be obtained in five grades. They are generally designed for use without liquid phase, but to achieve some special selectivity they may be coated with liquid. Without liquid phase the packing can be used up to 250 °C. This type of packing will probably appear in an even greater assortment in the near future.

The modification of adsorbents and the appearance of new types have all happened in the past few years, but they hold great promise. The trend is to develop suitable "tailor-made" adsorbents for the analysis of high boiling compounds and for special separation tasks.

With respect to the comparison and evaluation of GSC and GLC the opinions in the literature differ. The drawbacks of GSC are mainly associated with the difficulties encountered in the preparation and standardization of the adsorbents. The production of adsorbents is not an accurately reproducible process and there are differences not only between the same material from different firms, but even between different batches from the same firm. In the development of the adsorbent surface not only its chemical identity but also the reproducibility of its energy distribution (surface activity) are of decisive importance. There is no way for the unambiguous characterization of the adsorbent (see Chapter 3), while in GLC the nature of the liquid phase is unequivocally determined by its purity and application temperature. In the case of adsorbents the sample components may undergo catalytic or pyrolytic decomposition. In spite of these drawbacks adsorbent packings play an important role in gas chromatography and their importance is increasing.

5.2.2 Liquid coated packings (GLC)

a) The solid support

In gas–liquid chromatography the packing is a liquid phase on an inert solid support. The solid support must comply with two contradictory requirements: 1. it should provide for a thin, satisfactorily dispersed liquid film, which is equivalent to saying that the support should have a large active surface and strong affinity towards the liquid; 2. it should be inert towards the components of the sample. When choosing the support a compromise must be reached between these two demands; the inert support will give a lower efficiency, nevertheless in the case of certain compounds no active support can be employed because of tailing or some other effect. With active supports their detrimental effect might be reduced with the help of various methods.

A great variety of solids has been applied as support, from high surface area adsorbents to non-porous glass beads. The ideal support should meet the following requirements: 1. mechanical firmness to avoid dust formation; 2. particles with regular, preferably spherical shape to reduce the pressure drop; 3. particles with almost identical dimensions (narrow mesh fraction) to ensure highest efficiency; 4. pore structure should be open and uniform; 5. good wetting; 6. a homogeneous, chemically inert surface towards polar substances.

No such ideal support has been known so far, but recent types indicate a considerable improvement over the older types. Supports made of diatomaceous earth are the most commonly used and are available in a great variety. The quality of the support has a considerable influence on separation efficiency. Thus, in the description of a solid destined for use as support the surface characteristics and the structural parameters are most important. Surface characteristics include the size and chemical nature of the surface, while porosity and pore distribution are the main structural parameters. The surface characteristics determine primarily the participation of the support in the separation process. The structural parameters affect liquid distribution and the ease by which the molecules to be separated may reach the stationary phase, so these are characteristic mainly of column efficiency. No sharp distinction can be made between surface and structural parameters; the two together produce the properties required of the support.

Diatomaceous earth supports

Diatomaceous earth supports are silicious minerals with little other mineral contamination. Their chemical composition differs depending on their geographical origin; in general they contain 70–90% SiO_2, 3–10% Al_2O_3 with smaller quantities of Fe_2O_3, CaO, MgO, Na_2O and K_2O and silanol (Si—OH) and siloxane (Si—O—Si) groups on their surface. Two basic types of supports, namely a pink and a white type, may be prepared from diatomaceous earth depending on the conditions of preparation. The pink quality (P) corresponds as a matter of fact to the material of refractory bricks, it is prepared by crushing, pressing and calcination at temperatures >900 °C. The white quality (W) is produced by the addition of a small quantity of caustic soda and calcination at a high temperature. The chemical composition of the two types is almost identical, but there are essential differences in their physical properties, such as specific surface, pore distribution, and packed density. The characteristic physical properties of diatomaceous earth based supports are shown in Table 5.1. Owing to their surface and structural differences the pink and white supports will show a different gas chromatographic behaviour. The pink support is mechanically strong, with little dust formation and may be coated with 30–40 w% of liquid. Because of its surface activity the separation of polar substances, such as alcohols, ketones, etc. involves considerable tailing, even with a thick liquid coat. Owing to its acidic pH it induces isomerization of certain compounds (e.g. terpenes) or in others dehydration (e.g. unsaturated primary alcohols). For the analysis of non-polar compounds it provides a considerably more efficient packing than the W type. The white support is less firm, but has a more

Table 5.1

Properties of diatomite supports*

Property	*Pink (P)*	*White (W)*	*Chromosorb G*	*Chromosorb A*
pH	6–7	8–9	8.5	7.1
Volume weight (packed density), g/cm³	0.47	0.24	0.58	0.48
Void volume	0.80	0.90	—	—
Specific surface, m²/g	4.0	1.0	0.5	2.7
Specific surface, m²/cm³	1.88	0.24	0.29	1.3
Pore volume, cm³/g (0.005–100 μ)	1.0	2.78	—	—
Pore volume, cm³/g (greater than 5 μ)	0.23	2.2	—	—
Pore volume, cm³/g (smaller than 5 μ)	0.9	0.6	—	—

* From ref. 26 and Johns–Manville, Technical Bulletin, FF-133A.

highly inert surface and may be employed also for the analysis of polar substances.

The pink type includes the more widely known commercial grades such as Firebrick C–22, Chromosorb P, Gas-Chrom R, Phasesep P, Sterchamol, Anakrom P, etc. To the white type belong Celite 545, Chromosorb W, Gas-Chrom S, Phasesep W, Supasorb etc.

Comparison of the physical properties of supports, study of their efficiency and of the optimum conditions of their applicability was really begun in the 1960's. Some excellent reports have been published on the subject in the past few years[20–25] of which particularly Ottenstein's review[26] should be mentioned. On the basis of these works more and more new support types were evolved to meet the ever increasing demands which the support has to satisfy. A characteristic example here is Chromosorb G which has an even lower surface area than the W type, is very hard, has an alkaline pH and is an efficient support applicable to polar substances. It may be coated with a maximum of 10–15% liquid, but is generally used with 5–8%. Another new type, Chromosorb A, has properties which are between those of the P and W types, may be coated with 25% of liquid and is recommended mainly for preparative work. The main parameters of these two new types of supports are shown in Table 5.1. Similar improved qualities are marketed by other trade names. Some universal supports of diatomaceous earth type have been reported which are claimed to combine the advantages of the older P and W types. The efficient new support types are generally produced by the addition of higher quantities of caustic soda or some other alkali and calcination at higher temperatures with the final addition of other components.

Several methods have been applied to eliminate the adsorption effect of the solid support, i.e. to deactivate its surface. These methods can be classified into four groups: 1. acid or alkaline washing; 2. neutralization of the surface silanol groups; 3. coating the most highly active centres with a small quantity of liquid; 4. coating the support with a solid layer.

Acid washing is one of the oldest methods, its effect is not clear, and highly dubious. Nevertheless there are on the market some acid washed supports.

Alkaline washing is necessary when supports with acidic pH are to be used for the analysis of acid sensitive compounds.

The most generally used deactivation method is the silanization of the surface, that is to say, the conversion of the silanol group into an ether group. This is performed by treating the surface with dimethyldichlorosilane (DMCS) or hexamethyl-disilazane (HMDS).[22, 27] The surface area of the silanized support is somewhat reduced, but even polar substances will produce symmetrical peaks; silanized W may be considered the most highly inert support. However, in the case of strongly polar compounds silanizing alone is not sufficient; the small residual adsorption may be eliminated by coating with 0.1% of a strongly polar liquid, e.g. polyethyleneglycol (PEG).[27] Commercial supports are practically without exception also available in silanized qualities, mainly in HMDS treated form.

Saturation of the surface silanol group may also be achieved by coating with a small quantity of a liquid which has appropriate functional groups. For example Averill[28] obtained symmetrical peaks even for alcohols and ketones by mixing to the liquid phase 0.1–0.2% of a strongly polar liquid (Alkaterge T, Span 80, etc.).

Surface treatment depends also on the type of compound to be analyzed. For the analysis of basic nitrogen compounds the support must first be coated with a few per cent of KOH or K_2CO_3 prior to coating with the liquid phase.[29] In the analysis of fatty acids the addition of a few per cent of phosphoric acid is necessary to avoid tailing.

Another method of surface modification is coating with a solid substance. There were some early attempts to coat the support with silver, but with no avail. Later polymer coatings were tried; on Teflon coated supports the adsorption effect is very low and this combination is suitable for the analysis of highly polar substances.[13] A similar effect can be achieved with polyvinylpyrolidone coatings,[30] however the efficiency of polymer coated packings is considerably lower.

Recently supports have been modified with a solid porous layer. Kirkland[13] obtained highly selective packings with various liquid phases by coating a diatomaceous earth support with the same colloidal silica gel layer which had been mentioned among the adsorbents. Another similar solution is the coating of the support with graphitized carbon black, coupled with various liquid phases. A detailed comparison of the methods for the deactivation of supports is to be found in a recent paper by Ottenstein.[32]

Non-diatomaceous earth supports

Besides diatomaceous earth many other porous and non-porous solids were tried as supports.[31]

Among the *porous supports* are the adsorbents (primarily alumina and silica gel) already described. Owing to their high surface area the earlier adsorbents may be used with ample liquid coating and for special tasks only, however the modified adsorbents are often also satisfactory supports.

The tetrafluoroethylene polymers, marketed by various trade names (Teflon 1, Teflon 6, Haloport F, Chromosorb T, Fluoropak 80, Columpak, etc.) are the most important porous supports. Depending on the method of their preparation these polymers have different structures and properties, with specific surface between 0.2 and 12 m^2/g and pore diameters between 0.5 and 600 μm, but without any clear correlation between their physical parameters and efficiencies.[33, 34] The polytetrafluoroethylene supports all have inert surfaces which make them suitable for the analysis of such highly polar substances as water, ammonia, amines, alcohols, etc., which would produce considerable tailing on diatomaceous earth supports. Because of their inert surface, coating with the liquid phase is however less uniform and the efficiency of the support depends also on the quality of the liquid phase. Packings with polytetrafluoroethylene supports are considerably less efficient than those with diatomaceous earth,[33, 35] due partly to the difficulties of adding the liquid phase and partly to the softness of the material. Only at low temperature can a uniform, compact packing be prepared out of the soft, resilient particles. The efficiency of these packings may be increased with low liquid loading (2–5 w%) and very careful handling.[36] In the analysis of highly polar substances they satisfactorily complement the diatomaceous earth supports.

The use of polymer coating has already been mentioned in connection with the modification of supports.[13] They have the advantage of simplifying the preparation of the packing without a significant improvement in efficiency.

Polyethylene is also used occasionally as a support.

Of the porous supports the Tide type of detergent composed of an inorganic base with about 20% of alkyl-aryl sulphonate detergent has some practical importance. It is marketed in two forms, one is the original material in which the alkyl-aryl sulphonate serves as a partition phase, while in the other type the inorganic material obtained after the extraction of the sulphonate is used as support.[37, 38] This support is primarily suitable for the analysis of strongly basic compounds.

Recently porous glass powders (Vycor, etc.) with rather high surface areas (100–170 m^2/g) and about 25 Å average pore size have appeared on the market.[14]

A number of various porous materials have been tried as supports (sand, unglazed porcelain, carborundum, tin phosphate, etc.) without gaining general acceptance.[39]

As *non-porous supports* various metals were tried unsuccessfully. In this group the glass microbeads[40, 41] are the most important. These are spherical beads with diameters ranging from 0.05 to 0.2 mm; because of their regular shape they give packings with considerably higher gas permeabilities than those produced from irregularly shaped porous supports. The specific surface of the glass beads is small, essentially equal to the geometrical surface, about 0.02–0.04 m^2/g and the beads require only a very little liquid for coating. An efficient packing can be prepared with 0.02–0.5 w% of liquid; more liquid would produce islets at the contact points of the beads resulting in an increased liquid film resistance (i.e. the C_L term in the rate equation). As a result of the small quantity of liquid only small samples can be employed. Columns packed with glass beads are characterized by a high phase ratio ($\beta = V_G/V_L$) and are particularly suitable for the analysis of high boiling samples.[42] The advantage of these packings is that the analysis temperature is 100–200 °C lower than the boiling points of the components. This is of particular importance in the analysis of heat sensitive substances. Literature data on the efficiency of the glass bead packings are not unambiguous, though in general their efficiency lags behind that of the diatomaceous earth supports. It should however be mentioned that a comparison of different packings under identical conditions is not satisfactory. With different packings operation optima are achieved under different conditions and for a realistic comparison the optimum values must be chosen.

In the past few years considerable work has been devoted to the development of glass bead packings with the primary aim to increase the surface areas of the beads. The easiest way to increase the surface area is etching with alkali or hydrogen fluoride.[43] Supports with even higher efficiency can be prepared by coating the beads with a thin layer of a porous solid. Halász and Horváth[15, 44] have coated glass beads with finely powdered diatomaceous earth and metal oxides. And Kirkland[13, 45] modified the glass beads with diatomaceous earth, alumina and silica sols. These supports are able to take up 10–15% of liquid producing thereby supports with efficiencies equal to those prepared from diatomaceous earth. This is a very important trend in the improvement of supports, as it unites the advantages of both types, ensuring an adequate surface for the distribution of the liquid phase without hindering mass transfer in the gas phase (low C_g value) by an internal pore structure.

Several publications have appeared lately on the application possibilities and modifications of glass bead supports, including the effect of the chemical composition of glass and of surface etching.[46–48]

From the point of view of packing efficiency the shape and particle size of the support are also highly important. From particles with a regular spherical shape packings with higher permeabilities can be prepared. Column efficiency rises with decreasing particle size, while the pressure drop increases. In packed columns usually particles with

0.1–0.3 mm diameters are used. Highest efficiency is achieved when the diameters are nearly equal (narrow mesh fraction). Such narrow mesh fractions are also commercially available with 0.02–0.04 mm particle size distribution.

The mechanical firmness of the support is another important aspect when considering the efficiency of the packing. When the support crumbles during handling or column operation the fines will considerably increase the pressure drop and have a detrimental effect on efficiency.

b) Role and choice of the liquid phase

The fact that a great variety of liquids with different properties can be used as the stationary phase, and that the appropriate choice of the liquid phase provides packings suitable for the most diverse separation tasks, greatly contributed to the wide-spread application of gas–liquid chromatography. In recent years we have witnessed a steep rise in the number of liquids used in gas chromatography and today about 200 liquids applicable as stationary phases are commercially available. This high number of liquid phases is an immediate indication of the difficulties encountered in the choice of the appropriate liquid and even at present this choice is mainly performed by trial and error or on the basis of empirical data and correlations. Much work has been devoted in the past few years to the clarification of the interaction between the liquid phase and the solute and to the theoretical foundations of the choice of the liquid phase. For the most common analytical problems a range of suitable liquids has been established, though not even in this field has an agreement been reached in the literature. For the solutions of problems which are encountered in practical work a brief survey of the interaction between the liquid and the solute and of the choice of the appropriate liquid phase are still necessary.

There are three fundamental aspects to be considered in the choice of the liquid phase:

1. The volatility or vapour pressure of the liquid should be negligible at the operating temperature and the liquid should possess an adequate heat stability.

2. No irreversible reaction should take place between the sample components and the liquid phase.

3. It should be a powerful and at the same time a selective solvent for the components.

The first two stipulations permit a preliminary selection. For each liquid its maximum operation temperature is given, usually separately for isothermal and programmed temperature analysis. The permissible vapour pressure at the maximum temperature is about 0.1 mm Hg, this value is strongly dependent on detector sensitivity, as the evaporating liquid increases the noise level of the detector. Consequently with a given liquid phase the maximum applicable temperature is lower with the ionization detector than for example with the thermal conductivity cell. From the point of view of volatility the vapour pressure *vs.* temperature curve is not fully informative, for in general adsorption on the support reduces the vapour pressure of the liquid. The temperature limits in the literature should be critically viewed, as in practice generally lower temperatures are advisable.[49, 50] As well as affecting the detector, evaporation of the liquid also influences the reproducibility of the analytical results and column life. The heat stability of the liquid phase is often reduced by the catalytic effect of the support and contaminants in the carrier gas.

The stipulation that the liquid phase must be inert towards the components of the sample demands that no irreversible chemical reaction should take place between them. In addition the liquid should have a low viscosity and should wet the support to form a uniform film.

Separation in the column is determined by the solvent power (the value of the partition coefficient k) of the liquid phase and its selectivity with respect to the components under investigation (the value of the separation factor $\alpha = k_2/k_1$). In Chapter 3 the number of theoretical plates necessary for the separation of a given pair of components has been discussed. It appears from Eq. 3.42 that the number of plates depends on the value of α and the capacity ratio k' which for a given column is the function of the partition coefficient k. When the value of k is low, i.e. the component is poorly soluble, a higher number of plates will be required. This means that the value of α which characterizes the selectivity of the liquid phase is in itself not sufficient for the evaluation of the latter. Choice of the liquid phase on the basis of selectivity is even more complicated in the analysis of multicomponent mixtures, when the simultaneous separation of different components is required.

Selectivity, in fact, refers to the relative retention of two components, without providing any information on the mechanism of separation. In various systems separation is effected according to the boiling points, molecular weights or structures of the components. To speak about a "selective" stationary phase is meaningless, as every liquid phase is more or less selective, with the degree of selectivity greatly depending on the pair of components which have to be separated.[51]

The relative volatility of the components to be separated, i.e. the separation factor, is determined by the interactions between the liquid phase and the solutes, i.e. the van der Waals cohesion forces. These cohesion forces may be divided into a) London dispersion forces, b) Debye induction forces and c) Keesom orientation forces.

Dispersion forces are due to the attraction of the dipoles which arise from the momentary arrangement of the elementary charges. These forces will act between all types of molecules and in the separation of non-polar substances, e.g. saturated hydrocarbons, they are the only acting forces. The induction force is the result of the interaction between permanent dipoles and induced dipoles.

The orientation forces have their origin in the interaction of two permanent dipoles, with the hydrogen bond as their most important form. The hydrogen bond is weaker than the chemical bond, but stronger than the weak dispersion or induction forces. In the separation of polar components these forces participate to different degrees, depending on the functional groups and structure of the molecules. By polarity we understand the effect of the electrical field in the immediate vicinity of the molecule which depends on the number, nature and arrangement of the atoms, on the type of bond and the groups giving rise to polarity.

Various attempts have been made to classify the liquid phases in gas chromatography, but as no sharp demarcation line can be drawn between the properties of the various liquids, classifications of this type are only of an informative character (see e.g. Chapter 1).

In general the choice of the liquid phase is based on the polarities of the substances to be separated and of the liquid phase. The higher the polarity of the liquid the more it will retain polar components compared to non-polar substances with the same boiling points. Rohrschneider[52] investigated the relative polarity of various liquid phases on the basis of their retention of different types of hydrocarbons. Taking arbitrarily the polarity of β,β-oxy-dipropionitrile as 100 and the polarity of squalane as zero, he plotted the logarithms of the relative retention volumes with reference to n-butane on the relative polarity scale. Maier and Kárpáthy have carried out a similar classification.[53] These types of graphic representations provide means for the selection of a liquid phase with appropriate polarity for the separation of the components of a sample, but the method has no general validity, especially in the case of highly polar compounds with different functional groups. Littlewood[54] proved from the thorough evaluation of both his own and literature data that in addition to the specific retention of alkanes the relative retention of the subsequent members of the homologous series and also the probability of hydrogen bond formation with the liquid must be considered. With the exception of a few liquids the specific retention values can be estimated from these three parameters with about 10–20% accuracy. In his later works Rohrschneider[51] suggested empirically determined polarity factors for the characterization of the liquid phase.

The application of mixed phase[52, 53] is based on the comparative investigations of liquids of different polarities. A liquid phase of optimum polarity for the separation of a given pair of components may be prepared by mixing in appropriate proportions two liquids of different polarities or two liquid coated supports.[55, 56] Porter *et al.* worked out a computer program for the determination of the optimum mixed phases for various analytical tasks.[57]

In recent years a number of new liquid phases have been developed and used, mainly on the basis of the investigations mentioned. Up to a temperature limit of 200–250 °C many liquids of different polarities may be applied, these are mainly of the polyglycol and polyester types. Today some liquid phases with operating temperatures up to 300–400 °C are available, of which Apiezon greases, various silicone elastomers and asphaltenes have gained the greatest importance. A new class of packings are supports coated with inorganic salts and eutectics which can be applied to the analysis of high boiling and inorganic compounds.[3, 58, 59] A recent publication by Geiss *et al.* deals with the comparison of inorganic salt stationary phases.[60]

In addition to these stationary phases the so-called "specific" stationary phases form a class in themselves as phases with high selectivity for certain types of compounds, such as silver nitrate dissolved in ethylene glycol, benzyl cyanide or high boiling nitriles for olefins,[61] the effect of which is due to formation of a temporary complex

Table 5.2

Some common liquid phases

Solvents: Acetone = A; Benzene = B; Chloroform = C; Methanol = M; Toluene = T; Water = W.
Polarity: non-polar = —; fairly polar = +; strongly polar = ++.

Liquid phase	*Sol-vent*	*Maximum temperature, °C*	*Polar-ity*	*Applications*
Tri-isobutylene	A	30	+	Saturated and unsaturated C_1–C_5 hydrocarbons
Dimethylsulpholane	M	50	++	Saturated and unsaturated C_1–C_6 hydrocarbons
n-Hexadecane (n-cetane)	B	50	—	C_1–C_6 hydrocarbons, halogen derivatives
β, β-Oxy-dipropionitrile	M	70	++	C_1–C_7 paraffins, olefins, cyclo-paraffins, aromatics, alcohols, ketones, esters
Paraffin oil	T	100	—	Hydrocarbons, chlorine compounds, sulphides
Carbowax 400 (polyethylene glycol)	M	120	++	C_1–C_5 alcohols, ethers, ketones, amines
Tricresyl phosphate (tritolyl phosphate)	C	125	++	Aromatics, halogen derivatives, oxygen compounds
Carbowax 600	M	140	++	Oxygen compounds, halogen derivatives, nitrogen compounds
Squalane (hexamethyltetracosane)	T	150	—	Hydrocarbons, halogen derivatives, sulphides
Dinonyl phthalate*	A	150	+	Hydrocarbons, halogen derivatives, oxygen compounds
Carbowax 1500*	M	150	++	Aromatics, oxygen compounds, halogen derivatives, nitrogen compounds, sulphur compounds
Carbowax 6000	M	200	++	Aromatics, oxygen compounds, halogen derivatives, nitrogen compounds, sulphur compounds
Carbowax 20 M*	M	200	++	As above, plus polyfunctional alcohols

Ucon LB (polypropylene glycol)	M	200	+	Aromatics, alcohols, ketones, essential oils, amines
Ucon HB (polyalkylene glycol)	M	200	+	As above
Silicone oil DC 550*	A	200	—	Esters, aldehydes, hydrocarbons, boranes
Benton 34	T	200	++	Aromatics
Polyesters of succinic acid (e.g. LAC 296)	C	200–240	+	Esters of fatty acids, ethers, essential oils, and amino-acid esters
Polyesters of adipic acid (e.g. Resoflex)	C	200–240	+	
Silicone elastomer, XE 6A XE 60 (cyano)	A	250	+	Phenols, aromatics, terpenes, steroids
Apiezon M	T	275	—	Higher alcohols, fatty acid esters, essential oils
Apiezon L*	T	300	—	Higher oxygen compounds, fatty acids, nitrogen compounds, steroids, metal organic compounds
Silicone elastomer, SE 30 (dimethyl)	T	300	—	Alkaloids, steroids, nitriles, hydrocarbons, inorganic and metal organic compounds
Silicone elastomer, SE 52 (methyl-phenyl)	T	300	+	Alkaloids, steroids, carbohydrates
Silicone grease (vacuum)	T	350	—	Fatty acid esters, halogen compounds, inorganic compounds
Poly-phenyl tar	T	400	—	Polycyclic aromatics
Inorganic salts and salt eutectics (e.g. LiCl)	W	400–500	—	Inorganic and metal organic compounds, metal halides

* According to the survey of the Data Sub-Committee of the Gas Chromatography Discussion Group the most widely used liquid phases (Chapter 1, ref. 70, p. 360).

between the silver ion and the olefin. For the retention of aromatics, fluorene picrate or quite simply polyethyleneglycol may be used. In the analysis of amines and amino acid esters the heavy metal salts of fatty acids, e.g. zinc stearate or nickel oleate are used.

Other effects besides the interaction of the liquid phase and the solute may interfere, partly by affecting the selectivity of the stationary phase and partly by hindering qualitative identification through their influence on the retention parameters. These effects are temperature, adsorption on the support and on the liquid surface.

The separation factor α changes slightly with the temperature, and in certain cases even the sequence of elution is altered, so selectivity is temperature dependent. This fact may be utilized in high efficiency columns for the separation of closely related components.[62]

The surface of the support possesses a certain binding power, (especially in the case of polar components), even with relatively high liquid loading. Changes in liquid loading alter the relative influence of these factors, so the selectivity of the packing can be varied.[63]

Martin[64] was the first to point out that in the case of a polar liquid phase the relative retention of the components depends on the quantity and surface of the liquid phase. He proved the correlation between this effect and the adsorption on the surface of the liquid phase which is particularly significant in the case of highly polar liquid phases with non-polar solutes on a high surface area support with little liquid loading. Since then adsorption on the liquid surface has repeatedly been confirmed.[65, 66] With the optimum choice of liquid loading a separation can be achieved by the combined effects of solution and surface adsorption, which would otherwise be impossible with the same liquid phase.

There are still many problems to be solved in connection with the choice and classification of the liquid phases, from the point of view of the theoretical questions of solution and of satisfactory empirical methods and rules. For the practising gas chromatographer these problems are less significant, as many suggestions have been put forward in the literature on the use of a great variety of liquid phases in the analysis of all types of compounds. In practice the liquid phase is chosen with the help of literature data from the catalogues of the manufacturing firms which also quote the maximum operating temperatures and the general fields of applicability. The most widely used liquid phases are summed up in Table 5.2.

Quantity of the liquid phase. The quantity of the liquid phase is usually given in weight per cent of the support. In spite of its widespread use this method is not adequate, partly because of the considerable differences in the volume weights of the supports, thus with a constant weight per cent of liquid phase in a given column length this lat- may contain different quantities of liquids. The other difficulty arises from the difference between the specific surfaces of the supports. A given weight per cent of the liquid phase may give coats of different thickness, as supports with low surface areas may take up not more than 10–15 w% of liquid, while supports with high specific surfaces (e.g. Chromosorb P) will produce considerable tailing with low liquid loading. For comparisons it is more expedient to refer the quantity of the liquid phase to the unit volume. The optimum quantity of the liquid phase depends primarily on the specific surface of the support, but also on the nature of both the liquid phase and the components under investigation. Optimum liquid loading on high specific surface supports is 10–20 w%, on low specific surface supports 5–8 w%, on glass beads 0.1–0.5 w%. For commercial supports the optimum load range is usually cited. The influence of the quantity of liquid on column operation will be discussed later.

5.2.3 Preparation of packed columns

Preparation of the packing. The adsorbent of gas–solid chromatography and the solid support of gas–liquid chromatography are prepared in practically the same way, so their preparation will be discussed together.

The solid substance is crushed or ground and sieved to obtain the desired mesh fraction. After sieving, the fines adhering to the particles must be removed, as these will have a detrimental effect on both the pressure drop through the packing and efficiency. The dust is removed by blowing air through the screen or by wet sedimentation (flotation). Next the solid is dried, this operation cannot be omitted even when the dust is removed by some other method than wet sedimentation. The conditions for the removal of the adsorbed moisture differ depending on the nature of the material; active charcoal and molecular sieves are heated under vacuum at 300–350 °C for 3–4 hours, silica gel and alumina under atmospheric pressure at 150–180 °C for 3–4 hours, other solid supports are dried at 150 °C for 1–2 hours.

Adsorbents and solid supports of different particle sizes and mesh fractions are commercially available, these too should be dried before use. The adsorbents are allowed to cool in the desiccator, but it is advisable to pack the column with the still warm material.

Coating of the support with the liquid phase. Before coating, the support may be pretreated with washing and/or silanizing. Acid washing is followed by rinsing with ample deionized water and subsequent drying. Silanizing is carried out by washing with a hot solution of dimethyldichlorosilane or by boiling hexamethyl-disilazane with the solid under reflux. Silanizing should be followed by washing with toluene, methanol and acetone. Between each phase in the washing process it is advisable to remove the previous solvent and the adsorbed air by vacuum suction. These operations are however only required in certain analyses. After the preparatory operations the fines are carefully removed by blowing hot nitrogen through the packing.

The prepared support and the liquid for loading are weighed, so that the weight of the latter should be 5–20% of the weight of the support. The liquid is then diluted with a rather large quantity of a low boiling solvent. The selection of the solvent will depend on the liquid phase; the most frequently used solvents are: pentane, light petroleum, methanol, acetone, chloroform, methylene chloride, ether and toluene.

Various methods may be applied to the coating of the support, of which stirring is the most frequently used. The support and the diluted liquid are mixed in a porcelain dish when the quantity of the solvent should be sufficient to cover the entire support. It is advisable to add the support in small portions under slow stirring to the liquid. The thin slurry is slowly stirred to provide for a uniform contact between the support and the liquid phase, as vigorous stirring may cause the attrition of the support. The slurry is then filtered or the solvent removed without filtering by evaporation on the waterbath or under an infrared lamp till the odour of the solvent is no longer noticeable. The finished packing should feel dry and non-tacky. Another method uses the coating of the support in the column by the frontal technique.[67] A liquid–solvent mixture of given proportion is led through the packed column till the composition of the emerging fluid is the same as that of the introduced mixture. This method provides for uniform loading, but the determination of the actually used quantity of the loading liquid is cumbersome. A method suggested recently is to percolate the liquid through the support in a separate heated column,[68] the amount of liquid deposited is directly proportional to the concentration in the solution. This method provides for highly uniform and reproducible loading. In the two latter methods the solvent is removed by blowing hot nitrogen through the column. Coating within the column can be applied to restore the liquid phase after prolonged use.

The packing operation. Before introducing the packing into the column it is advisable once more to remove the fines. For packing, one end of the column is closed with a glass wool pad or by wire mesh plug and the other connected through rubber tubing to a funnel. The material is introduced in small portions through the funnel and the packing in the column compressed by gentle shaking or with the help of a vibrator, when both too weak and too strong vibration may be detrimental to the density of the packing. For spiral columns the method suggested is to pack the straight tube and to coil it after packing. But even coiled columns can be perfectly packed by attaching a vacuum pump to the outlet end and applying effective vibration.[69] Villalobos described a simple apparatus and method for the reproducible packing of the column.[70]

The column is packed till about 1 cm free space is left at the inlet end and the packing kept in place with a wire mesh plug. When the finished column is not immediately assembled it should be short-circuited with a flexible tubing.

The packing can be removed by blowing compressed air through the column with simultaneous vibration. The filling and removal operations have a detrimental effect on the packing, thus it is advisable to prepare as many columns as the number of packings needed and to store the packed columns, when the columns can be changed for the different tasks.

Stabilization of the packing. The liquid coated packings must be stabilized prior to use to remove the volatile solvent. This is usually performed in the gas chromatograph itself by passing a carrier stream through the column at a temperature of about 20–30 °C higher than the maximum temperature to be applied. Very careful stabilization is particularly important with the ionization detectors, as even slight contaminations in the carrier gas will affect their operation. The packing is stabilized when the recorder plots a stable straight base line.

Preparation of packed capillaries. The preparation of the recently evolved packed capillary columns requires a special technique.[2, 71] At present these columns are made mainly of glass. They are 1–2 m long, 2–4 mm internal diameter thick glass tubes loosely filled with 0.1–0.15 mm large solid support particles, then drawn at a temperature of 600–700 °C to about 30–50 times of their original length and coiled into a spiral in an appropriate device. In this way a loosely filled capillary with 0.2–0.5 mm internal diameter is obtained. The support is coated by passing through the column a 1–5 w% solution of the liquid phase in an appropriate solvent.

5.3 Open tubular columns

5.3.1 General features

Martin raised the possibility of using empty (open tubular) columns with liquid coated walls as far back as 1956, but their practical application followed the theoretical and practical work of Golay[72, 73] which was published in 1957–58. In subsequent years many investigators became interested in the development of open tubular columns and in the study of their operating conditions.[74–81] As a result the application of open tubular columns has gained wide acceptance and since the 1960's they may be considered standard parts of the apparatus whose operation complements the field of application of packed columns. The preparation and application of open tubular columns has been described in two excellent monographs.[82, 83]

In the case of the open tubular column the internal wall of the tube is coated with a thin liquid layer. The permeability of these columns surpasses considerably that of the packed columns, thus very long, highly efficient columns can be prepared. While in the case of packed columns with column lengths which are still practicable from the point of view of pressure drop and with the new highly efficient packings the number of theoretical plates is of the order of 10^4, with open tubular columns theoretical plate numbers of the order of 10^5 or even 10^6 may be achieved. It has been shown in Chapter 3 that the number of theoretical plates is not unambiguously characteristic of the separation power of the column and especially for early eluted components a higher number of theoretical plates is required for a given separation in the case of open tubular columns than of packed columns. Nevertheless, especially in the separation of heavier components a separation efficiency may be achieved with open tubular columns which is impossible with packed columns.

The high efficiency of open tubular columns may be put to various uses; they are less demanding in the choice of the liquid phase than packed columns. For example the separation of *o*-, *m*- and *p*-xylene can be performed with a packed column only with special liquid phases (e.g. 7,8-benzoquinone) while these isomers can be separated in an open tubular column with squalane as the stationary phase.[84] With the same liquid phase analysis time is considerably shorter in an open tubular column and analysis can be performed at a lower temperature than in a packed column. The open tubular column has special assets in the analysis of liquid samples of a wide boiling range, when even the heavy components will appear as sharp symmetrical peaks on the chromatogram, thus offering a possibility of separation of wide boiling range hydrocarbons or other complex mixtures.

Columns of 30–100 m length are the most frequently used, but for more complicated analyses columns several hundreds of meters long are not unusual. Shortening of the column results in reduced efficiency, but for certain tasks 1–30 m columns will give quite satisfactory results. With short open tubular columns the analysis time is very short and multicomponent mixtures may be analyzed within a few seconds.[77, 79]

The operation of open tubular columns can be modified in a highly versatile manner. By changing the length and diameter of the column and with the proper choice of carrier gas rate and temperature, conditions can be chosen for the solution of highly diverse problems.

Besides the difficulties associated with the liquid coating of the column walls, the application problems of the open tubular columns are linked to the small sample quantities which are necessitated by the low capacity of the columns. Only a 1/100th to 1/1000th fraction of the sample used with packed columns can be injected into the small diameter (0.2–0.4 mm) open tubular columns. Sample injection is performed by the splitters described in Chapter 4 when the reliability of analysis will depend on the stable and linear operation of the latter. The other problem arises in connection with the detector; because of the small sample quantities only highly sensitive ionization detectors can be used. These problems have been more or less solved in recent years and with the latest types of apparatus accurate and reproducible analyses can be performed. The latest types of open tubular columns have a larger diameter or a modified surface allowing the injection of considerably larger samples without stream splitting or the application of less sensitive detectors.

5.3.2 Columns with liquid coated walls

The classic capillary column worked out by Golay was prepared by coating the even internal surface of the tube with a liquid layer. The term "capillary column" is misleading and is no longer used, as the capillary effect does not contribute to separation. The first columns were prepared from tubes with 0.2–0.4 mm internal diameter. The tubes may be made of copper, aluminium, stainless steel, plastics (polyamide) or glass. Because of their oxidation at higher temperatures the copper capillaries have only a limited applicability. Aluminium tubes are the easiest prepared metal tubes and are quite adequate also from the point of wetting, though stainless steel capillaries are the most popular.[81] Quite efficient columns can be prepared from polyamide type plastics[72] (nylon, perlon) and glass. Glass capillaries are quite easy to fabricate in the laboratory with the device constructed and described by Desty *et al.*[85]

There are some early papers on the investigation of the applicability of open tubular columns with larger diameters,[74, 75] but their detailed study, including operation conditions, dates from the 1960's. Jentzsch and Hövermann[86, 87] and Ettre *et al.*[88] prepared open tubular columns with 0.5–1.5 mm internal diameter and studied their optimum operating conditions. With the usual capillaries of about 0.3 mm internal diameter the pressure drop is too great and only a low carrier gas rate (1–5 ml/min) can be realized; with larger diameters the permeability of the column ($K = r^2/8$) is considerably higher and even with low inlet pressures adequate gas rates can be achieved, resulting in a considerable decrease of analysis time. Another important feature of the columns with greater diameters is the increase of the maximum applicable sample quantity ten to one hundredfold and their operation with 0.1–1 μl directly injected liquid sample without a stream splitter. Because of the higher sample quantities, instead of the ionization detector the thermal conductivity cell may also be used, while with an ionization detector these columns are suitable for trace analysis.

Larger diameters are of course associated with reduced column efficiency, but even 1 mm internal diameter columns might be more efficient than the packed columns. The small pressure drop accompanying larger diameters allows the preparation of very long (200–300 m) columns.

The greatest problem encountered in the preparation of open tubular columns with liquid coated walls is the production of a uniform, thin liquid film. The usual tube materials when untreated can be coated only with non-polar or only slightly polar liquids. Because of their high surface tension polar liquids will form islets and not a continuous layer on the wall, resulting partly in tailing due to the wall effect and partly in considerably reduced efficiency. When polar components are analyzed with a non-polar liquid phase the interaction with the tube wall may also result in considerable tailing. Consequently without modification the open tubular columns have been used mainly with non-polar (squalane) or only slightly polar (polypropylene glycol) liquid phases and then primarily for the analysis of hydrocarbons. In recent years extensive work has been done on the determination of the wetting properties and surface tension of various liquids on different metal, glass and plastic tubes. From these studies important conclusions were drawn on the effect of the roughness of the internal tube wall, the quality of the liquid and of the diluent and of the temperature and presence of active substances on the production of the film.[89]

Coating of the tube wall. After coiling and assembling the fixtures, but prior to coating, the tube of appropriate dimensions must be perfectly degreased by rinsing with light petroleum or diethyl ether.

The liquid phase is applied after dilution with a volatile solvent (ether-pentane, methylene chloride, chloroform, etc.). Both the liquid phase and the diluent must be freshly distilled, or dust and any other floating particles must be completely removed by filtration through a G4 sintered glass filter.

Golay coated his column with liquid[72] by filling the entire tube with a 1% methylene chloride solution of polyethylene glycol, and drawing the tube slowly through an oven heated to 100 °C. At this temperature the diluent evaporated and a thin liquid layer was left on the wall. With this method a uniform film is obtained, with the disadvantage that the tube can be wound only after coating and the procedure is therefore inapplicable to long columns.

The method introduced by Dijkstra and de Goey[90] for the coating of the empty tube is at present generally used. Depending on the quantity of liquid intended for coating a 5–20% solution of it is prepared and part (about 5–10% of its length) of the column is filled with this solution. This liquid "plug" is made to travel through the column by a slow constant gas stream introduced at one end. The gas rate varies between 0.5 and 10 cm/sec, but has to be kept constant. The introduction of such a low quantity of gas is no easy task; the micro-electrolytic cell which has been suggested by Kaiser[91] has proved to be the most satisfactory for this purpose. The thickness of the liquid layer will depend on the rate at which the gas pushes the plug forward. After the liquid plug has passed through the column this latter has to be stabilized, that is the residual solvent and any non-adhering liquid phase have to be removed. This is performed by first passing a purge gas (usually N_2) at the temperature of coating at increasing rate through the column followed by a temperature rise up to a maximum of about 20–30 °C higher than maximum analysis temperature. Stabilization usually takes 4–12 hours.

The thickness of the liquid film depends partly on the material of the tube and partly on the experimental conditions of coating. For metal tubes the usual layer thickness is 0.3–2 μm, for glass tubes 0.2–0.5 μm. It is almost impossible to determine the real film thickness, as even in the case of smooth tubes there might arise certain effects which will cause an unevenness of the liquid film. The only practical way to establish the quality of coating is gas chromatographic analysis itself. An adequately stabilized open tubular column operated within a given temperature range may have a long life with constant column characteristics for months. Because of the poor adherence of the liquid phase the temperature limit of open tubular columns is lower than that of packed columns.

5.3.3 Modified open tubular columns

Several methods have been worked out to modify the internal wall of open tubular columns, partly with the view to apply polar liquids and partly to raise column capacity by the use of thicker liquid layers. These modified versions are the results of recent investigations and their appearance has greatly contributed to the spreading of open tubular columns.

Deactivation of the tube wall. Averill[28] has proved that as with solid supports, the adsorption activity of the empty tube may also be considerably reduced by small quantities of highly polar liquids (corrosion inhibitors). By adding 0.1–0.5% of some such substance to the liquid phase the tailing of the polar components can be partly avoided, or alternatively the tube may be coated also with polar liquids. Zlatkis and Walker[91] modified the internal wall of a copper tube with a fine layer of silver, gold, platinum and mercury. Kiselev *et al.*[92] modified the surface of a glass tube by treatment with trimethyl-chlorosilane.

Production of greater surface areas. Greater surface areas are most easily achieved in glass columns. By treating the internal wall of the glass tube with an ammonium hydroxide[93] or NaOH[94] solution at 100–200 °C a porous silica layer is formed which may be used either for gas–solid chromatography or as a support for higher (1–10%) liquid loads. In a similar manner an active alumina layer can be produced on an aluminium tube.[95]

Coating of the wall with a porous solid layer. There have been some early attempts to coat the internal wall of the tube with a solid layer, but the practical application of the method is based on the work of Halász *et al.*[96–98] They coated open tubular columns of various diameters with thin layers (1–100 μ) of different types of porous solids (graphitized carbon black, alumina, silica gel, Sterchamol, iron oxide). The layer was precipitated from the suspension of the finely ground (1 μ) solid. Several methods have been described recently for the effective deposition of the solid substance and for increasing the thickness of the porous layer.[99, 100] Without a liquid phase these columns can be used in GSC, with a liquid phase in GLC. The solid layer has practically no effect on the permeability of the column, but provides a means of making open tubular columns with a higher capacity for uniform film thicknesses of various liquids. As already discussed in Section 5.3.1 higher capacity is an asset from the point of view of both sample injection and detection. With uniform thin liquid layers it is possible to prepare highly efficient columns.

5.4 Back-flush and multicolumn arrangements

It is in general impossible to find a perfect solution for the analysis of wide boiling range samples or of those containing different types of components with a single column. The sample may often contain high boiling components or contaminations which prolong analysis time, and on the other hand by escaping elution they may reduce the activity of the column. To solve problems of this type various column arrangements for the serial or parallel, simultaneous or alternative operation of two or more columns were elaborated. These arrangements will considerably reduce analysis time, while at the same time they open a way to the one-step analysis of multicomponent, complex mixtures. Most commercial equipment includes a special switch for the implementation of the different column arrangements; these switches are obtainable as ancillary devices.

5.4.1 Back-flush

The analysis of wide boiling range samples is usually rather time consuming and for the components eluted last asymmetric, elongated peaks are obtained. Quite often a separation of the heavy components is not even required. In these cases the method of back-flushing,

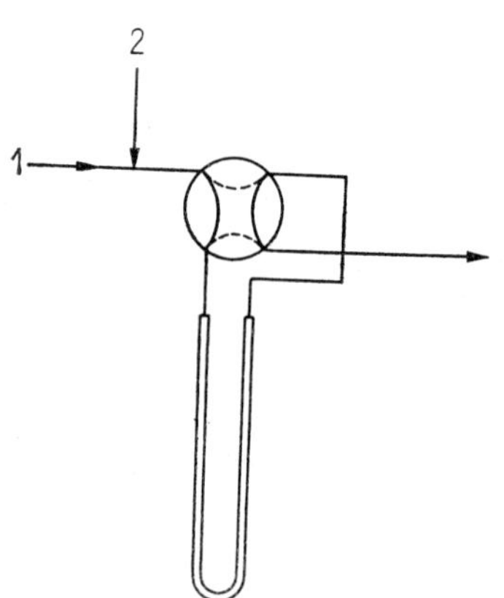

Fig. 5.2. Schematic diagram of column back-flush. 1. Carrier gas inlet; 2. sample injection

as shown diagrammatically in Fig. 5.2, may give very satisfactory results. After the injection of the sample the carrier gas is led through the column till the components to be determined emerge individually with the carrier gas. Now the flow of the carrier gas through the column is reversed by turning a valve, so that the heavier components still in the column are flushed out in the opposite direction. The back-flush method was mainly applied to the analysis of hydrocarbon fractions, e.g. gasoline.[101, 102] In this way the heavy components are not separated and may appear together in one or perhaps several large peaks. Villalobos *et al.*[103, 104] have studied in detail the theoretical problems and accuracy of the back-flush method. However, a more efficient separation of the heavy components may be achieved by connecting a second column into the back-flush branch. From fore-flush and back-flush retention data columns for various separation tasks may be designed. The back-flush method is also applied to the arrangement of two or more columns and with appropriate switches to open tubular columns.[105] To eliminate peak distortion and broadening which accompany the back-flush method Deans[106] describes a procedure applicable to two columns with the sample by-passing the switch.

In cases when the determination of the heavy components is not required these are directly blown off during the back-flush operation.

A dual column back-flush method was suggested for the detailed standard assay of natural gases.[107] Various back-flush methods are applied mainly in process chromatographs where a reduction of analysis time is particularly important.[108] A comparative evaluation of the back-flush methods is to be found in Eppert's recent publication.[109]

5.4.2 Two or more alternately operating columns

To achieve a complete separation of wide boiling range samples or those containing different types of compounds often several stationary phases are needed. To eliminate column change two or more columns may be mounted in the apparatus and may be operated alternately (Fig. 5.3). Columns containing two different stationary phases (polar and non-polar types) are also desirable for qualitative identification (see Chapter 7). The columns are usually mounted in a common thermostat when the maximum temperature of application is determined by the more volatile liquid phase. One of the columns can be mounted outside the thermostat, e.g. in a separate cooling vessel. This method has been used in the analysis of samples containing components of different types,[110] but has the disadvantage of requiring separate sample injection into each column, with the added complication of a difficult eluting of certain components from the different columns.

5.4.3 Two or more columns in series

In these arrangements in general the columns initially connected in series are disconnected after a certain period of time and only the lightest

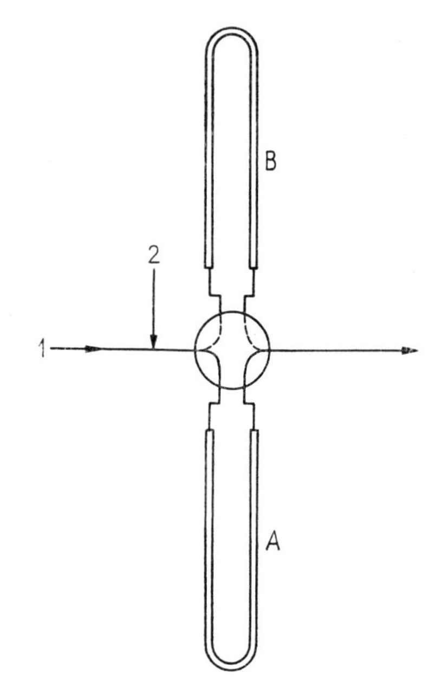

Fig. 5.3. Alternative dual column arrangement. 1. Carrier gas inlet; 2. sample injection; A, B — columns

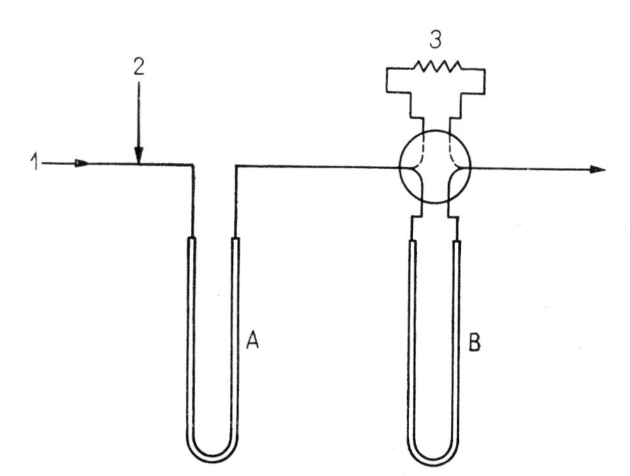

Fig. 5.4. Two columns in series with the intermittent by-pass of column B. 1. Carrier gas inlet; 2. sample injection; 3. choke

components travel through all the columns. The diagrammatic scheme of the simplest arrangement of this type is shown in Fig. 5.4 which may be used e.g. for the analysis of samples containing permanent gases and hydrocarbons. Column A is packed with a liquid loaded support, column B with a 5A molecular sieve. After the permanent gases have reached column B, while the hydrocarbons are still in column A, column B is by-passed by the carrier gas which first elutes the hydrocarbons from column A. After the end of this operation the carrier gas stream is again connected to column B in which the permanent gases are separated. The pressure drop in the choke (capillary) parallel to column B is the same as the pressure drop in column B. This method has the advantage that the adsorbent is not saturated with the heavy components and a single sample injection is sufficient. Two columns with different packings may be operated with a single switch in the following variations: A, B, AB, BA. The columns may be switched in and out to give the desired variation even during analysis thereby providing conditions for highly versatile applications. Columns in series may be operated by two or more detectors joined to the individual columns. Several columns may be arranged partly in series and partly parallel such as the arrangement of Huran *et al.* for the detailed analysis of motor exhaust gases.[111]

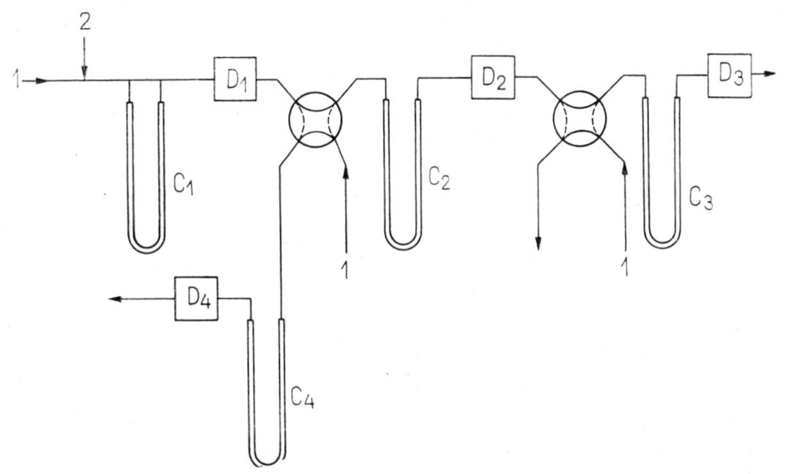

Fig. 5.5. Schematic diagram of four column analysis.[112] *C_1–C_4 – Columns; D_1–D_4 – detectors. 1. Carrier gas inlet; 2. sample injection*

Bloch has described the arrangement of four columns with four detectors,[112] where three columns were in series with intermittent disconnection, while the fourth column connected after the first served the separation of the heaviest components. The arrangement of the apparatus is shown diagrammatically in Fig. 5.5 and is suitable e.g. to the analysis of samples containing permanent gases and C_1–C_7 hydrocarbons (70 components) from a single sample injection.

A special solution of operating in series columns is the so-called heart cut method when an arbitrary "cut" from the components, that is fractions, which have been partially separated or concentrated in the first column is led into a second column for more thorough separation. The possibilities and laboratory realization of the method are described by Deans.[113]

5.4.4 Two or more parallel columns

A simple diagrammatic scheme of the parallel arrangement of two columns is shown in Fig. 5.6. The sample travels through both columns for which the stationary phases are chosen so that the best separation of the light components is effected in one column and that of the heavy components in the other column. One packing is e.g. a molecular sieve for the separation of permanent gases, the other active charcoal for hydrocarbon vapours. The length of the two columns is chosen in such a way that the emerging components shall not interfere with one another, that is, they shall appear in succession on the chromatogram. The simplest way to ensure this is to include an adjustable choke in one branch. The method has the disadvantage that heavy components may reach both packings and may be slowly eluted or irreversibly bound, for example, CO_2 on the molecular sieve. With two or more parallel columns with identical packings but different lengths even wide boiling range fractions may be rapidly analyzed. The heavy components are separated on the shorter column, the lighter fractions on the longer column.[114]

Two parallel columns are usually operated with separate sample injectors and detectors and are connected to separate branches. Figure 4.25 showed examples of this type for a differential arrangement used in programmed temperature (A) and for the application of two columns of different types (B). Two parallel columns with different packings, but operating with a single sample injector and dual channel recording are particularly suited to the qualitative identification of components in complex samples.[115] In the case of wide boiling range samples the apparatus can be operated with programmed temperature. In the latest equipment there is a possibility of operating three

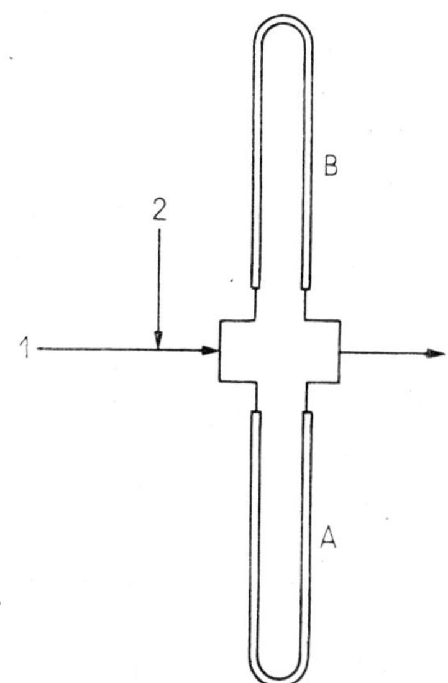

Fig. 5.6. Two columns in parallel. 1. Carrier gas inlet; 2. sample injection; A, B—columns

and even four parallel columns, or more correctly separate analytical units.

In the case of certain problems the optimum analytical conditions may be ensured by the alternate series and parallel operation of two columns. This method is highly suitable to the analysis of natural gases.[116]

5.4.5 Pre-columns

Quite often the sample to be analyzed contains such heavy components that their elution from the packing may require a considerable period of time. The sample may also contain contaminants reacting with the packing. Certain components (e.g. water) in the sample produce particularly asymmetric peaks which may overlap with the peak of the component under investigation, thereby frustrating all attempts at analysis. To eliminate these difficulties so-called pre-columns have been used which are intended to retain the interfering components. The pre-columns may be mounted in the thermostat before the column, or adjoining the sample injector on the front of the apparatus. The simplest arrangement of the pre-columns is in gas analysis, when they are joined to the gas sampling valve (see Chapter 4).

The pre-columns may operate in three different ways. One way is by binding heavy components or certain types of compounds to a special packing in the pre-column and thus preventing them reaching the separating column. In group analysis the pre-column selectively retains certain types of compounds which will reach the separating column after having been separated from the other components. In the pre-column physical adsorption, the formation of unstable addition products or irreversible chemical reactions may take place. In cases of physical adsorption or the formation of addition products the pre-column may usually be regenerated by back-flushing, should however a chemical reaction occur in the column, the packing will have to be changed after a certain number of analyses.

For the comparative test of crude oil samples the C_8 and heavier components were retained by Apiezon L grease in a pre-column and the light fractions were analyzed.[117] For the retention of heavy impurities a pre-column packed with active charcoal might be quite satisfactory. For the removal of water a pre-column with highly polar packing (glycerol or polypropylene glycol) was suggested.[118]

Various types of pre-columns were used for the retention of certain types of compounds. One of the most widely used methods for the retention of normal hydrocarbons is a pre-column packed with 5 A molecular sieves.[119, 120] The determination of normal paraffins is feasible up to C_{40} with this method and the pre-column may be used in conjunction with an open tubular column.[121] Olefins may be retained by a pre-column packed with sulphuric acid loaded silica gel.[122] Various pre-column packings have been described for the retention of olefins, acetylenes, bromides, iodides and aldehydes.[123] A column packed with mercury perchlorate dispersed on a solid support was found highly efficient for the retention of unsaturated compounds; for detailed examinations this pre-column was used in various arrangements.[110]

The other role of the pre-columns is the conversion of certain compounds into others which will be easier to separate or to detect. As a matter of fact, this method may be considered the first application of the microreactor technique which has become popular so rapidly in the recent years. Here packed pre-columns were primarily used to remove the water from the sample, i.e. to determine the water content. In a pre-column packed with calcium hydride water is reduced to hydrogen; when the pre-column is packed with calcium carbide, water will form acetylene with the latter and acetylene will appear as a sharp peak on the chromatogram from which it may be accurately determined.[125] This method is highly sensitive and may be applied to a wide concentration range.

The third field of application of the pre-columns is the concentration of small quantities of contaminants. Before the appearance of the ionization detectors the thermal conductivity cells lacked the sensitivity required for trace analysis. For the detection of contaminants in the order of ppm a preliminary concentration was necessary which was usually performed[126] on an adsorbent filled pre-column at −40 to −80 °C. With the highly sensitive ionization detectors it is possible to perform direct trace analysis without preliminary concentration. However, quite often the peak of the component which is present in a high concentration will overlap with the peak of the contaminating component under investigation and for an adequate separation the partial or total preliminary removal of the main component will be necessary.

5.5 Factors affecting column operation

A great variety of factors influence column operation. Reports on a great deal of practical and theoretical work concerning the study of these factors have been published in the literature. Some of this work, up to the beginning of the 1960's, dealt mainly with the study of the dependence of relative peak broadening or H value, as being characteristic of column efficiency on the various factors and parameters. The H value alone is however insufficient to describe column operation (see Chapter 3) and particularly great differences are obtained when packed and open tubular columns are compared. The main objective of papers published after 1961–62 on this subject was to devise correlations by which column operation may be described and to carry out comparative tests with special emphasis on the comparison and evaluation of columns of different types. The influence of the different parameters on column performance was subjected to detailed study and valuable data were obtained for the design of an optimum column for a given analytical task and for the choice of the optimum operation conditions.

It should however be pointed out that these problems are far from being solved. At present there are no correlations for the calculation of the efficiency and other parameters of the column from the available data, neither can the influence of the various parameters be predicted. Essentially the design of the column and the choice of the operation conditions is still a matter of trial and error. Since in recent years a very great volume of data and experience has been gathered in this field, these empirical methods are obviously more reliable today than they were five or ten years ago. Quite often the data and findings in the literature contain contradictions regarding the influence of various parameters. This is primarily due to the fact that they attempt to draw general conclusions from the results of experiments which have been carried out within a relatively narrow range and by neglecting several parameters. The other problem arises because quite often one and the same parameter may have several effects and these may differ for different factors, so that the final effect appears as their resultant. Finally it is highly important that the influence of the parameters depends also on the component under investigation and for the readily and slowly eluted peaks the influences are often entirely different.

We may therefore say that today we are not yet in a position to present an unambiguous evaluation of the effects of parameters and factors on column operation, but are in need of a system which would investigate the necessary correlations and find quantitative formulas for their description. In the following section we present a brief survey of the effect of factors and parameters on column operation, and this mainly from the point of view of practical applications.

5.5.1 Column efficiency

As already discussed in Chapter 3 the efficiency of the gas chromatography column can be described by the broadening of the gas chromatographic peak which for a given column may be the number of theoretical plates, N, or the height equivalent of a theoretical plate,

H (or more correctly the relative peak broadening). When studying column efficiency usually the changes in the value of H are given as functions of the various parameters. In the discussion of the rate theories in Chapter 3 the various correlations for the description of H were also mentioned. From the van Deemter and the modified rate equations the effect of the factors in the equations, and of the operating conditions which affect the latter, on the value of H may be studied. In the following we shall investigate changes in column efficiency on the basis of the original equation (3.7) and the modified van Deemter equation (3.9). In addition to the effect of the factors in the equations for H, there are still other factors which may influence the value of H, for example the quantity and composition of the sample, and the nature of the solid support. Bohemen[127] and later other authors have pointed out that the construction and operation conditions of the gas chromatograph may have a considerable influence on the value of H. All those parameters which result in a decrease of the value of H will raise column efficiency, thus it is of fundamental importance for the practising gas chromatographer to be acquainted with their effect.

The factors influencing column efficiency may be classified into three groups for the sake of easier understanding: a) general apparatus parameters; b) column parameters; c) operation conditions. In our scrutiny of the various factors we shall first deal with findings pertaining to the packed column and point out as far as possible the differences between packed and open tubular columns.

A detailed theoretical and experimental study of the operation of packed capillary columns is presented in a recent paper by Landault and Guiochon.[128, 129]

a) General apparatus parameters

Construction of the sample injector. The dead volume of the sample injector has a considerable influence on peak broadening; rapid sample injection may be realized with an injector of minimum volume and uniform, rapid carrier gas sweep. In the case of open tubular columns the construction and operation of the stream splitter is equally important.

Joints and connections. The dead volume of the connections between the sample injector and the column, as well as between the column and the detector, in the case of packed columns the dead volumes at the inlet and outlet openings, may all cause peak broadening.

Detector. Peak broadening greatly depends on the dead volume and response time of the detector, mainly with the thermal conductivity cell. In the case of ionization detectors (e.g. FID) this effect is negligibly low, where however the time constant of the recorder might have a considerable influence.

b) Column parameters

The diameter of the column. The diameter of the column is not included in the equation of H and has only an indirect effect on its value. With increasing diameter the uniformity and density (compactness) of the packing diminishes, but this effect is small and may be eliminated by careful packing. Thc quantity of the sample which can still be injected increases quadratically with the diameter of the column, but this is of importance in the case of preparative columns only. Practice has shown a decrease in the value of H with decreasing column diameter. The reason for this is uniform compact packing, and the continuous rapid injection of the sample which can be realized with small diameter columns. The new, highly efficient columns are prepared, instead of the 6–8 mm internal diameter of the old columns, with 2–3 mm diameter and operate with very small samples.

Open tubular columns have the highest efficiency; the lowest value of H can be achieved with 0.2–0.3 mm internal diameters. The open tubular columns with greater internal diameters have higher H values, but for other reasons such as pressure drop, and analysis time, they may nevertheless often be used advantageously.

Column length. Concerning the length of the column the data in the literature are rather contradictory. The rate equation is based on the assumption that the value of H is independent of the column length, which means that the efficiency of the column is directly proportional to its length. This has been experimentally confirmed by some authors, while others found that with greater column length this increase is not linear; moreover above a certain length the increase in efficiency will be only slight. This may have its reason partly in the difficulties encountered when a long column has to be filled evenly and densely, though in the commonly used columns of 2–4 m length such problems do not arise. The other reason for diminished efficiency may be sought in the greater pressure drop and the related change in flow rate of longer columns. This effect can be corrected by reducing the ratio of the inlet to the outlet pressure, that is by choking the emerging gas. It is nevertheless clear that a greater column length is not necessarily associated with higher efficiency; for every separation there is an optimum column length which depends on the separation in hand and on the operation conditions, primarily on the value of the capacity ratio, k'.

The situation is similar in the case of the open tubular column where primarily the greater pressure drop limits the increase of the length.

Column shape. In packed coiled columns the uniformity of the packing may also depend on the dimensions of the spiral. A narrow spiral usually raises the value of H, so the ratio of tube diameter to coil diameter should be at least 1:10. In the case of columns with greater internal diameters the influence of coiling or bending on the increase of H is greater.[130]

The phase ratio of the column. The phase ratio as defined by Eq. 2.43

$$\beta = \frac{V_G}{V_L} \qquad 2.43$$

is the ratio of the free gas volume to the volume of the stationary phase in the column. The quotient of the partition coefficient, k, and the phase ratio, β, determines the capacity ratio, k', of the column which plays an important role in the H value of the rate equation and in the characterization of different types of columns. In packed columns the phase ratio, β, depends on the parameters of the solid support and on the density of the liquid phase. In GSC the true density of the adsorbent is the determining factor in the value of β. In GLC with a given support and a given liquid phase the value of β will decrease exponentially with the increase of the quantity of the liquid loading[131] and will hardly change above approximately 20 w%. The liquid quantity in weight per cent of the support is not unequivocally defined and is not suitable for the direct comparison of different supports. Depending on the specific gravity of the support for a given column volume great differences in the quantity of the liquid, and thus in the value of β, may be obtained. In packed columns with low loading (about 1 w%) β is roughly 200, with 10 w% loading roughly 10 and decreases to 5 with loadings of about 20 w%. With the most commonly used loadings between 5 and 15 w% the value of β varies between 5 and 50.

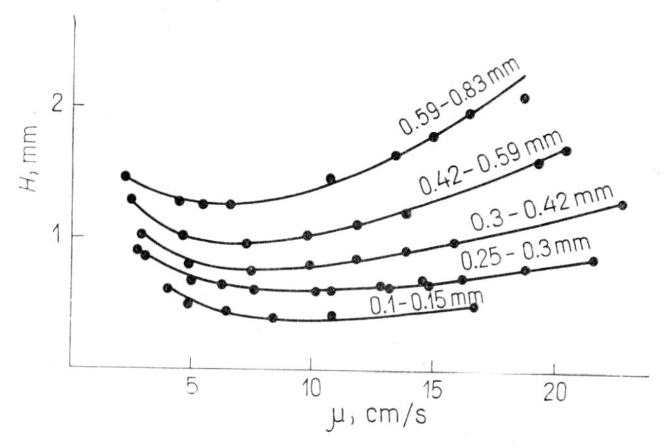

Fig. 5.7. Effect of particle size of the support on the value of H

In the case of open tubular columns the value of β is considerably higher, generally between 200 and 2000 depending on the column diameter and the thickness of the liquid layer.

Nature of the solid support. The parameters of the solid support and its effect on the separation process were extensively discussed in Section 5.2.2. From the point of view of column efficiency the most important requirement is a negligible adsorption activity of the support coupled with mechanical firmness to avoid attrition and greater pressure drop due to the latter. The surface and porous structure of the support must be such as to permit its coating with a uniform liquid film. A support with evenly distributed and large pores provides for the even distribution of the liquid which has a significant influence on the mass transfer factor and thus on the value of H.

Particle size of the solid support. Much work has been devoted to the study of the optimum particle size of the solid support.[132–134] Column efficiency will increase with decreasing particle size, i.e. the value of H will be lower, while the pressure drop considerably increases thus limiting the reduction of the particle size. Figure 5.7 shows the values of H measured with various mesh fractions *vs.* linear gas velocity. It appears from the figure that with decreasing particle size the value of H will also decrease considerably. Flattening of the curves in the range of small particle sizes indicates that at small particle sizes the influence of the flow rate is significantly less and the optimum range wider.

Investigation of packings prepared from different mixtures of narrow mesh fractions showed a considerable increase in the value of H with the extension of the particle size range. According to the van Deemter equation the size of the particles influences mainly the first term in the equation ($\mathbf{A} = 2\lambda d_p$, where d_p is the diameter of the particle), i.e. the value of **A** is directly proportional to the particle size. The coefficient λ which is characteristic of the non-uniformity of the voids between the particles also increases with the extension of the particle size range because of the growing irregularity of the void distribution. As shown in Fig. 5.7 the terms **B** and **C** also change with the particle size, moreover investigation of fractions of different mesh ranges have indicated their dependence also on the latter. Changes in the term **B** and in the gas mass transfer term $\mathbf{C}_g$, of the term **C** are related to the distribution of the free volumes, while the change in the liquid phase mass transfer part $\mathbf{C}_l$ of the term $\mathbf{C}_l$ may be the result of the greater thickness of the liquid film. The relative effect of $\mathbf{C}_g$ and $\mathbf{C}_l$ will depend on the particle size, but mainly on the distribution of the liquid film. The most frequently used mesh fractions today are between 80 and 100 (0.15–0.18 mm), between 100 and 120 (0.11–0.15 mm), though occasionally smaller mesh fractions (< 0.1 mm) are also used.

Besides the particle size the shape of the particles also influences the development of gas paths and thereby the separation power of the column. Particles with regular, preferably spherical, shapes may greatly improve both the efficiency and permeability of the column.[135]

Summing up, it may be said that from the aspect of column efficiency small particles with a narrow mesh size range, with a regular, preferably spherical, shape are the most favourable solid supports.

Nature of the liquid phase. The points to be observed in the choice of the liquid phase were discussed in Section 5.2.2. The properties of the liquid phase affect column efficiency through the partition coefficient, k, which appears in the factor k' of the rate equations (Eq. 2.51)

$$k' = \frac{k}{\beta} = \frac{kV_L}{V_G} \qquad 2.51$$

Thus for a given phase ratio β, k' is directly proportional to the distribution coefficient, k. It was found that generally when k is high the concentration difference between the two phases is greater and mass transfer rate will be higher. The influence of k on the value of H can however only be discussed together with the phase ratio β, therefore it will be treated together with the effect of the liquid quantity.

Viscosity should be mentioned next in connection with the properties of the liquid phase. In the term **C** of the van Deemter equation liquid diffusivity is in the denominator. Duffield and Rogers have shown[136] that viscosity is inversely proportional to liquid diffusivity, thus the value of the term $\mathbf{C}_l$, and consequently the value of H, will decrease with decreasing viscosity. Viscosity is however not the decisive factor in the choice of the liquid phase, instead the partition coefficient k, which characterizes solvent power, and the separation factor, α, which characterizes the selectivity of the liquid, are here the important factors. Their effect will be discussed together with the other important parameters of column operation, resolution power, and the required number of theoretical plates.

Quantity of the liquid phase. There are several fairly different literature data on the effect of the quantity of the liquid. Earlier work studied the influence of liquid quantity mainly in the range of 15–40 w%. Keulemans and Kwantes[137] obtained the most efficient separation with a liquid load of 15–20 w%. When plotting the value of H *vs.* the quantity of liquid de Wet and Pretorius[138] obtained a minimum curve with H_{min} at about 30 w% of liquid load. The investigations of Cheshire and Scott,[132] Purnell[133] and also of Duffield and Rogers[136] have unambiguously proved that with decreasing liquid quantity the value of H will significantly diminish; this fact is related to the decrease of mass transfer resistance in the liquid phase. The other advantage of low liquid loading is the flattening of the H *vs.* u hyperbolas indicating that a higher carrier gas rate may be applied without endangering efficiency. These results provided the basis for the application of highly efficient packed columns with low liquid load. The reduction of the quantity of liquid is however limited by the specific surface and kind of solid support. On a support with a greater specific surface even larger quantities of liquids may form a thin film, while due to the liquid islets formed on supports with smaller specific surfaces diffusion in the liquid phase will require a longer period of time. This effect is conspicuous when we compare diatomite and glass bead supports (see Section 5.2.2). In the case of solid supports with a porous structure the permissible lowest quantity of liquid will depend on the adsorption activity of the support which in cases of unsatisfactory loading will result in tailing. Besides the nature and pretreatment of the solid support the optimum quantity of liquid load will also depend on the nature of the liquid, the nature of the components in the sample and the temperature of analysis.

The effect of the quantity of the liquid phase on column efficiency may be discussed from an overall aspect on the basis of the rate equation by inspecting the film thickness, d_f, and the capacity ratio, k', which contains the partition coefficient, k, this being characteristic of the interaction between the liquid phase and the component.

The term **C** in the van Deemter equation stands for the mass transfer resistance in the liquid phase and is the same as the term $\mathbf{C}_l$ in the modified equations:

$$\mathbf{C}_l = \frac{2}{3} \frac{k'}{(1 + k')^2} \frac{d_f^2}{D_l} \qquad 5.1$$

Figure 5.8 shows the changes in the quotient $k'/(1 + k')^2$ (Eq. 5.1) *vs.* k', from which it can be seen that when $k' < 1$ then $k'/(1 + k')^2$ and consequently $\mathbf{C}_l$ will increase with increasing values of k', reaching a maxi-

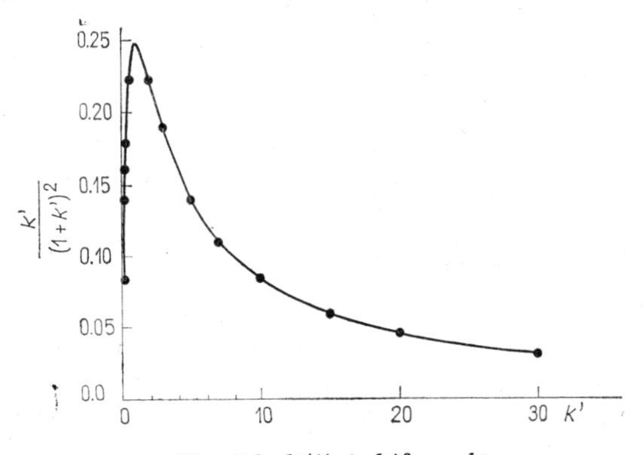

Fig. 5.8. $k'/(1 + k')^2$ *vs.* k'

mum of 0.25 at $k' = 1$. When $k' > 1$ then $k'/(1 + k')^2$ decreases exponentially with increasing values of k' tending asymptotically to zero.

It follows from Eq. 2.51 that k' is directly proportional to the volume

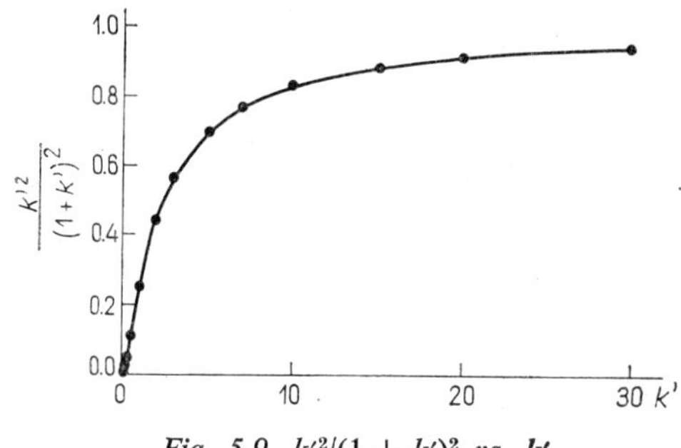

Fig. 5.9. $k'^2/(1 + k')^2$ *vs.* k'

of the liquid phase and with $k' > 1$ its increase results in the decrease of the term $\mathbf{C}_l$. On the other hand $\mathbf{C}_l$ is directly proportional to the square of the effective film thickness, d_f. Because of the relatively small effect of k', as a result of the two opposing influences $\mathbf{C}_l$ will decrease with decreasing liquid loads.

In the modified rate equations there are several complementary terms for the mass transfer resistance in the gaseous phase. Of these the most important and generally accepted is the term $\mathbf{C}_g$ in Eq. 3.11, namely

$$\mathbf{C}_g u = c_a \frac{k'^2}{(1 + k')^2} \frac{d_g^2}{D_g} u \qquad 3.11$$

Figure 5.9 shows the changes of the quotient $k'^2/(1 + k')^2$ *vs.* k'. It appears that with increasing values of k' the $\mathbf{C}_g$ term increases corresponding to the increase of $k'^2/(1 + k')^2$, tending to the limiting value of 1 and indicating that the $\mathbf{C}_g$ term gains in importance at high k' values, while at $k' = 0$ its value is zero, like the value of $\mathbf{C}_l$.

Changes in the minimum value of H are best illustrated by the measurements of Dal Nogare and Chiu.[131] Figure 5.10 shows H_{min} *vs.* k' for various phase ratios. At low k' values and in columns with high values of β, that is, in the case of low liquid loading, mass transfer resistance in the gas phase dominates, while at low values of β, that is higher liquid loads, diffusion in the liquid phase is the primary rate controlling process and according to Fig. 5.8 the value of H passes through a maximum at $k' = 1$. With increasing values of k' diffusion in the liquid phase approaches zero asymptotically and the value of

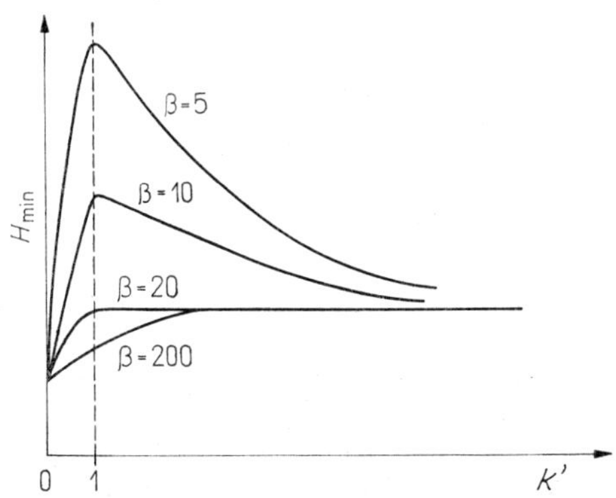

Fig. 5.10. H_{min} *vs.* k' *for various phase ratios*

H is determined by the term $\mathbf{C}_g$ which, as shown in Fig. 5.9, also approaches a limiting value; thus diffusion resistance in the gas phase will be the rate controlling process. In the range $1 < k' < \infty$ the diffusion resistance of the liquid phase is added to the diffusion resistance in the gas phase and the relative effect of these two resistances will depend on the values of k' and β.

In open tubular columns the effect of the liquid load is again highly complex and in the choice of the quantity of liquid practical difficulties must be borne in mind. In open tubular columns mass transfer resistance is, according to Golay's equation (Eq. 3.25), the sum of the following two terms:

$$\mathbf{C} = \mathbf{C}_l + \mathbf{C}_g = \frac{2}{3} \frac{k'}{(1+k')^2} \frac{d_f^2}{D_l} + \frac{1 + 6k' + 11k'^2}{24(1+k')^2} \frac{r^2}{D_g} \qquad 5.2$$

The term $\mathbf{C}_l$ is the same as in the case of packed columns (Eq. 5.1) and changes similarly with k' and d_f. Figure 5.11 shows the changes of the $\frac{1 + 6k' + 11k'^2}{24(1+k')^2}$ part of the term $\mathbf{C}_g$ *vs.* k'. Further, at low values of k' this quotient increases steeply, approaching asymptotically a limiting value. Accordingly a decrease in the value of k' results in lower values of the term $\mathbf{C}_g$.

In open tubular columns therefore the value of H decreases with decreasing liquid load. When choosing the quantity of the liquid phase

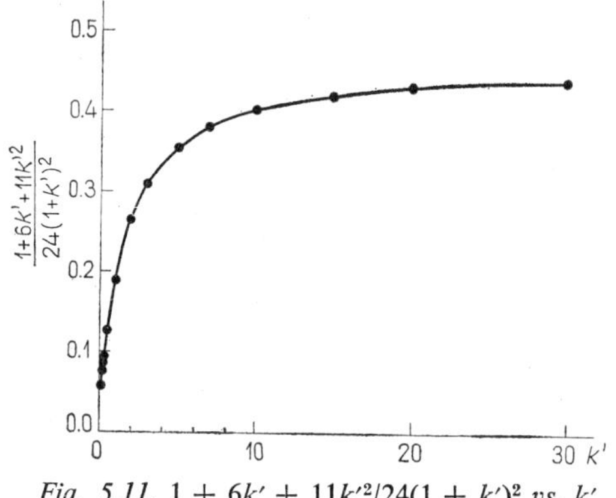

Fig. 5.11. $1 + 6k' + 11k'^2/24(1 + k')^2$ *vs.* k'

it should not be forgotten that with decreasing values of k' the number of theoretical plates, N_{ne}, necessary for a given separation will increase (see Section 5.5.2). The film thickness in open tubular columns varies in practice between 0.5 and 2.5 μ.

c) Operating conditions

Carrier gas velocity. The effect of carrier gas velocity on the value of H was discussed in Chapter 3 in connection with the van Deemter equation. Since with increasing linear gas velocity the term **B** decreases, while the term **C** increases, the value of H will pass through minimum (Fig. 3.2), and H_{min} can be found at the optimum gas velocity. From the equation for packed (Eq. 3.7) and open tubular columns (Eq. 3.23) the optimum average linear gas velocity u, for minimum values of H is for both columns

$$u_{opt} = \sqrt{\frac{\mathbf{B}}{\mathbf{C}_g + \mathbf{C}_l}} \qquad 5.3$$

Theoretically u_{opt} may be calculated by substitution of the corresponding values of $\mathbf{C}_g$ and $\mathbf{C}_l$ into Eq. 5.3. However it is impossible to carry out this calculation in practice, as these terms contain factors which cannot be measured directly (d_f, d_g). Similar calculations were carried out for open tubular columns by neglecting the term $\mathbf{C}_l$ and this work has been reported by Littlewood.[139] The value of the optimum gas velocity depends on the mass transfer resistance. When mass transfer resistance is low higher velocities may be achieved. As both term **B** and term **C** contain data characteristic of a given component (D_g, k') the optimum gas velocity will be different for each component. This means that in the case of samples with different components the optimum gas velocity for one component will not necessarily correspond to an optimum value e.g. for a later eluted component. Optimum gas velocity is also temperature dependent; with increasing column temperature the value of u_{opt} usually increases. The gas velocity which is optimum from the aspect of column efficiency (H_{min}) is not identical with the optimum gas velocity for separation; this latter is about 1.5–2 times higher.[140] At velocities higher than u_{opt} the term **B** will be negligibly small and the value of H will be controlled by the mass transfer resistance. Analysis time considerably decreases with increasing gas velocity, so that quite often higher than optimum gas velocities are applied at the expense of efficiency, and this mainly in open tubular columns.

Nature of the carrier gas. In gas–solid chromatography the carrier gas may also have a significant effect on the separation process. Adsorbents with high specific surfaces will adsorb the carrier gas too, so that some of the adsorption centres will be covered by the molecules of the carrier gas, and the adsorption of the components will change depending on the adsorption capacity of the carrier gas. In a short period of time a heavier carrier gas with higher adsorption capacity (e.g. nitrogen) may produce sharper peaks than hydrogen or helium.[142] The properties of the carrier gas are explicitly included in the terms **B** and $\mathbf{C}_g$ (namely in the diffusion coefficient D_g) of the rate equation for both packed and open tubular columns. D_g appears in the numerator of the term **B** and in the denominator of term $\mathbf{C}_g$, thus it has an inverse effect on the two terms. It has been shown that the relative influences of the terms **B** and $\mathbf{C}_g$ depend on the gas velocity; at low gas velocities ($< u_{opt}$) the term **B** has a considerable importance, while at high gas velocities it can be neglected. Thus at low gas velocities it would be desirable to lower the value of D_g which can be achieved by the application of carrier gases with higher densities (nitrogen, argon). However at the high gas velocities in practical gas chromatography the term $\mathbf{C}_g$ will dominate, its value decreases with increasing D_g, or with low density carrier gases (hydrogen, helium). The nature of the carrier gas affects not only column efficiency, but also the pressure drop in the column. With increasing viscosity of the carrier gas the pressure drop in a given column increases roughly linearly; with a low viscosity carrier gas (e.g. hydrogen) higher velocities may be applied with a given constant pressure drop.

Effect of pressure. The data in the literature on the effect of the average pressure and pressure drop in the column are rather contradictory. The gas diffusion coefficient, D_g, increases with increasing pressure, together with the partition coefficient, k, resulting in higher values of k'. As these factors appear in several terms of the rate equation, the effect of pressure will change in accordance with earlier considerations, with the velocity applied and the relative effects of the terms $\mathbf{C}_g$ and $\mathbf{C}_l$. It may be stated that in general the value of H will decrease somewhat with increasing pressure. Similar results were obtained when column efficiency was investigated at reduced pressures.[142]

The role of the pressure drop in the column was discussed in connection with Eq. 3.17. When the pressure drop is not negligible, changes in gas velocity with expansion will lead to zone broadening, that is, to higher values of H. This effect can be accounted for by the application of the pressure correction factor, j. As the column will operate with optimum efficiency at a given gas velocity, in the case of large pressure drops a velocity differing from the optimum value will prevail in a considerable part of the column. This finding is included in many handbooks. To ensure maximum column efficiency the application of a high inlet pressure and lower inlet to outlet pressure ratios was suggested, which can be achieved by choking the emerging gas. Recent detailed studies have however proved that the effect of pressure drop in the column on the value of H is considerably less as had been assumed in earlier work.[143, 144] Pressure drop in the column is a factor of practical importance, as in commercial gas chromatographs the inlet pressure cannot exceed a given value (usually 3–6 atm).

Effect of temperature. Though the temperature is not explicitly included in the rate equation, its change has a significant effect on several terms of the equation. The effect of temperature is extremely complex and different depending on operating conditions and on the components. The most important effect of the temperature is manifest in the change of the partition coefficient k, which appears partly in the change of the retention volume, partly through the factor k' in the change of the value of H. Retention volume decreases exponentially with increasing temperature and there is a well-known correlation between the logarithm of the retention volume and the reciprocal value of the absolute temperature. Harris and Habgood[145] have subjected to detailed study the effect of temperature on the efficiency of columns. Their main findings can be briefly summed up as follows: of the terms in the rate equation the temperature affects primarily the diffusion coefficients D_g and D_l and the capacity ratio, k'.

The gas diffusion coefficient D_g increases with the 1.81th power of the absolute temperature. The liquid diffusion coefficient D_l also increases with the temperature at a rate depending on the temperature coefficient of the viscosity of the liquid. This effect is particularly significant in the case of high viscosity liquids.

As the heat expansion of the liquid is negligible, β will be constant and the capacity ratio k' will change, in accordance with the change of the partition coefficient k, exponentially with the reciprocal value of the absolute temperature, or in other words will considerably decrease with increasing temperature. Changes in these factors will affect in different ways the various terms in the rate equation. At low gas velocities, when the term **B** has an important role, the value of the latter will increase with the temperature corresponding to the increase of D_g. At velocities of practical importance column efficiency is determined by the changes of the terms $\mathbf{C}_g$ and $\mathbf{C}_l$. In term $\mathbf{C}_g$ the effect of k' is negligible compared to the increase of D_g in the denominator, resulting in a slight decrease of $\mathbf{C}_g$ with the temperature. Changes in the term $\mathbf{C}_l$ will depend partly on the change in D_l and partly on the change of the quotient $k'/(1 + k')^2$ (see Fig. 5.8). When $k' < 1$, the quotient will decrease with increasing temperature. When $k' > 1$, the quotient will first increase and later decrease with increasing temperature. As a result of the various influences the function $\mathbf{C}_l$ *vs.* T may differ for the various components in a given column, e.g. in the case of propane $\mathbf{C}_l$ will decrease considerably, in the case of octane it will somewhat increase with increasing temperature. When mass transfer in the gas phase is the controlling process, efficiency for every component will increase with increasing temperature.

Temperature has an important influence on gas velocity too; in general u_{opt} increases and H_{min} decreases with increasing temperature.

Increasing temperature however raises the number of theoretical plates necessary for a given separation (see below) and thus optimum temperature will be chosen on the basis of other considerations.

Effect of sample size. The size of the sample which can be injected depends on the capacity of the column, in GLC on the quantity of the liquid phase. Keulemans[146] gave the following definition for the maximum quantity of sample, v_s:

$$v_s = c\, v_{eff} \sqrt{N} \qquad 5.4$$

where c is a proportionality factor (for plug-like injection into packed columns $c \approx 0.02$), N is the number of theoretical plates and v_{eff} the effective volume of the theoretical plate:

$$v_{eff} = \frac{V_G + kV_L}{N} \qquad 5.5$$

where V_G is the gas volume of the column, and V_L the volume of the liquid phase.

Both theoretical and experimental studies have indicated a considerable increase in column efficiency with decreasing sample quantities. The shape of the peak also reacts sensitively to larger samples, large samples will produce broad asymmetric peaks. According to Bohemen and Purnell[133] the value of H increases linearly with the size of the sample; the degree of this increase is different for the various components. De Wet and Pretorius[138] found that changes in the value of H with the size of the sample depend on the partition coefficient k; with high partition coefficients and high liquid loading the value of H will only increase slightly.

Since the value of H decreases with decreasing sample sizes, that is column efficiency improves, it is expedient to work with the smallest possible sample. Reductions in the quantity of the sample are limited primarily by the sensitivity of the detector. To eliminate the effect of the sample size, the measurements of the value of H, by which the columns are evaluated and compared, should be extrapolated for precise investigations to a zero quantity and this H_0 value used.

Effect of the method of sample injection. The method of sample injection has a considerable effect on the shape of the elution peak and thus on the value of H. At injection the sample occupies in gas or vapour form a certain volume. Formerly liquid samples were evaporated in a heated head and this vapour volume was swept by the carrier gas into the column. In general it is advisable to adjust

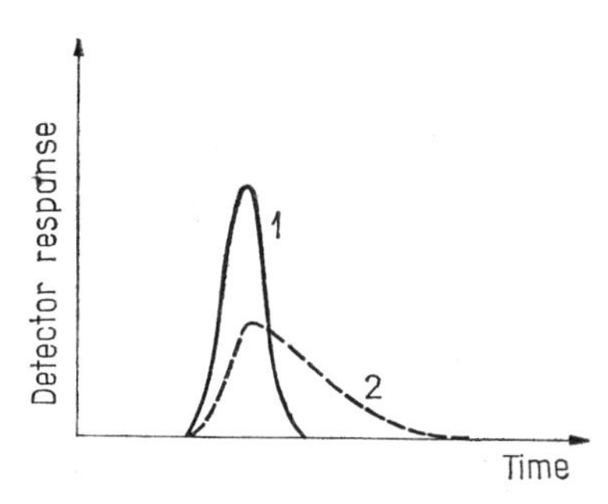

Fig. 5.12. Effect of the method of sample injection on the shape of the peak. 1. Plug-like injection; 2. exponential injection

the temperature of the evaporating head so that it should be 30–50 °C above the boiling point of the heaviest component. Theoretically there are two extreme methods of sample injection, namely "plug-like" injection and continuous, so-called exponential injection into the carrier gas. Figure 5.12 compares these two different methods. In practical gas chromatography injection approximates more the "plug" type.

Recently injection by evaporation is often replaced by direct on-column injection of the liquid. The condition of the applicability of this method is the smallness of the sample. There are no appropriate data available for the objective comparison of the method of vapour phase and liquid phase injections.

It is of fundamental importance in both types of sample injection that this process snould be rapid and uniform. Inadequate sample injection will result in a considerable increase of the value of H.

5.5.2 Evaluation and comparison of the operation of columns

The number of theoretical plates, N, or the height of the theoretical plates, H, which characterizes column efficiency is alone insufficient to describe the separation process or the operation of the column, neither does it offer a way of comparing columns of different types. The operation of columns is studied and compared by utilizing the quantities discussed in Section 3.4, namely on the basis of the resolving power, R_s, the number of theoretical plates necessary for a given separation, N_{ne}, or the necessary column length, L_{ne}, the necessary (or minimum) analysis time, t_{ne}, and the pressure drop, ΔP_{ne}.

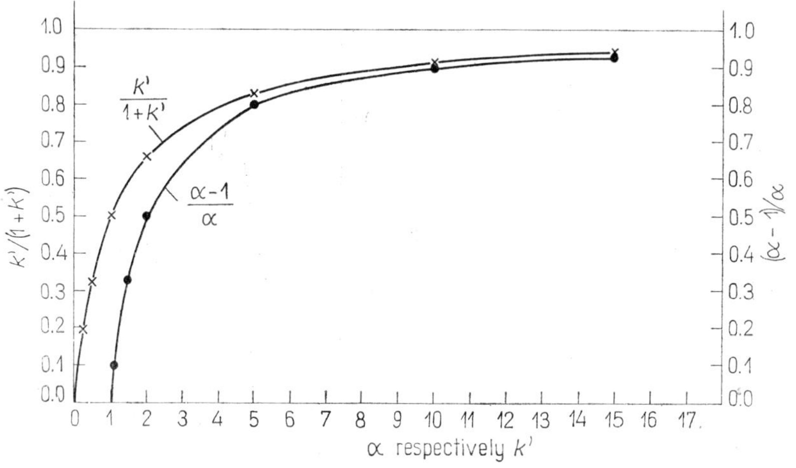

Fig. 5.13. $(\alpha-1)/\alpha$ *vs.* α *and* $k'/(1+k')$ *vs.* k'

Resolution. The equation generally used for the resolution R_s (defined by Eq. 3.36) is obtained from Eq. 3.42:

$$R_s = \frac{\sqrt{N}}{4}\left(\frac{\alpha-1}{\alpha}\right)\left(\frac{k'}{1+k'}\right) \quad 5.6$$

where N and k' refer to the later eluted component of the component pair to be separated.

Under isothermal conditions α and k' are constant for a given column, so resolving power will change with the number of theoretical plates only. Figure 5.13 shows the changes of the factors $\frac{\alpha-1}{\alpha}$ and $\frac{k'}{1+k'}$ *vs.* α and k' respectively. The two curves are similar, both terms first increase steeply and then asymptotically approach the limiting value 1, indicating that an increase of both α and k' will raise the resolving power of the column. The effect of the change is particularly significant in cases of low α, ($\alpha < 3$) and low k', ($k' < 3$).

The values of α and k' depend primarily on the nature of the liquid phase, but also on the magnitude (k') and ratio of the partition coefficients (α). In a given system, temperature is one of the most im-

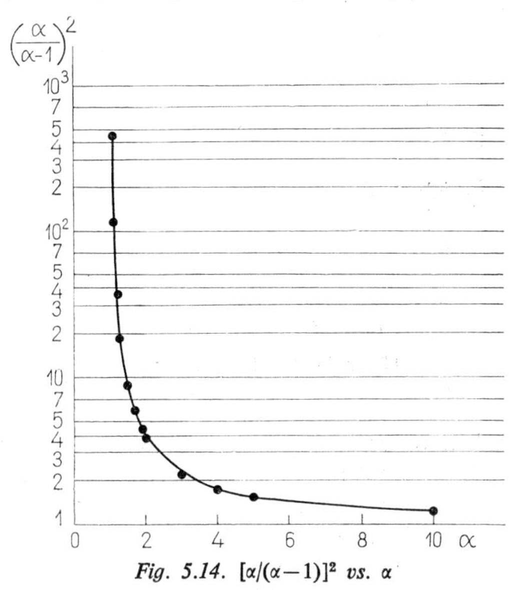

Fig. 5.14. $[\alpha/(\alpha-1)]^2$ *vs.* α

portant operating parameters. The value of k' increases with decreasing temperature, as discussed in the preceding paragraph. The value of α will change to a degree depending on the system under investigation, but in general it will also increase with decreasing temperature. The result is that at lower temperatures fewer theoretical plates or a shorter column will be needed for the same separation efficiency. Karger and Cooke[147, 148] have investigated in detail the effect of operating conditions on the resolution for "normalized" analysis time kept at a constant value. Their method and experimental results are a considerable help in establishing optimum column operation.

Required plate number. The number of theoretical plates necessary for a given separation with resolution R_s may be calculated from Eq. 3.42:

$$N_{ne} = 16R_s^2 \left(\frac{\alpha}{\alpha - 1}\right)^2 \left(\frac{1 + k'}{k'}\right)^2 \qquad 3.42$$

When the desired resolution is fixed (usually at $R_s = 1.5$), then the necessary number of plates will be determined by the values of α and k'. Figure 5.14 shows the changes in the factor $\left(\frac{\alpha}{\alpha - 1}\right)^2$ as a

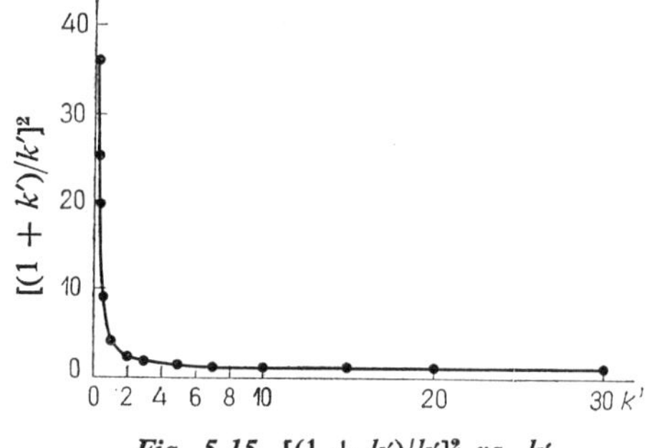

Fig. 5.15. $[(1 + k')/k']^2$ *vs.* k'

function of α. The curve is very steep in the range of low values of α ($\alpha < 2$), so with a slight increase of α (e.g. by reducing the temperature) the necessary number of theoretical plates will considerably diminish. It appears from Fig. 5.15 that the factor $[(1 + k')/k']^2$ also decreases steeply with increasing values of k', but when k' has reached a value of about 3, the change will be only slight and tend asymptotically towards 1. The overall effect of α and k' is illustrated in Fig. 5.16 showing the necessary number of theoretical plates as a function of k' at a resolution $R_s = 1.5$ for different α values.

Data from the curves calculated from Eq. 3.42 are summed up in Table 5.3. It appears from Fig. 5.16 that at low k' values ($k' < 3$) N_{ne} will increase significantly with decreasing values of k', and when k' is greater than 3, N_{ne} will depend almost exclusively on α.

Equation 3.42 offers a possibility of comparing packed and open tubular columns. The essential difference between the two column types is inherent in the phase ratio β and the permeability value, K. For packed columns the value of β is in practice mostly between 5 and 30, for open tubular columns between 100 and 1000, appearing as a 10- or 100-fold difference in the capacity ratio, k'. Thus, for example according to Table 5.3 when the separation factor $\alpha = 1.1$ with a $k' = 2$ packed column of 9801 theoretical plates the same separation can be achieved as with an open tubular column of $k' = 0.2$

Table 5.3

Changes in the necessary number of theoretical plates (N_{ne}) for a given separation with different α and k' values (Eq. 3.42) $R_s = 1.5$

k' \ α	1.05	1.1	1.5	2.0	3.0
0.1	1,920,996	527,076	39,204	17,424	9,801
0.3	298,103	81,893	6,091	2,707	1,523
0.5	142,884	39,204	2,916	1,296	729
1.0	63,504	17,424	1,296	576	324
2.0	35,519	9,801	729	324	182
3.0	28,228	7,754	577	256	144
5.0	22,861	6,273	467	207	117
7.0	20,734	5,689	423	188	106
10.0	19,210	5,271	392	173	97

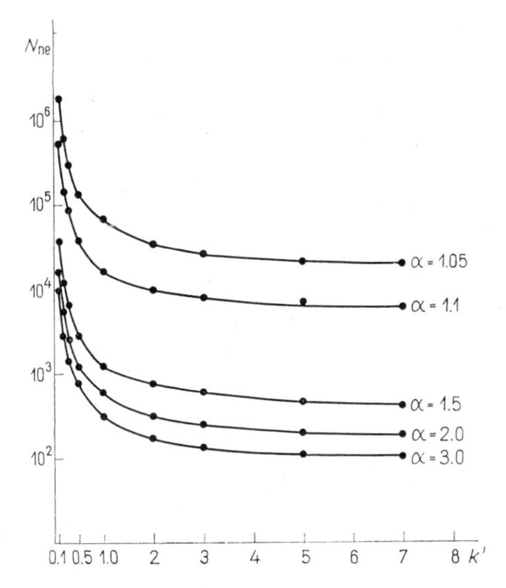

Fig. 5.16. The number of theoretical plates necessary for separation (N_{ne}) vs. the capacity ratio at different values of α

with 156,816 theoretical plates. Differences of this order occur however only for poorly soluble (the value of k is small), rapidly eluted components, as already discussed in Chapter 3. For readily soluble components the value of k' increases and correspondingly the value of N_{ne} for open tubular columns will rapidly decrease. The greater number of theoretical plates required from open tubular columns may however easily be realized in practice by increasing the length of the column because of the higher permeability or smaller pressure drop of the open tube and ultimately a more efficient separation can be achieved with open tubular columns than with packed columns. The latest packed capillary columns occupy a place between the two basic types from the point of view of both their β values and their permeabilities.

The column may be characterized by the column length instead of the necessary number of theoretical plates for a given separation according to Eq. 3.46:

$$L_{ne} = HN_{ne} \tag{3.46}$$

Analysis time. In the evaluation of column operation the necessary analysis time is a very important factor. According to Eq. 3.53 the necessary analysis time for a resolution R_s is

$$t_{ne} = 16R_s^2 \left(\frac{\alpha}{\alpha - 1}\right)^2 \frac{(1 + k')^3}{k'^2} \frac{H}{\bar{u}} \tag{3.53}$$

The effect of the change in the value of α was discussed in relation to Fig. 5.14. The influence of k' on analysis time may be studied with the help of Fig. 5.17. The value of the quotient $(1 + k')^3/k'^2$ decreases rapidly with increasing values of k' and passes through a

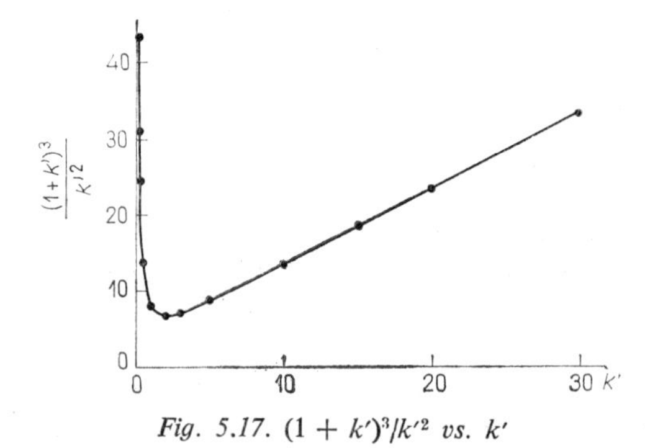

Fig. 5.17. $(1 + k')^3/k'^2$ *vs.* k'

minimum at $k' = 2$, after which it shows a linear increase. For a given column the analysis time depends on the average carrier gas velocity, $\bar{u}$. Short analysis times can be achieved with high values of $\bar{u}$. It has been shown in Section 5.5.1 that the value of H increases with increasing values of $\bar{u}$ which can be compensated by increasing the column length to obtain the desired separation. The increase of gas velocity is however limited by the greater pressure drop which may be written as follows:

$$\Delta p_{ne} = \frac{\eta}{j' K} \bar{u} H N_{ne} \quad 5.7$$

where η is the viscosity of the carrier gas and j' is the modified pressure drop correction factor:

$$j' = \frac{(p_i/p_o) + 1}{2} j \quad 5.8$$

where j is the pressure drop correction factor of James and Martin (see Eq. 2.7).

In Eq. 5.7, $HN_{ne} = L_{ne}$ that is a certain L_{ne} column length and Δp_{ne} pressure drop will belong to every given separation.

When the permeability of the column is low and a short column must be used this shall operate with a gas velocity corresponding to the minimum of the H *vs.* $\bar{u}$ curve, i.e. at H_{min}.

Should the permeability of the column permit the application of higher $\bar{u}$ values and greater lengths then it is recommended to choose the gas velocity $\bar{u}$ from the correlation $H/\bar{u}$ *vs.* $\bar{u}$ which is also a minimum curve. The minimum value of $H/\bar{u}$ belongs to higher velocities than the corresponding H_{min} value.

Columns of different types and construction are usually compared at a constant temperature or under operation conditions corresponding to constant values of k'. Comparisons of this kind are however greatly dependent on the system under investigation. It may be argued that there is no column type which would be the "best" column for the solution of all problems. The various types of columns complement each other quite satisfactorily from many aspects, and the choice of the optimum column depends on the problem in hand.

The choice of columns and of the optimum operating conditions is still more or less subject to trial and error, though the correlations discussed in the foregoing offer valuable help in this work. In these investigations essentially the correlation H *vs.* $\bar{u}$ has to be determined experimentally, so that the separation parameters can be calculated from the equations.

Starting from the work of Karger and Cook,[147, 148] Guiochon[140] subjected the optimum operation of the column and the influence of the various parameters to a detailed study. From his work some important conclusions with direct practical value can be drawn.

The operating conditions should be chosen in such a way that the necessary resolution shall be achieved within a given period of time. Should resolution prove to be too low or too high, analysis time can be altered as desired by an appropriate change in column length. The decrease of column length is about four times the decrease in resolution.

Analysis time depends heavily on the desired resolution. Thus in accurate quantitative analysis 3–4 times longer analysis times are required for the resolution $R_s = 1.5$ than for the less sharp resolution $R_s = 1$. In general the required resolution can be achieved in an optimum period of time, when the k' value of the packed columns is about 3 and that of open tubular columns about 1.5 with a carrier gas velocity in the range of 1.4–2 times the optimum value obtained from the H *vs.* $\bar{u}$ curve.

References

1. I. Halász, K. Hartmann and E. Heine, Chapter 1, ref. 70, p. 38
2. I. Halász and E. Heine, *Anal. Chem.*, **37,** 495 (1965)
3. C. G. Scott and C. G. S. Philips, Chapter 1, ref. 70, p. 266
4. P. T. Eggertsen, H. S. Knight and S. Groennings, *Anal. Chem.*, **28,** 303 (1956)
5. A. V. Kiselev and K. D. Scherbakova, Chapter 1, ref. 62, p. 198
6. A. V. Kiselev and K. D. Scherbakova, Chapter 1, ref. 65, p. 345
7. A. V. Kiselev, Chapter 1, ref. 70, p. 238
8. A. V. Kiselev *et al., Anal. Chem.*, **36,** 1526 (1964)
9. K. D. Scherbakova and W. K. Tschurikins, Chapter 1, ref. 71, Suppl. Vol., p. 77
10. C. G. Scott, Chapter 1, ref. 66, p. 36
11. A. V. Kiselev and Y. I. Yaskin, Chapter 1, ref. 71, Suppl. Vol., p. 43
12. J. J. Kirkland, *Anal. Chem.*, **35,** 1292 (1963)
13. J. J. Kirkland, Chapter 1, ref. 70, p. 285
14. H. L. Macdonell *et al., Anal. Chem.*, **35,** 1253 (1963)

15. I. HALÁSZ and CS. HORVÁTH, *Anal. Chem.*, **36,** 1178 (1964)
16. A. G. ALTENAU and L. B. ROGERS, *Anal. Chem.*, **36,** 1726 (1964)
17. S. DAL NOGARE, *Anal. Chem.*, **37,** 1450 (1965)
18. O. L. HOLLIS, *Anal. Chem.*, **38,** 309 (1966)
19. O. L. HOLLIS and W. V. HAYES, Chapter 1, ref. 73, p. 8
20. W. J. BAKER, E. W. LEE and R. F. WALL, Chapter 1, ref. 59, p. 21
21. R. STASZEWSKI and J. JANÁK, *Coll. Czech. Chem. Commun.*, **27,** 532 (1962)
22. D. T. SAWYER and J. K. BARR, *Anal. Chem.*, **34,** 1518 (1962)
23. R. G. SCHOLZ and W. W. BRANDT, Chapter 1, ref. 67, p. 7
24. G. BLANDENET and J. E. ROBIN, *J. Gas Chromatog.*, **2,** 225 (1964)
25. N. C. SAHA and J. C. GIDDINGS, *Anal. Chem.*, **37,** 822 (1965)
26. D. M. OTTENSTEIN, *J. Gas Chromatog.*, **1,** 11 (1963)
27. J. BOHEMEN *et al.*, *J. Chem. Soc.*, 2444 (1960)
28. W. AVERILL, Chapter 1, ref. 67, p. 1
29. E. D. SMITH and R. D. RADFORD, *Anal. Chem.*, **33,** 1160 (1961)
30. B. A. KNIGHTS, *J. Gas Chromatog.*, **2,** 160 (1964)
31. G. EWALD and H. ZECH, *Z. anal. Chem.*, **215,** 8 (1966)
32. D. M. OTTENSTEIN, *J. Gas Chromatog.*, **6,** 129 (1968)
33. K. J. BOMBAUGH and W. C. BULL, *Anal. Chem.*, **34,** 1237 (1962)
34. J. J. KIRKLAND, Chapter 1, ref. 68, p. 77
35. L. S. ETTRE, *J. Chromatog.*, **4,** 166 (1960)
36. C. LANDAULT and G. GUIOCHON, *J. Chromatog.*, **9,** 133 (1962)
37. D. H. DESTY and C. L. HARBOURN, *Anal. Chem.*, **31,** 1965 (1959)
38. A. W. DECORA and G. J. DINEEN, *Anal. Chem.*, **32,** 164 (1960)
39. Y. R. NAVES, *J. Soc. Cosmetic Chemists*, **9,** 101 (1958)
40. A. B. LITTLEWOOD, Chapter 1, ref. 60, p. 23
41. C. HISHTA, I. P. MESSERLY and R. F. RESCHKE, *Anal. Chem.*, **32,** 1730 (1960)
42. L. SZEPESY and J. SIMON, Chapter 1, ref. 61, p. 435
43. R. W. OHLINE and R. JOJELA, *Anal. Chem.*, **36,** 1681 (1964)
44. I. HALÁSZ and CS. HORVÁTH, *Anal. Chem.*, **36,** 2226 (1964)
45. J. J. KIRKLAND, *Anal. Chem.*, **37,** 1458 (1965)
46. A. M. FILBERT and M. L. HAIR, *J. Gas Chromatog.*, **6,** 150 (1968)
47. A. M. FILBERT and M. L. HAIR, *J. Gas Chromatog.*, **6,** 218 (1968)
48. C. HISHTA and J. BOMSTEIN, *J. Gas Chromatog.*, **5,** 395 (1968)
49. S. T. PRESTON, *J. Gas Chromatog.*, **1,** (3) 8 (1963)
50. J. TAKÁCS, J. BALLA and L. MÁZOR, *J. Chromatog.*, **16,** 218 (1964)
51. L. ROHRSCHNEIDER, *Z. anal. Chem.*, **211,** 18 (1965)
52. L. ROHRSCHNEIDER, *Z. anal. Chem.*, **170,** 256 (1959)
53. H. J. MAIER and O. C. KÁRPÁTHY, *J. Chromatog.*, **8,** 308 (1962)
54. A. B. LITTLEWOOD, *J. Gas Chromatog.*, **1,** (11), 16 (1963)
55. G. P. HILDEBRAND and C. N. REILLEY, *Anal. Chem.*, **36,** 47 (1964)
56. H. RAUSCHMANN, *Z. anal. Chem.*, **211,** 32 (1963)
57. R. S. PORTER *et al.*, *Anal. Chem.*, **36,** 260 (1964)
58. W. W. HANNEMAN, *J. Gas Chromatog.*, **1,** (12) 18 (1963)
59. F. M. ZADO and R. S. JUVET, Chapter 1, ref. 73, p. 283
60. F. GEISS, B. VERSINO and H. SCHLITT, *Chromatographia*, **1,** 9 (1968)
61. A. ZLATKIS, G. S. CHAO and H. R. KAUFMAN, *Anal. Chem.*, **36,** 2354 (1964)
62. A. G. POLGÁR, J. J. HOLST and S. GROENNINGS, *Anal. Chem.*, **34,** 1226 (1962)

63. E. D. SMITH and J. L. JOHNSON, *Anal. Chem.*, **35**, 1204 (1963)
64. R. L. MARTIN, *Anal. Chem.*, **33**, 347 (1961); **35**, 116 (1963)
65. R. L. PECSOK, A. YLLANA and A. ABDUL-KARIM, *Anal. Chem.*, **36**, 452 (1964)
66. D. E. MARTIRE, *Anal. Chem.*, **38**, 244 (1966)
67. E. D. SMITH, *Anal. Chem.*, **32**, 1049 (1960)
68. J. F. PARCHER and P. URONE, *J. Gas Chromatog.*, **2**, 184 (1964)
69. L. SZEPESY, J. SIMON and P. SIMON, *Magy. Kém. Folyóirat*, **67**, 27 (1961)
70. R. VILLALOBOS, *J. Gas Chromatog.*, **6**, 367 (1968)
71. C. LANDAULT and G. GUIOCHON, Chapter 1, ref. 70, p. 121
72. M. J. E. GOLAY, Chapter 1, ref. 59, p. 1
73. M. J. E. GOLAY, Chapter 1, ref. 60, p. 36
74. A. ZLATKIS and H. R. KAUFMAN, *Nature*, **184**, 4101 (1959)
75. R. P. W. SCOTT, *Nature*, **183**, 1753 (1959)
76. R. P. W. SCOTT and G. S. HAZELDEAN, Chapter 1, ref. 63, p. 144
77. D. H. DESTY and A. GOLDUP, Chapter 1, ref. 63, p. 162
78. D. H. DESTY, A. GOLDUP and B. H. F. WHYMAN, *J. Inst. Petroleum*, **45**, 287 (1959)
79. D. H. DESTY, A. GOLDUP and W. T. SWANTON, Chapter 1, ref. 67, p. 105
80. I. HALÁSZ and G. SCHREYER, *Z. anal. Chem.*, **181**, 367 (1961)
81. R. D. CONDON, *Anal. Chem.*, **31**, 1717 (1959)
82. R. KAISER "*Chromatographie in der Gasphase*, II. Teil. Kapillar-Chromatographie", Bibliograph. Institut, Mannheim (1962)
83. L. S. ETTRE, "*Open Tubular Columns in Gas Chromatography*", Plenum Press, New York (1965)
84. I. HALÁSZ and G. SCHREYER, *Chemie Ing. Techn.*, **32**, 675 (1960)
85. D. H. DESTY, J. N. HARESNAPE and B. H. F. WHYMAN, *Anal. Chem.*, **32**, 203 (1960)
86. D. JENTZSCH and W. HOVERMANN, Chapter 1, ref. 66, p. 204
87. D. JENTZSCH and W. HOVERMANN, *J. Chromatog.*, **11**, 440 (1963)
88. L. S. ETTRE, E. W. CIEPLINSKI and W. AVERILL, *J. Gas Chromatog.*, **1**, (2) 7 (1963)
89. F. FARRE-RIUS, J. HENNIKER and G. GUIOCHON, *Nature*, **196**, 63 (1962)
90. G. DIJKSTRA and J. DE GOEY, Chapter 1, ref. 60, p. 58
91. A. ZLATKIS and J. O. WALKER, *J. Gas Chromatog.*, **1**, (5) 9 (1963)
92. A. V. KISELEV *et al.*, Chapter 1, ref. 65, p. 345
93. W. LEIPNITZ and M. MOHNKE, *Chem. Tech.*, **14**, 753 (1962)
94. F. A. BRUNER and G. P. CARTONI, *Anal. Chem.*, **36**, 1522 (1964)
95. D. L. PETITJEAN and C. J. LEFTAULT, *J. Gas Chromatog.*, **1**, (3) 18 (1963)
96. I. HALÁSZ and E. E. WEGNER, *Brennstoff-Chem.*, **42**, 261 (1961)
97. I. HALÁSZ, Chapter 1, ref. 66, p. 133
98. I. HALÁSZ and CS. HORVÁTH, *Anal. Chem.*, **35**, 499 (1963)
99. R. KAISER, *Chromatographia*, **1**, 34 (1968)
100. D. W. GRANT, *J. Gas Chromatog.*, **6**, 18 (1968)
101. W. A. DIETZ, Chapter 1, ref. 59, p. 87
102. R. A. MEYER, Chapter 1, ref. 59, p. 93
103. R. VILLALOBOS, R. O. BRACE and T. JOHNS, Chapter 1, ref. 64, p. 39

104. R. Villalobos and G. S. Turner, Chapter 1, ref. 68, p. 105
105. D. J. McEwen, *Anal. Chem.*, **36,** 279 (1964)
106. D. R. Deans, *J. Chromatog.*, **18,** 477 (1965)
107. ASTM D 1945—64.; ASTM Standard Part 18, 634 (1965)
108. G. Goretti, A. Liberti and G. Nota, Chapter 1, ref. 73a, p. 1
109. G. Eppert, *J. Gas Chromatog.*, **6,** 361 (1968)
110. B. H. Mitzner and P. Gitoneas, *Anal. Chem.*, **34,** 589 (1962)
111. R. W. Hurn, J. D. Chase and K. J. Hughes, *Ann. N. Y. Acad. Sci.*, **72,** 675 (1959)
112. M. G. Bloch, Chapter 1, ref. 64, p. 133
113. D. R. Deans, *Chromatographia*, **1,** 18 (1968)
114. I. Lysyj, Chapter 1, ref. 67, p. 443
115. C. Merritt and J. T. Walsh, *Anal. Chem.*, **34,** 908 (1962)
116. B. Kolb and F. Wiedeking, *Chromatographia*, **1,** 98 (1968)
117. R. L. Martin and J. C. Winters, *Anal. Chem.*, **31,** 1954 (1959)
118. Aerograph Research Notes, Wilkens Instr. Co. Spring Issue, 1960
119. N. Brenner and U. J. Coates, *Nature*, **181,** 1401 (1958)
120. F. T. Egertsen and S. Groennings, *Anal. Chem.*, **33,** 1147 (1961)
121. E. R. Adlard and B. T. Whitham, Chapter 1, ref. 67, p. 371
122. R. L. Martin, *Anal. Chem.*, **32,** 336 (1960)
123. W. F. Trottman-Dickenson, Perkin-Elmer Instr. News, **12,** (3) 12 (1961)
124. C. R. Ferrin, J. O. Chase and R. W. Hurn, Chapter 1, ref. 67, p. 423
125. L. Szepesy and J. Simon, Chapter 1, ref. 69, p. 190
126. N. Brenner and L. S. Ettre, *Anal. Chem.*, **31,** 1815 (1959)
127. J. Bohemen, *J. Chem. Soc.*, 2630 (1961)
128. C. Landault and G. Guiochon, *Chromatographia*, **1,** 119 (1968)
129. C. Landault and G. Guiochon, *Chromatographia*, **1,** 277 (1968)
130. J. C. Giddings, *J. Chromatog.*, **3,** 520 (1960)
131. S. Dal Nogare and Jen Chiu, *Anal. Chem.*, **34,** 890 (1962)
132. J. W. Cheshire and R. P. W. Scott, *J. Instr. Petroleum*, **44,** 74 (1958)
133. J. Bohemen and J. H. Purnell, Chapter 1, ref. 63, p. 6
134. J. H. Purnell, *Ann. N. Y. Acad. Sci.*, **72,** 592 (1959)
135. M. Schröter and E. Leibnitz, Chapter 1, ref. 65, p. 199
136. J. J. Duffield and L. B. Rogers, *Anal. Chem.*, **32,** 340 (1960)
137. A. J. Keulemans and A. Kwantes, Chapter 1, ref. 58, p. 15
138. W. J. de Wet and V. Pretorius, *Anal. Chem.*, **32,** 169 (1960)
139. A. B. Littlewood, "*Gas Chromatography: Principles, techniques, and applications*", Academic Press, London, 173 (1962)
140. G. Guiochon, *Anal. Chem.*, **38,** 1020 (1966)
141. S. A. Green and H. E. Roy, *Anal. Chem.*, **29,** 569 (1957)
142. A. C. Locke and W. W. Brandt, Chapter 1, ref. 66, p. 55
143. J. C. Giddings, *Anal. Chem.*, **36,** 721 (1964)
144. J. C. Sternberg, *Anal. Chem.*, **36,** 921 (1964)
145. W. E. Harris and H. W. Habgood, *Talanta*, **11,** 115 (1964)
146. A. J. Keulemans, "*Gas Chromatography*", Reinhold, New York 199 (1951)
147. B. L. Karger and W. D. Cooke, *Anal. Chem.*, **36,** 985 (1964)
148. B. L. Karger and W. D. Cooke, *Anal. Chem.*, **36,** 991 (1964)

6. Qualitative and Quantitative Analysis

6.1 Qualitative evaluation of the chromatogram

The qualitative evaluation of the chromatogram is in fact the identification of the peaks on the chromatogram, or rather, of the components corresponding to the different peaks. The fundamental condition for the success of qualitative evaluation is the adequate separation of the components or, in other words, every peak must correspond to a single component. The fulfilment of this requirement may be ascertained in the case of relatively simple samples from the shape of the peaks, while for more complex samples containing isomers and compounds of different types separation on several columns or by ancillary methods, to be described later, may be necessary.[1]

The qualitative identification of components is one of the most difficult problems of gas chromatography. The retention data, which represent the time, or after appropriate correction, the carrier gas volume corresponding to the maximum peak of the component were discussed in the chapter on the fundamentals of gas chromatography. Under given conditions the retention volume is an unequivocal function of the partition coefficient, that is of a thermodynamic parameter which characterizes the interaction between the stationary phase and the component. The retention volume, though it may provide some valuable information, is nevertheless inadequate for the unequivocal identification of the components in a complex, multicomponent mixture. The retention data are useful for the identification of a sample whose components are known or which consists of members of a homologous series, but are insufficient for the identification of unknown components. In these cases auxiliary methods are needed for the qualitative analysis of the gas chromatographically separated components, i.e. selective detectors, spectrometric methods, paper or thin layer chromatography and various other analytical methods have to be applied. These two types of qualitative analyses will be treated separately.

6.1.1 Qualitative evaluation from gas chromatographic retention data

The fundamental condition for applying retention data to qualitative identification is to have a great many retention data of adequate accuracy available for the various stationary phases and components. Many investigators and a number of committees have been engaged ever since the early days of gas chromatography in the classification, evaluation and generalization of retention data. Essentially, retention data may be listed as any of the following parameters:

a) specific retention volume;
b) relative retention;
c) R_{x9} or theoretical nonane method;
d) retention index.

a) Specific retention volume

Depending on the column parameters and on operating conditions the retention volume may vary significantly. The specific retention volume, V_g, (Chapter 2, Eq. 2.47) was introduced by Littlewood *et al.*[2] to eliminate the influence of the temperature, flow rate, pressure drop and of the amount of stationary phase. Ambrose and Purnell[3] have introduced the following Antoine type equation to account for the temperature dependence of the retention volume:

$$\log V_g = A + \frac{B}{t + C} \qquad 6.1$$

where t is the temperature of the column in °C and the constants A, B and C can be determined from the plot of the equation by trial and error.

Specific retention volume is a theoretically exact value and has an important role in the determination of physical chemical data from gas chromatographic measurements, but is not used for practical qualitative identifications, partly because its determination is cumbersome and partly because it furnishes reliable values only under accurately controlled conditions and this accuracy is not necessarily ensured by most commercial equipment.

In general absolute retention data are applicable only to the identification of components within the same homologous series. For a given column and at constant temperature the logarithm of the retention volume is a linear function of the number of carbon atoms,[4]

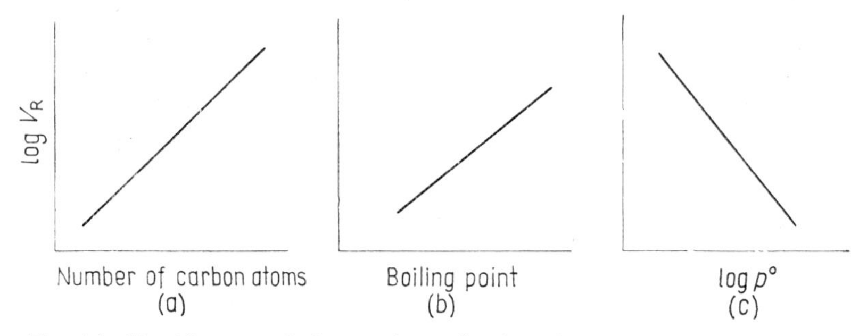

Fig. 6.1. Identification of the members of a homologous series: (a) from the number of carbon atoms; (b) from the boiling point; and (c) from the vapour pressure

as shown in Fig. 6.1(a). Since within a homologous series the boiling points rise smoothly with molecular weight, there will be a linear correlation between the retention volume and the boiling point, i.e. the logarithm of the vapour pressure[5] (Figs 6.1(b) and 6.1(c)). Components belonging to different homologous series may be identified from retention data obtained on two or more stationary phases, of different polarities. When the logarithms of the retention volumes measured on different stationary phases are plotted, the members of the homologous series will lie on parallel straight lines thereby providing means for certain identification,[6] as shown also by Fig. 6.2.

b) Relative retention

Errors arising from the inaccurate measurement of column parameters and operation conditions may be eliminated by measuring the retention relative to a similar component and introducing a correction for the dead volume of the column. The reference component has to be chosen in such a way that relative retention should not be more than 3–4. Relative retention is independent of the flow rate and the pressure drop and in the case of components of a similar type will depend only slightly on the temperature. Relative retention is theoretically independent also of the quantity of the liquid phase, but in the case of polar components certain deviations may occur. The choice of the reference standard is extremely important and several substances available in adequate purity have been suggested

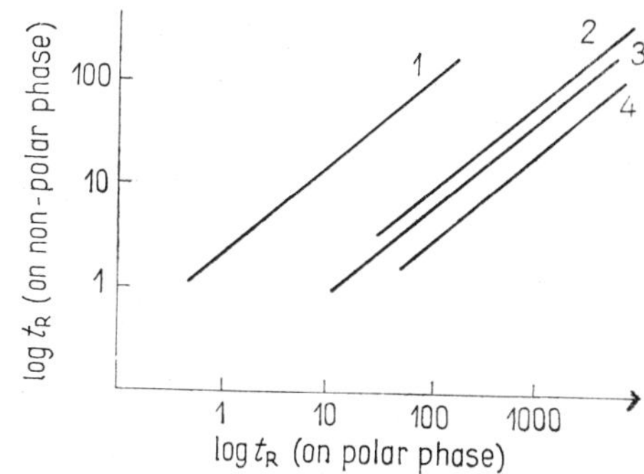

Fig. 6.2. Qualitative identification from retention data on two different stationary phases. 1. Alkanes; 2. esters; 3. ketones; 4. alcohols

and applied to this purpose, such as e.g. n-butane, n-pentane, benzene, p-xylene, naphthalene, methyl ethyl ketone, cyclohexanone, butyric acid, camphor, etc.

Provided the reference standard does not interfere with the components of the sample it is added to the latter before injection or chromatographed under identical conditions. As substances belonging to different groups of compounds may furnish identical retention volumes, the components can be unequivocally identified only from a comparison of relative retention volumes on two columns with different polarities. When the logarithms of the relative retentions are plotted in a rectangular co-ordinate system parallel lines are obtained for the various homologous series.[7] In determining relative retention to specify the temperature is sufficient, though its value changes less with the temperature than that of the absolute data. Relative retention data might be applied directly to the identification of the component provided the necessary tabular compilations are available. There are several detailed data compilations in the literature[8] and the *Journal of Chromatography* and the *Journal of Gas Chromatography* regularly publish retention tables.

The greatest drawback of the simple relative retention method is the lack of a universal reference standard and the literature data refer to different standards which are difficult to compare and cannot always be used.

c) R_{x9} or theoretical nonane method

For relative retention data with reference to a single standard substance Smith[9] has introduced the theoretical nonane value. The theoretical nonane value may be determined as follows: 1. log t (retention time) *vs.* number of carbon atoms is plotted from a chromatogram of n-paraffins. 2. On the diagram the point is marked where the straight line best fitting the experimentally determined points intersects the vertical line corresponding to the carbon atom number = 9. The experimental data are referred to the standard retention time represented by this "theoretical" nonane value. 3. The sample under investigation should be analyzed together with a n-paraffin. The relative retention of the unknown substance with reference to the n-paraffin used (R_{xa}) can be calculated in the usual way from their common chromatogram. The relative retention with reference to theoretical nonane will be:

$$R_{x9} = R_{xa} \cdot R_{N9} \qquad 6.2$$

where R_{N9} is the relative retention of the n-paraffin with N carbon atoms with reference to the theoretical nonane. On a polar stationary phase it might be more expedient to use a theoretical n-pentanol standard which can be determined from the n-alcohol series in the same way.

The problems of the applicability and reproducibility of the theoretical nonane method were subjected to detailed scrutiny by Evans and Smith.[10, 11, 12] The accuracy of the determination depends on the distance of the primary standard n-paraffin from n-nonane. The other problem of the method is the difficulty of describing the temperature dependence of the results.

d) Retention index (I)

The retention index system elaborated by Kováts refers the retention data not to a fixed standard, but to a series of standards.[13–16] The retention index of any arbitrary component can be determined by logarithmic interpolation from the values of two standard compounds.

In the retention index system the n-paraffins with even carbon atom numbers serve as reference standards with ethane as the first of the series. It was established later that the n-paraffins with odd carbon atom numbers are just as suitable, so that the accuracy of the system can be improved and extended to components with less than two carbon atoms.[17] For the sake of simplicity the points corresponding to n-paraffins on the scale are given as 100 times the number of carbon atoms. The equation which defines the modified retention index is:

$$I = 100 \frac{\log V_R^0 (\text{unknown}) - \log V_R^0 (\text{n-C}_N)}{\log V_R^0 (\text{n-C}_{N+1}) - \log V_R^0 (\text{n-C}_N)} + 100.N \qquad 6.3$$

where V_R^0 is the net retention volume, n-C_N the n-paraffin with N carbon atoms, n-C_{N+1} the n-paraffin with $N + 1$ carbon atoms and according to definition:

$$V_R^0 (\text{n-C}_N) \leqslant V_R^0 (\text{unknown}) \leqslant V_R^0 (\text{n-C}_{N+1})$$

It should be pointed out here that in the original publications and in literature in general the equation for n-paraffins with even numbers of carbon atoms is quoted.

As the n-paraffins used as reference substances are analyzed either separately or mixed with the sample, but always under identical conditions, instead of the net retention volume the retention time corrected for the air peak or the corresponding distances on the chromatogram may also be used for the calculation of the retention index. The basis of the retention index and the use of the logarithmic scale are in fact given by the correlation illustrated in Fig. 6.1, according to which the logarithm of the retention volume (or time) changes within a homologous series and at a given temperature linearly with the number of carbon atoms, so that the retention index has a linear scale which simplifies its treatment. The method has the further advantage that the I value provides a good indication of the retention of the component, e.g. the figure 610 indicates the elution of the component after the elution of n-hexane.

Figure 6.3 shows a simple diagrammatic method for determining the retention index. The corrected retention volume or time is determined from the chromatogram of the mixture containing the sample and an appropriate n-paraffin (A). The standard n-paraffins should be chosen in such a way that they shall appear as peaks clearly separated from the peaks of the sample components. When these values are plotted on the logarithmic scale, the retentions of the n-paraffins will be separated by equal distances (B). On the linear retention index scale (C) the points corresponding to the number of carbon atoms of the n-paraffins are fixed (100 *N*) and the distance between two

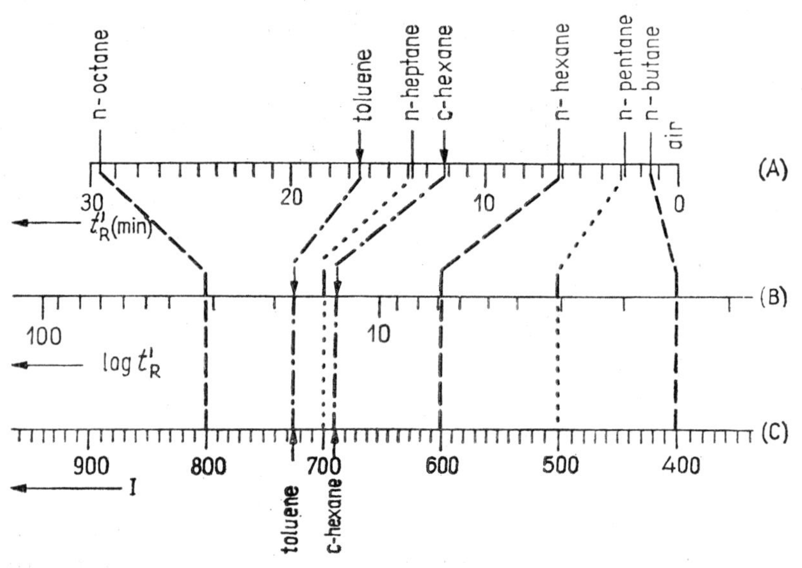

Fig. 6.3. Graphic determination of the retention index.[17] *(A) Adjusted retention times from the chromatograms of the components under investigation (cyclohexane and toluene) and of n-paraffins. (B) Retention times plotted on a logarithmic scale. (C) Retention times of n-paraffins, on a linear scale the distance between two neighbouring n-paraffins divided into 100 equal parts*

neighbouring carbon atom numbers is divided into 100 equal parts. In this way the retention indices of the unknown component or components can be directly read off. To simplify these rather lengthy calculations Hupe[18] has constructed a nomogram for the determination of the retention index.

For n-paraffins and for non-polar components in general the retention index changes linearly, but only slightly, with the temperature indicating that temperature constancy has no decisive influence. Polar components may on the other hand display a considerable temperature dependence. The retention index system has the great advantage that the n-paraffins chosen as standards are easily available in adequate purity, cover the entire range of the retention data and are in linear correlation with most of the stationary phases corresponding to Fig. 6.1(a). In certain cases, however, secondary standards (e.g. alcohols or fatty acid methyl esters) may be needed. If the relative retentions of the secondary standards with n-paraffins as the reference substance are known, the retention index of the unknown component is obtained by calculation.[12]

In the relevant tables the retention indices include the temperature of measurement and the stationary phase; e.g.

$$I_{150}^{\text{Apiezon}}$$

It is helpful to list the temperature coefficient of the retention index referring to 10 °C temperature changes, e.g.

$$\partial I/10\,^{\circ}\mathrm{C} = 5.1$$

including the temperature range of the test.

From their own experiments and the working up of the retention data of other authors Kováts *et al.* have proposed some general rules for the use of the retention index in identifications and calculations.

From data measured on a single stationary phase the following rules can be deduced: 1. The retention index of a member of any homologous series increases by about 100 units when the number of carbon atoms increases by one; the accurate value depends on the polarity of the stationary phase. 2. The retention index measured on a non-polar phase has a simple relationship to the boiling point of the component, so that the approximate boiling point can be determined in this way. 3. On a non-polar phase the difference between the retention indices of isomeric compounds (δI) can be calculated from the difference between the boiling points (δt):

$$\delta I \approx 5\delta t \qquad 6.4$$

4. Identical substituents on compounds with similar structures raise the retention indices in the same degree.

Valuable data on molecular structure may be obtained from measurements with two or more stationary phases. Kováts used in his measurements Apiezon L and Emulphor 0 as stationary phases, but his measurements could be extended to any stationary phase.[19] Here the following general statements are valid: 1. The retention index of non-polar compounds (paraffins) is practically constant irrespective of the stationary phase. 2. On non-polar stationary phases the retention index of any compound is almost the same. 3. The difference in the retention indices (ΔI) of some substance on stationary phases of different polarities is characteristic of the molecular structure of the substance. 4. The functional groups, double bonds and rings in the compound provide separate increments to be added to the retention index. The retention index of some compound may in most cases be calculated additively from these increments. This method has been applied to the identification of various organic compounds.

Schomburg's work[20, 21, 22] contains some highly detailed data on the application of the retention indices for the study of the structure of various types of hydrocarbons and carboxylic esters. Changes in the intermolecular dispersion forces due to the incorporation of functional groups into the initial molecular skeleton may be followed from the differences between the retention indices.[23]

The use of retention indices is a very great help in the analysis of steroids. Knights[24] identified compounds with unknown structures from the ΔI values obtained from the retention indices which he measured on six different types of columns. More information may be gained by reacting the functional groups (hydroxyl, carbonyl, etc.) and from the determinations of the new $\Delta I(r)$ values. Several publications appeared lately dealing in great detail with the identification of the separated components on the basis of the retention index, mainly of hydrocarbons. The retention indices published in the literature represent a considerable help to the practising gas chromatographer.[25, 27] Retention indices have great possibilities in the solution of problems connected with the elucidation of molecular structures as well as in qualitative analysis.

In addition to the qualitative identification of the components and its role in the study of molecular structure the retention index system is further applicable to the description of the separation characteristics of the various stationary phases, thus to their unambiguous characterization.[14]

The system used for qualitative identifications has to meet demands of high accuracy. Detailed studies in recent years have revealed that the accuracy of retention indices surpasses the accuracy which may be achieved with the other relative retention data. The reproducibility for hydrocarbons is highly satisfactory, about 0.3 units, which means that 300 values can be interpolated between two neighbouring n-paraffins. The scattering for polar compounds is greater, but even here an approximately 2% reproducibility can be achieved.

The following factors have the greatest influence on the accuracy of the retention index: 1. the adsorption of the support, mainly in the case of polar substances; 2. difference between the stationary phase qualities and ageing of the liquid phase (e.g. due to oxidation); 3. inaccuracies in the measurement and control of the temperature and of the carrier gas rate; 4. effect of sample quantity, particularly in the case of non-linear partition isotherm; 5. effect of sample composition in the case of larger samples.

To comply with the requirements of items 4 and 5 the smallest permissible sample quantity should be used. In addition it is of course highly important that the column shall have an adequate separation power, since the overlapping of the peaks results also in a shift of the retention indices. The retention index system was first described in 1958, but the method has gained general acceptance only in recent years. In Europe the use of the retention index system began to spread at the beginning of the 1960's, but in the USA it has only been used quite recently. Extensive work on the generalization of the retention index system and on the collection of extensive data was carried out particularly in France (GAMS) and in the Netherlands.[17]

To promote the uniform denotation and general use of gas chromatographic data the Data Sub-Committee of the Gas Chromatography Discussion Group worked out at the Vth Gas Chromatography Symposium in Brighton the following recommendations for the measurement and quotation of retention data:[24]

1. Relative retention data should refer to a chromatogram containing an internal standard.

2. As far as possible the internal standards should be n-paraffins.

3. If necessary other internal standards may also be used, but their relative retentions with reference to n-paraffins must be known.

4. The retention data must be corrected in accordance with the dead volume of the apparatus.

5. The relative retention of the unknown component with reference to the n-paraffin with N carbon atoms should be given the symbol R_{XN} and this should be considered the primary unit of relative retention.

6. Relative retention data should be quoted as R_{x9} or, better, as the retention index I.

7. For every datum the slope b, of the straight line log V *vs.* number of carbon atoms, constructed from the data of at least three n-paraffins must be quoted. The equation of the slope is the following:

$$b = \frac{\log V_{N+1} - \log V_N}{N} \qquad 6.5$$

or by substituting Eq. 6.5 into Eq. 6.3:

$$I = 100\left[\frac{\log V_{XN}}{b} + N\right] \qquad 6.6$$

8. In addition to these data the operation parameters, as far as possible the temperature, carrier gas rate, average pressure, solid support, percentage of the liquid phase, the quantity of the sample and the type of detector used, should also be quoted.

For the general acceptance and applicability of the retention index system the factors which may reduce accuracy, primarily uniformity and reproducibility of column packings, must be solved. In addition the retention indices of a great number of compounds on various stationary phases must still be determined and classified. The application of retention indices in programmed temperature gas chromatography (Chapter 7) also needs further investigations.

6.1.2 Qualitative identification by auxiliary methods

In general it is impossible to identify unequivocally the components of samples containing unknown compounds or compounds of various types from their retention data and therefore different auxiliary methods must be applied. Auxiliary methods are often necessary to decide whether the peak represents a single component only. The many diverse methods used for qualitative identification may be classified as follows:

a) Gas chromatographic methods

The peaks are identified from the retention data of pure substances or model mixtures.

The component is identified from the increase of the peak after the addition of the pure substance to the sample under investigation.

The components are converted into their derivatives and compared with known derivatives.

Certain types of compounds are selectively extracted from the sample by the methods described in the section dealing with pre-columns (Chapter 5).

Specific detectors or detector combinations are used. Certain detectors are capable of sensing some compound types selectively and with high sensitivity. The parallel operation of two or more different detectors may provide detailed information on components of various types in the sample. These possibilities were discussed in the section dealing with the operation of detectors (Section 4.5). The most important specific detectors are: electron capture detector (for halogens, metal organic compounds, etc.), sodium ionization detector (for phosphorus compounds, halogen derivatives), the micro-coulometer (for sulphur, phosphorus, chlorine compounds), gas density balance (for determination of molecular weight), radioactive detectors (for the detection of labelled compounds), etc. The most frequently used detector combinations are: flame ionization and electron capture detectors; flame ionization and sodium ionization detectors; flame ionization detector and micro-coulometer, etc.

b) Other chromatographic methods

Liquid chromatographic methods may be suitable in general for the preliminary concentration of the samples, or perhaps for the preliminary separation of certain types of compounds.

For the identification of the components emerging from the gas chromatographic column thin layer chromatography (TLC) or paper chromatography has been used. In one method the column effluent is condensed by cooling or led on to cooled TLC plates and then run.[28] The column effluent may also be led directly on a mechanically moved thin layer plate which is connected in parallel to the detector, so that each peak will produce a separate spot.[29] In the multidimensional chromatographic method described by Janák *et al.*[30] the thin layer plate or the chromatographic paper strip is moved with logarithmic speed so that the members of a homologous series will appear at equal distances. Unknown components may be identified through their derivatives or with different developing agents.

c) Spectrometric methods

The spectrometric methods furnish the most detailed information for the identification of unknown components. The application of spectrometers as secondary detectors was described in Section 4.5. Because of the small quantity of the column effluent the mass spectrometer alone is suitable for the direct analysis of the sample, though recently infrared spectrometers have also been applied under special conditions. The application of infrared, ultraviolet and nuclear magnetic resonance methods requires generally the collection of a larger quantity (10–100 μg) of the components. For qualitative identification especially the rapid mass spectrometers are now joined directly to both packed and open tubular columns.

The mass spectrometer provides detailed information on molecular structure and weight, the interpretation of the spectra can however be quite difficult. Infrared spectrometry furnishes useful information on the structure of the molecule which then may be complemented by ultraviolet and nuclear magnetic resonance methods. For an unambiguous identification often the combination of all these methods may be required. The high price of the spectrometers and the need of highly qualified operators restrict the wider application of these methods.

d) Use of chemical reagents

Chemical methods are used partly to remove certain types of compounds prior to the injection of the sample and partly for qualitative analysis of the column effluent. Bassette and Whittnak[31] e.g. removed the carbonyl compounds, sulphides, acids and amines by preliminary sodium bisulphite and mercury chloride treatment and carried out the identification by comparison with the chromatogram of the sample not pretreated. Certain types of compounds may also be removed by the preliminary selective extraction of the sample.[32] Hoff and Feit[33] have elaborated the so-called Syringe Reaction Technique for the investigation of compounds with various functional groups. By mixing the diluted vapour of the sample under investigation with various reagents in the sample injector syringe certain components will be selectively bound or converted. They described methods for the detection of aldehydes, alcohols, ketones, ethers, unsaturated bonds, aromatic hydrocarbons and paraffins. Walsh and Merritt[34] and also Casu and Cavalotti[35] led the column effluent into specific reagents. By splitting the effluent into five streams and allowing these to bubble through the reagents they were able to detect nine types of functional groups from their specific colour reactions. This method is also suitable for the detection of different components with different functional groups giving a single peak on the chromatogram. A relatively large sample is needed; the limit of detectability by colour reactions is between 20 and 100 μg.

Methods which involve the transformation of certain types of compounds into derivatives suitable for gas chromatographic analysis or identification also belong to this group. Such transformations can be performed by first reacting the sample with an appropriate reagent, by the application of a pre-column with appropriate packing or on the column itself. Preliminary catalytic or chemical transformation belong in fact to microreactor techniques though in practice no sharp distinctions can be made. The investigation of alcohols,[36] amino acids,[37] carboxylic acids,[38] aldehydes and ketones,[39] esters[40] etc. were performed by means of various transformations. Crippen and Smith[1] have presented a review on the transformation of various functional groups.

An excellent survey of the chemical methods applied to qualitative identifications is presented together with a large number of literature citations in a recent publication by Littlewood.[41]

e) Application of the microreactor technique

The coupling of microreactors with the gas chromatograph has extended the application possibilities of gas chromatography in many ways. These methods will be described in detail in the Chapter on special techniques; here we shall point out only some application possibilities in qualitative identification. The compounds may be transformed or degraded by catalytic or thermal reactions and from the analysis of the products the initial substance may be identified. The method which applies thermal degradation alone is called pyrolysis gas chromatography and will be discussed in Chapter 7.

The microreactors are generally mounted before the sample injector and the reaction products reach the column directly. There are some methods in which the reactor follows the column and the products of the separated components are then determined after separation on a second column. The most frequently applied reactions are: hydrogenation, hydrocracking (hydrogenolysis), dehydrogenation and oxidation.

Catalytic hydrogenation serves the determination of unsaturated bonds, mainly in the analysis of hydrocarbons[42, 43] and unsaturated fatty acid methyl esters.[44] Dehydrogenation too is suitable for the analysis of hydrocarbon mixtures,[42] e.g. for the transformation of naphthenes into aromatic compounds. Catalytic hydrogenolysis was introduced by Thompson *et al.* for the determination of sulphur compounds in petroleum fractions.[45] The method was later extended to the analysis of halogen, oxygen and nitrogen compounds[46] and also to the removal of other heteroatoms[47] (phosphorus, sulphur, metals). These methods may be applied to both aliphatic and cyclic compounds; their common characteristic is cleavage and hydrogenation at the heteroatom, after which the original compound can be identified from the one or two hydrocarbons produced.

As an improvement of these methods Beroza[48–50] worked out the so-called "carbon skeleton chromatography" for determining the structure of various compounds. Here the reaction takes place on the solid support coated with a Pd catalyst and depending on the experimental conditions hydrogenation, hydrogenolysis or dehydrogenation will proceed at 300–400 °C. In this way the unsaturated bonds will be saturated and depending on the position of the heteroatom hydrocarbons with the same or lower number of carbon atoms than the initial compound will be formed. In addition to hydrocarbons, this method may also be applied to the analysis of alcohols, aldehydes, ketones, acids, esters, ethers, amines, amides, sulphides, halogen compounds, etc.

One way to identify the effluent component is by elementary analysis. Elementary analysis by gas chromatography is an important microanalytical method. For qualitative identification the separated component emerging from the column is led through a microreactor, where it decomposes into its elementary constituents and the decomposition products are then separated by a second column. Carbon content is determined[51] after transformation into CO_2 on a copper oxide or cobalt oxide catalyst at 700–1000 °C. The hydrogen content can be oxidized to water and the water reacted with calcium carbide to form acetylene[52] or reduced on calcium hydride or an iron catalyst to hydrogen. The nitrogen content is determined by oxidation to nitric oxide which might then be reduced to nitrogen.[53] Oxygen is determined in the form of carbon monoxide.[54] Organic sulphur is oxidized to SO_2 or SO_3 or reduced to H_2S.[55] Halogens can be determined after oxidation in the form of Cl_2 or Br_2 or after reduction in the form of the corresponding hydrogen halide.[56]

Pyrolysis gas chromatography is an important new branch of the identification methods and is used mainly for the analysis of high polymers, non-volatile solids or high boiling liquids. After pyrolysis at a high temperature the decomposition products are led directly into the column. Under appropriately chosen constant operation conditions pyrolysis will yield decomposition products which are characteristic of the structure of the substance under investigation. In simpler cases the individual decomposition products which are characteristic of the sample components (e.g. in the case of polymers the monomers) might be determined even quantitatively, in the case of more complicated samples the chromatogram of pyrolysis (pyrogram) is like a fingerprint of the investigated sample. The method and applications of pyrolysis gas chromatography will be discussed in detail in Chapter 7. Pyrolysis gas chromatography has been applied to the qualitative identification of a great variety of substances, as well as to the study of their structure. Its most important applications are the analysis of polymers, elastomers and synthetic rubbers[57–60] of barbiturates,[61, 62] phenothiazines,[63] purine and pyrimidine derivatives,[64] dyes,[65] proteins and amino acids,[66] various hydrocarbons[67, 68] etc. In the February 1967 issue of the *Journal of Gas Chromatography* several reviews were published on the application of pyrolysis gas chromatography to the identification and study of the structure of compounds. Beroza and Coad[69] have published a review on reaction gas chromatography with 447 references.

f) Collection of samples

The direct methods of identification are in some cases inadequate and for further tests a quantity of the separated components has to be collected. Except for mass spectrometry, the other spectrometric methods require preparation of the sample as well. The collected material may then be tested also by pyrolysis gas chromatography, by gas chromatography of some derivatives or by other physical and microchemical methods. Depending on the test method the necessary quantity varies between 0.01 and 100 mg. This quantity can be produced with the usual analytical columns from several sample injections.

6.2 Quantitative analysis

The quantitative evaluation of gas chromatographic analysis may be performed either from the height of the peaks on the chromatogram or from the area below the peaks. The response of integral detectors (the volume of gas collected in a burette, the result of automatic titration, etc.) provides directly a measure of the quantity of the components and the composition of the sample can be calculated by means of simple conversion factors.

When differential detectors are used quantitative evaluation will consist of two steps, with the determination of peak heights or areas below the peaks as the first task and the calculation of the quantity of components, that is of the composition of the sample from the above data, as the second. Both peak height and the area under the peak respond sensitively to the factors which affect detector operation (temperature, rate of carrier gas, pressure, etc.). The basic condition for successful quantitative analysis is a high degree of constancy of the operating conditions, as already pointed out in Chapter 4 in connection with the various parts of the apparatus. The accuracy of quantitative analysis is significantly affected by the apparatus parameters, primarily by separation efficiency, by the characteristics of the detector and its ancillary devices and by the operation and characteristics of the recorder. Finally the practice and skill of the operator are quite important factors which may dominate in the injection of the sample and also in the various steps of quantitative evaluation.

Recently Emery has dealt in detail with the importance of quantitative evaluations and the development of evaluation methods.[70] The effect of changing operation conditions on quantitative analysis and improvements in the accuracy of analysis are discussed in a paper by Mikkelson.[71]

6.2.1 Peak height method

Measuring the height of the peak is the simplest method of quantitative evaluation. It is primarily suitable for the determination of rapidly eluted components producing high and narrow peaks, and less appropriate for components which give low and elongated peaks after a longer period of time. There is no direct and simple correlation between the peak height and the quantity of the component and for quantitative analysis calibration with the pure component is indispensable. The correlation peak height *vs.* concentration is generally linear in a narrow concentration range only, so calibration must extend to the entire concentration range involved, as extrapolation is unreliable. Peak height responds very sensitively to changes in operating conditions, especially to changes in temperature, flow rate, sample quantity and in the method of injection.[72]

Quantitative evaluation from the height of the peak will be more reliable when the relative height with reference to an internal standard is measured. As internal standard a known quantity of a substance is used which can be completely separated from the components of the sample, while its peak will lie in the vicinity of the peak of the component under investigation. The internal standard mixed with the sample may compensate for slight fluctuations in operation conditions which would otherwise affect the height of the individual peaks. Pollard and Hardy[73] achieved a 1.4% reproducibility in the analysis of chloromethanes by means of this method, but in general considerably higher discrepancies may occur.

The determination of peak height is a rapid and simple method of quantitative evaluation. In routine works analyses, mainly of gas analyses, in continuous process control and with process chromatographs this is the chief method of quantitative evaluation.

6.2.2 Peak area method

a) Determination of the area under the peak

The accuracy of quantitative analysis can be improved by the determination of the area under the peak. The area below any peak on the chromatogram can be mathematically expressed as the time integral of the detector response which is in quantitative correlation with the concentration of the component in the carrier gas or with the mass of the component which reaches the detector in unit time (see Section 4.5). Measurement of the area ensures equal sensitivity for the analysis of both readily and slowly eluted components.

The first step in quantitative evaluation is the determination of the peak areas. This may be performed by one of the following methods (the standard deviation σ in brackets denotes the reproducibility of the method):

1. The peaks are cut out with scissors and the paper is weighed. This is a lengthy method and rather inaccurate as it depends also on the quality of the paper and ruins the chromatogram (σ = 3–5%).
2. Calculation of the area from the peak height and width measured at half peak height.[74] This method is simple and fairly accurate for symmetric peaks (σ = 2.5–4%) but inapplicable to asymmetric peaks.
3. Multiplication of peak height by retention time is a simple and rapid method with an accuracy similar to that of the previous procedure.
4. Measuring the area with the planimeter is lengthy and cumbersome with the possibility of a large subjective error (σ = 4–6%).
5. Application of an automatic integrator; either mechanical (σ = 1–2%), electromechanical (σ = 1–2%) or electronic (σ = 0.2–0.5%). The highest speed and accuracy can be achieved with the electronic digital integrator (see Section 4.7). The latest types work independently of the recorder and print the sum of the values of the areas.

Detailed theoretical and practical comparison of the manual integrating methods is given in a recent publication of Ball *et al.*[75]

Determination of the areas under partly separated peaks is a problem in itself. In the case of sharp symmetric peaks the area might be halved by a vertical line drawn at the minimum as shown in Fig. 6.4(a). For broader asymmetric peaks approximations may be applied. Figure 6.4(b) illustrates the construction of approximating triangles. If the internal side of the peak is asymmetric an isosceles triangle should be constructed from the vertical drawn at the peak maximum and the external side of the peak. An adequate accuracy can be achieved in the case of overlapping peaks provided the minimum reaches at least the half height of the lower peak.

The approximate determination of peaks appearing as shoulders is illustrated in Fig. 6.4(c). Automatic integrators begin a new counting at the minima between the peaks or at inflection points, so that this provides only a rough approximation of the true quantities of partly separated components.

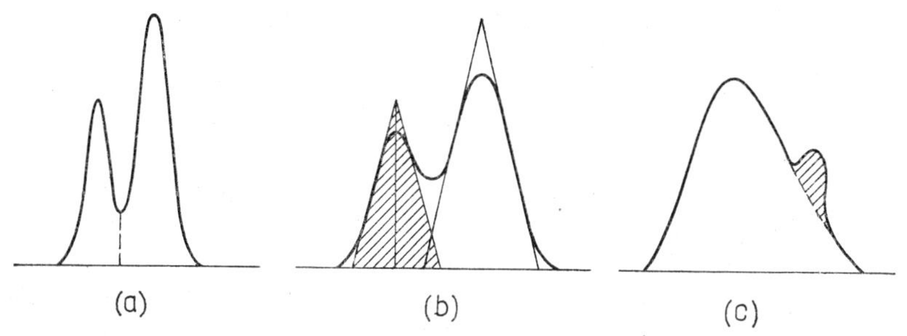

Fig. 6.4. Determination of the area of partially separated peaks

b) Calculation of the percentage composition of the sample

Normalization. This is the simplest method for calculating the composition of the sample and is nowadays in general use, especially in the analysis of multicomponent mixtures (e.g. hydrocarbon fractions). The method is based on the elution of the total sample and the indication of the quantity of the various components by the areas under the peaks. The percentage composition of the *i*th sample will be:

$$i\% = \frac{A_i}{\sum^{n} A_i} \times 100 \qquad 6.7$$

where A_i is the peak area of the component i, and the denominator the sum of all peaks areas.

The problems of the quantitative evaluation of detector response were discussed in connection with the various detectors in Section 4.5. In the following we shall describe the calculations with respect to the two most frequently used detectors, namely the thermal conductivity cell and the flame ionization detector.

Because of the difference in their thermal conductivity the quantities of the components are not exactly proportional to the detector response of the thermal conductivity cell. Many contradictory data were published on the proportionality between the percentage area and the molar or weight per cent composition of the sample. It was however confirmed later that the percentage areas are nearer to the weight per cent composition, but for really accurate analysis the specific correction factor of each substance has to be introduced[76-78] (see Chapter 4). Errors of 20–30% or even higher may be obtained in the molar per cent composition, especially in the case of high molecular weights as against 5–10% errors in weight per cent composition.[79]

The weight per cent composition of hydrocarbons calculated from the percentage area with the flame ionization detector may be determined with less than 5% relative error.[80-86]

Application of correction (normalization) factors. The use of correction factors which are characteristic of the substances under investigation is an indispensable condition of accurate quantitative analysis. The quantity of the component i in per cent will be:

$$i\% = \frac{f_i A_i}{\sum_{i}^{n} f_i A_i} \times 100 \qquad 6.8$$

where f_i is the correction factor of the component i and every peak area must be multiplied by the corresponding correction factor. These factors are usually given in mol/area or gram/area units. The correction factors are heavily dependent on the nature of the component, the construction of the detector, operating conditions, column packing[87] and also on the composition of the sample.[88] Kaiser in his book[89] shows a detailed compilation of the correction factors including their conversion. To reduce the effect of operating conditions relative correction factors were introduced with reference to some selected standard substances. Messner *et al.*[78] (see Chapter 4) calculated the relative molar responses (RMR) with benzene as reference substance (RMR = 100). The value of RMR changes linearly with the molecular weight within a homologous series, though considerable deviations may occur at high temperatures and in the case of heavy components.[87] For practical purposes usually relative correction factors per unit weight are used, primarily with benzene as reference substance. From the peak area of a sample of known composition the correction factor per unit weight will be:

$$f_i = \frac{w\%}{\%\ \text{area}} \qquad 6.9$$

To determine relative correction factors a known quantity of the standard substance is added to the mixture of accurately determined quantities of the components and the area values are determined from several samples. From these measurements the relative correction factor of the component *i* will be:

$$f_{ir} = \frac{w_i\% \times A_s}{w_s\% \times A_i} \qquad 6.10$$

where *s* is the reference standard substance (benzene). For determining correction factors high purity substances must be used and operation conditions must be kept accurately at constant values.

The correction factors quoted for the thermal conductivity cell refer usually to H_2 and He carrier gases. There is a fairly good agreement between some data from various sources in the literature, while others show considerable deviations. In addition to Kaiser's book a detailed compilation may be found in a recent paper by Dietz[90] for both the thermal conductivity cell and the flame ionization detector. When using the correction factors it must be remembered that they are quoted in different ways in the literature.

Correction factors must be used for accurate analysis also with flame ionization detectors. For hydrocarbons the relative sensitivity of the detector is almost constant (± 5%) with somewhat greater deviations for benzene and toluene. The deviations are however quite considerable in the case of other compounds, e.g. for alcohols relative sensitivity varies between 0.23 and 0.85, thus for compounds other than hydrocarbons the direct percentage area can involve a very large (20–50%) error. In general the relative correction factors are referred to n-heptane.

The great volume of published data offers little help in the solution of practical quantitative analytical tasks. As already mentioned the correction factors depend on a number of constructional and operation factors and the literature data are only very seldom directly useful. The applicability of the literature data to the given system must be checked in each case by preliminary analysis of pure substances. Sometimes it may be possible to use the literature data with an appropriate conversion factor determined from the measured data of a few components. For accurate analysis usually direct calibration will be necessary. It should be mentioned here that the correction or calibration factors change also with changing retention times, that is, with the ageing of the column[91] and depend further, especially in the case of components with different properties, also on the composition of the sample.

Application of an internal standard. Errors due to the apparatus and to operation conditions may be considerably reduced by the application of the internal standard method. Prior to analysis an accurately measured quantity of the standard is mixed with the measured quantity of the sample. The standard should be chosen in such a way that it should satisfactorily separate from the sample components, should be eluted near the peak or peaks under investigation, should be present in a concentration similar to that of the sample and should be as far as possible a chemically similar type of compound. The concentration of the component i will be:

$$i\% = \frac{A_i \times w_{is}}{A_{is}} \times 100 \qquad 6.11$$

where A_i is the peak area of component i, A_{is} is the area corresponding to the internal standard, and w_{is} is the weight per cent of the internal standard in the sample.

The internal standard method is particularly advantageous when not all components are eluted or, for example, are partly retained by a pre-column or the quantity of only some components is required, then with the application of correction factors or by careful calibration a very high degree of accuracy can be achieved.

Calibration by direct injection. A rather cumbersome but often used method of quantitative analysis consists in the recording of calibration curves obtained from the analysis of pure substances. The calibration curve represents the correlation between area and the quantity of the injected substance. In the case of ionization detectors the very small sample presents a special problem. The method has the advantage of being applicable to the entire concentration range involved, with the disadvantage of an accuracy highly dependent on operating conditions (primarily sample injection). In general the extrapolation of calibration curves will involve very considerable errors.

The other method of direct calibration is the injection of standard samples with known compositions, so that areas both smaller and greater than that of the peak under observation are obtained and performing linear interpolation between the areas for two different percentage compositions.

Calibration curves will change just as correction factors do, depending on the apparatus and operating conditions. Calibration will have to be checked at regular (usually weekly) intervals with the injection of a standard mixture of known composition.

Deans deals in a recent publication with the calculation methods applied to quantitative analyses and with the detailed comparison of the main sources of errors stressing again the importance of calibration with pure substances.[92]

References

1. R. C. Crippen and C. E. Smith, *J. Gas Chromatog.*, **3**, 37 (1965)
2. A. B. Littlewood, C. S. G. Phillips and D. T. Price, *J. Chem. Soc.*, 1480 (1955)
3. D. Ambrose and J. H. Purnell, Chapter 1, ref. 66, p. 369
4. A. T. James and A. J. P. Martin, *J. Appl. Chem.*, **6**, 105 (1956)
5. L. J. Sullivan, J. H. Lotz and C. B. Willingham, *Anal. Chem.*, **28**, 495 (1956)
6. J. S. Lewis, A. W. Patton and W. J. Kaye, *Anal. Chem.*, **28**, 370 (1956)
7. D. H. Desty and B. H. F. Whyman, *Anal. Chem.*, **29**, 320 (1957)
8. P. R. Scholly and N. Brenner, Chapter 1, ref. 64, p. 263
9. J. F. Smith, *Chem. and Ind.*, 1024 (1960)
10. M. B. Evans and J. F. Smith, *J. Chromatog.*, **5**, 300 (1961)
11. M. B. Evans and J. F. Smith. *J. Chromatog.*, **6**, 293 (1961)
12. M. B. Evans and J. F. Smith, *J. Chromatog.*, **9**, 147 (1962)
13. E. Kováts, *Helv. Chim. Acta*, **41**, 1915 (1958)
14. A. Wehrli and E. Kováts, *Helv. Chim. Acta*, **48**, 2709 (1959)
15. E. Kováts, *Z. Anal. Chem.* **181**, 351 (1961)
16. E. Kováts, *Helv. Chim. Acta*, **46**, 2705 (1963)
17. L. S. Ettre, *Anal. Chem.*, **36**, 31A (1964)
18. K. P. Hupe, *J. Gas Chromatog.*, **3**, 12 (1965)
19. G. Guiochon, *Anal. Chem.*, **36**, 1672 (1964)
20. G. Schomburg, *J. Chromatog.*, **14**, 157 (1964)
21. G. Schomburg, *J. Chromatog.*, **23**, 1 (1966)
22. G. Schomburg, *J. Chromatog.*, **23**, 18 (1966)

23. J. Jonas, J. Janák and M. Kratochvil, *J. Gas Chromatog.*, **4**, 332 (1966)
24. Data Sub-Committee: Chapter 1, ref. 70, p. 348
25. H. Widmer, *J. Gas Chromatog.*, **5**, 506 (1967)
26. G. Arich and S. Volpe, *J. Gas. Chromatog.*, **6**, 384 (1968)
27. A. Matukuma, Chapter 1, ref. 73a, p. 4
28. J. A. Attaway and R. W. Wolford, Chapter 1, ref. 70, p. 170
29. R. Kaiser, *Z. anal. Chem.*, **205**, 284 (1964)
30. J. Janák, I. Klimes and K. Hana, *J. Chromatog.*, **18**, 270 (1965)
31. R. Basette and C. H. Whittnak, *Anal. Chem.*, **32**, 1098 (1960)
32. R. Suffis and D. E. Dean, *Anal. Chem.*, **34**, 480 (1962)
33. J. E. Hoff and E. D. Feit, *Anal. Chem.*, **36**, 1002 (1964)
34. J. T. Walsh and C. Merritt, *Anal. Chem.*, **32**, 1378 (1960)
35. B. Casu and L. Cavalotti, *Anal. Chem.*, **34**, 1514 (1962)
36. J. K. Haken, *J. Gas Chromatog.*, **1**, (10) 30 (1963)
37. A. Zlatkis, J. F. Oro and A. P. Kimball, *Anal. Chem.*, **32**, 162 (1960)
38. J. R. Hunter, *J. Chromatog.*, **7**, 288 (1962)
39. J. W. Ralls, *Anal. Chem.*, **36**, 946 (1964)
40. J. Janák, J. Novak and J. Sulovsky, *Coll. Czech. Chem. Comm.*, **27**, 2541 (1962)
41. A. B. Littlewood, *Chromatographia*, **1**, 133 (1968)
42. R. Rowan, *Anal. Chem.*, **33**, 658 (1961)
43. C. E. Döhring and H. G. Hauthal, *Acta Chim. Hung.*, **37**, 125 (1963)
44. T. L. Mounts and H. J. Dutton, *Anal. Chem.*, **37**, 641 (1965)
45. C. J. Thompson *et al.*, *Anal. Chem.*, **32**, 424 (1960)
46. C. J. Thompson *et al.*, *J. Gas Chromatog.*, **4**, 1 (1966)
47. C. J. Thompson *et al.*, *Anal. Chem.*, **37**, 1042 (1965)
48. M. Beroza, *Nature*, **196**, 768 (1962)
49. M. Beroza and R. Sarmiento, *Anal. Chem.*, **35**, 1353 (1963)
50. M. Beroza and R. Sarmiento, *Anal. Chem.*, **37**, 1040 (1965)
51. A. Dijkstra *et al.*, *J. Gas Chromatog.*, **2**, 180 (1964)
52. A. A. Duswalt and W. W. Brandt, *Anal. Chem.*, **32**, 272 (1960)
53. M. L. Parsons *et al.*, *Anal. Chem.*, **35**, 842 (1963)
54. A. Gotz, *Z. Anal. Chem.*, **181**, 92 (1961)
55. H. P. Burchfield *et al.*, *J. Gas Chromatog.*, **3**, 28 (1965)
56. J. C. Mamaril and C. E. Meloan, *J. Chromatog.*, **17**, 23 (1965)
57. E. Groten, *Anal. Chem.*, **36**, 1206 (1964)
58. B. C. Cox and B. Ellis, *Anal. Chem.*, **36**, 90 (1964)
59. H. Feuerberg and H. Weigel, *Z. anal. Chem.*, **199**, 121 (1964)
60. W. Kupfer, *Z. anal. Chem.*, **192**, 219 (1963)
61. J. Janák, Chapter 1, ref. 63, p. 387
62. D. F. Nelson and P. L. Kirk, *Anal. Chem.*, **36**, 875 (1964)
63. C. R. Fontan, N. C. Jain and P. L. Kirk, *Mikrochim. Acta*, Nos 2–4, 326 (1964)
64. E. C. Jennings and K. P. Dimick, *Anal. Chem.*, **34**, 1543 (1962)
65. D. Braun and G. Vorendohre, *Farbe Lack*, **69**, 820 (1963)
66. L. N. Winter and D. W. Albro, *J. Gas Chromatog.*, **2**, 1 (1964)
67. C. A. Cramers and A. I. M. Keulemans, *J. Gas Chromatog.*, **5**, 58 (1967)
68. J. France and J. Blaha, *J. Chromatog.*, **6**, 396 (1961)

69. M. BEROZA and R. A. COAD, *J. Gas Chromatog.*, **4,** 199 (1966)
70. E. M. EMERY, *J. Gas Chromatog.*, **5,** 596 (1967)
71. L. MIKKELSEN, *J. Gas Chromatog.*, **5,** 601 (1967)
72. H. W. PATTON, J. S. LEWIS and W. J. KAYE, *Anal. Chem.*, **27,** 170 (1955)
73. F. H. POLLARD and C. J. HARDY, *Anal. Chim. Acta*, **16,** 135 (1957)
74. E. CREMER and R. MÜLLER, *Z. Elektrochem.*, **55,** 217 (1951)
75. D. L. BALL, W. E. HARRIS and H. W. HABGOOD, *J. Gas Chromatog.*, **5,** 613 (1967)
76. A. I. M. KEULEMANS, A. KWANTES and G. W. RYNDERS, *Anal. Chim. Acta.* **16,** 29 (1957)
77. D. M. ROSIE and R. L. GROB, *Anal. Chem.*, **29,** 1263 (1957)
78. A. F. MESSNER, D. M. ROSIE and P. A. ARGABRIGHT, *Anal. Chem.*, **31,** 230 (1959)
79. R. L. GROB *et al.*, *J. Chromatog.*, **3,** 545 (1960)
80. R. A. DEWER, *J. Chromatog.*, **6,** 312 (1961)
81 J. C. STERNBERG, W. S. GALLOWAY and T. L. JONES, Chapter 1, ref. 67, p. 231
82. L. S. ETTRE and H. N. CLAUDY, *Chem. in Canada*, **12,** (9) 32 (1960)
83. L. S. ETTRE and W. AVERILL, *Anal. Chem.*, **33,** 680 (1961)
84. L. S. ETTRE, Chapter 1, ref. 67, p. 307
85. L. S. ETTRE, *J. Chromatog.*, **8,** 525 (1962)
86. A. J. ANDREATCH and R. FEINLAND, *Anal. Chem.*, **32,** 1021 (1960)
87. H. P. KAUFMANN *et al.*, *Fette u. Seifen*, **64,** 501 (1962)
88. H. HOLZHAUSER, Chapter 1, ref. 71, p. 197
89. R. KAISER, "*Chromatographie in der Gasphase*, III. Teil. Tabellen zur Gas-Chromatographie," Bibliograph. Institut, Mannheim (1962)
90. W. A. DIETZ, *J. Gas Chromatog.*, **5,** 68 (1967)
91. A. SCHER and R. KULMAST, *Fette u. Seifen*, **67,** 745 (1965)
92. D. R. DEANS, *Chromatographia*, **1,** 187 (1968)

7. Special Techniques

7.1 Programmed temperature gas chromatography (PTGC)

Method. Gas chromatographic analysis under isothermal conditions fails to provide adequate separation for wide boiling range samples. At low temperatures the heavy components will appear only after a long analysis time in the form of broad peaks, difficult to evaluate, while at high temperatures the light components are not resolved. Samples of this type can be adequately analyzed by separation at two or three different temperatures. Analysis may be performed however from a single sample injection by gradually increasing the column temperature during analysis; this so-called programmed temperature method has contributed greatly to the wider application of gas chromatography. The wide-spread use and importance of programmed temperature may be assessed from the number of relevant publications. The literature survey of Harris and Habgood[1–3] includes 426 publications up to 1964 and there is also a monograph dealing with this subject.[4]

Column temperature can be programmed by different methods. Linear temperature programming is the most usual, but in addition simple ballistic programming, exponential programming (by the linear programming of heating performance) and various combinations of programmed temperature and isothermal operations may also be used.

Apparatus construction. Griffiths *et al.*[5] as far back as 1952 used gradually increased temperature in the analysis of alkyl chlorides. In the following years programmed temperature was applied by several authors, but the analytical usefulness of these methods was greatly limited by the poor reproducibility of the results due to the inadequacy of the apparatus. In the years 1958 and 1959 several papers were published on the proper construction of the apparatus and on the implementation of a reproducible temperature program.[6–10] At the beginning of the 1960's the first commercial equipment with programmed temperature appeared on the market. By the accurate and reproducible adjustment of the operation conditions an accuracy equal or higher than would correspond to isothermal conditions may be achieved particularly in the case of wide boiling range fractions. The most important requirements which a programmed temperature apparatus has to meet may be summed up as follows:

Carrier gas system. The viscosity and density of the carrier gas, and consequently the pressure drop in the column will change with increasing temperature. When using concentration sensitive detectors (e.g. thermal conductivity cells) the flow rate of the carrier gas must be kept at a constant value which necessitates the incorporation of a flow regulator.[8] With a mass sensitive detector (e.g. flame ionization detector) the apparatus may be operated by keeping the inlet pressure, that is the pressure drop, at a constant value.[11, 12] Some detailed work has been published recently on the method by which a constant carrier gas rate can be realized.[13]

Sample injection. A very carefully constructed sample injector with very small dead volume is needed for the rapid evaporation of the wide boiling range sample to reach the column without fractionation. Direct on-column sample injection proved to be highly efficient in this respect (see Section 4.2).

Thermostating. An appropriate heating system is the most important prerequisite for the method. The column, the sample injector and the detector have to be thermostated separately. The sample injector head and the detector are conveniently kept under isothermal conditions at the maximum temperature of the analysis or at a temperature approximating this. Thermostating of the columns should provide for rapid and accurate heating and also for rapid cooling for repeated analyses. Several methods have been proposed for heating the column, as for instance winding the heating wire on the column, or by induction heating of metal columns, but efficiently constructed air thermostats providing for rapid air change were found to be best (see Section 4.4). The temperature program can be varied within wide limits and the latest apparatus can be adjusted to values from 0.5 to 50 °C/min rates. Commercial designs use mainly linear temperature programming when the temperature regulating potentiometer is operated by a synchronous motor.[6, 8] In the latest apparatus various techniques are used including linear, gradual and combined temperature programming. In certain types programmed temperature is realized on the basis of a preliminary plotted arbitrary curve by the incorporation of a photoelectric system which reads the curve. With this method any linear, non-linear or gradual program can be realized which is particularly useful in the analysis of samples containing components which are not all members of the same homologous series.[14] In general the programmed temperature apparatus can be operated at high temperatures (400–500 °C). After analysis the thermostat and column can be rapidly cooled, usually by the automatic opening of the ventilation and the blowing of cold air through the thermostat. Cooling from 400–500 °C to 30 °C requires 5–10 minutes. Low temperature programming has further expanded the application possibilities of the programmed temperature technique.[15,16] The thermostat may be cooled to −75 °C with liquid CO_2; in the latest apparatus cooling with liquid air (−180 °C) can also be applied. This method is particularly suitable for wide boiling range samples which contain gases.

Column. Programmed temperature can be applied with both packed and open tubular columns. In gas–liquid chromatography the appropriate choice of the liquid phase presents the greatest problem, as here higher demands with respect to volatility have to be met than in the case of the isothermal method[17], and volatility is a specially important factor in trace analyses. With programmed temperature the maximum permissible temperature is lower than under isothermal conditions, or the packing has to be preconditioned at higher temperatures. The problem is best solved by a dual channel gas chromatograph, that is, with two parallel columns filled with the same packing. The sample is injected into one of the columns only, the other serves as a reference branch, so that the drift of the base line due to evaporation can be eliminated.[18] At high temperatures the choice of stationary liquid phases is greatly limited, as only a few liquids are stable at temperatures above 400 °C. When adsorbent packings are used no such problems will arise and the applicability of solid adsorbents can be considerably extended in programmed temperature gas chromatography, as at high temperatures even the heavy components are rapidly eluted with the production of symmetrical peaks. The recently evolved polymer gel packings are here very important, as they may be used without liquid coating and are applicable to a multitude of tasks (see Section 5.2).

The other parts of the apparatus are the same as those used in isothermal equipment.

Theory. It is fairly complicated to present a theoretical description of programmed temperature gas chromatography, as its theory is even less clarified than that of isothermal methods. Among others Habgood and Harris,[19] Giddings[20, 21] and Said[22] have carried out some important work in this field. A modified equation was proposed for the number of theoretical plates which characterizes column efficiency,[19] namely:

$$N = 16 \left(\frac{V_{TR}}{w} \right)^2 \qquad 7.1$$

where V_{TR} is the retention volume of the component under isothermal conditions at the temperature T_R and w is the width of the peak base. In the case of programmed temperature gas chromatography the unambiguous definition of the retention data is the most difficult task including the utilization of data obtained under isothermal conditions. Owing to changing temperature the dependence of the retention of the component on operating conditions is highly complex, and only approximate descriptions are known for these correlations even in the simplest case of a linear temperature program.[19–21] Consequently the identification of the components will present a far more complicated problem than in isothermal operations. Identification is based on the adjustment of accurately controlled and reproducible operation conditions, primarily on the control and reproducibility of the initial temperature, the heating rate, the final temperature and of the carrier gas rate. Initial temperature and carrier gas rate have a particularly significant effect on the analysis of light components, while the analysis of heavy components is primarily affected by the heating rate.

Several attempts to describe the retention data have been published. In the case of a linear temperature program Kováts's retention index system is approximately applicable with the modification of substituting into Eq. 6.3 instead of the logarithms of the retention volumes the values of the retention temperature.[23] This method however needs further improvement.[24]

The use of the relative elution temperature has been applied quite satisfactorily in practical gas chromatography;[25] this is described by the following equation:

$$T_{RE} = \frac{T_e}{T_{es}} \qquad 7.2$$

where T_e is the elution temperature of the component, and T_{es} the elution temperature of a chosen standard. The elution temperature is defined by the following correlation:

$$T_e = T_i + P_r \left(\frac{D}{R}\right) \tag{7.3}$$

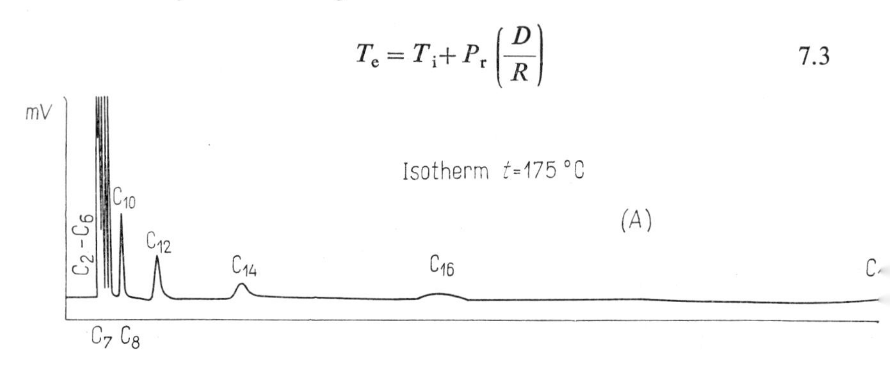

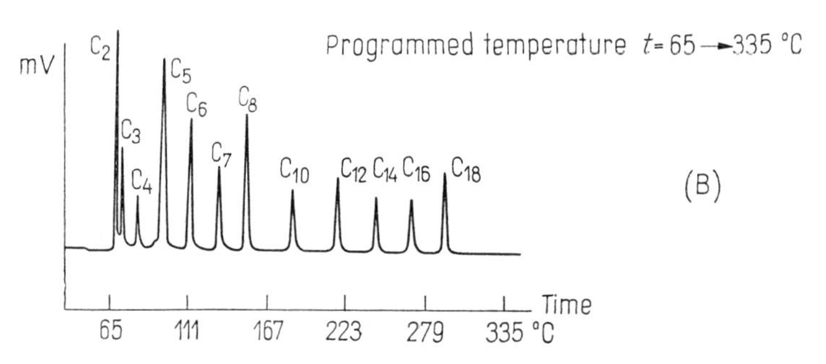

Fig. 7.1. Comparison of isothermal and programmed temperature gas chromatographic analysis of fatty acid methyl ester

where T_i is the initial temperature in °C, P_r the heating rate in °C/min, D the distance between sample injection and peak maximum in cm, R the chart speed of the recorder in cm/min.

Between certain limits relative elution temperature is independent of the initial temperature and of the heating rate.

Applications. Programmed temperature is most frequently used in the analysis of wide boiling range samples.[26] Figure 7.1 shows a comparison of isothermal and programmed temperature methods for the analysis of the same sample. Programmed temperature gas chromatography up to high temperatures has opened new perspectives in the analysis of natural organic substances and of many materials which hitherto were impossible to analyze.[27, 28] The methyl esters of normal fatty acids with C_3 to C_{36} carbon atoms can now be analyzed in a single process, so can various vitamins, hormones, lipids, etc. One of the most important applications of programmed temperature gas chromatography is the analysis of various hydrocarbon fractions up to C_{60}. The method is highly efficient for the analysis of wide boiling range petrol and Diesel oil fractions and the characterization of crude petroleum samples from the results of a single analysis.[29] Particularly high resolutions can be achieved with programmed temperature open tubular columns.[30] Schulz and Reitmeyer give details on the separation efficiency of complex hydrocarbon mixtures with the open tubular column and on the qualitative identification of the components from the retention indices.[31, 32]

Programmed temperature gas chromatographic analysis of samples whose composition is unknown will furnish valuable information on their boiling range and complexity.

Programmed temperature greatly extends the possibilities of gas analysis with solid adsorbent packings. One of its most important applications is the programming of molecular sieve columns for the analysis of samples containing permanent gases and heavier components.[33] Figure 7.2 presents such an example, when under isothermal conditions analysis would need several columns with different packings.

Programmed temperature is very valuable in the trace analysis of

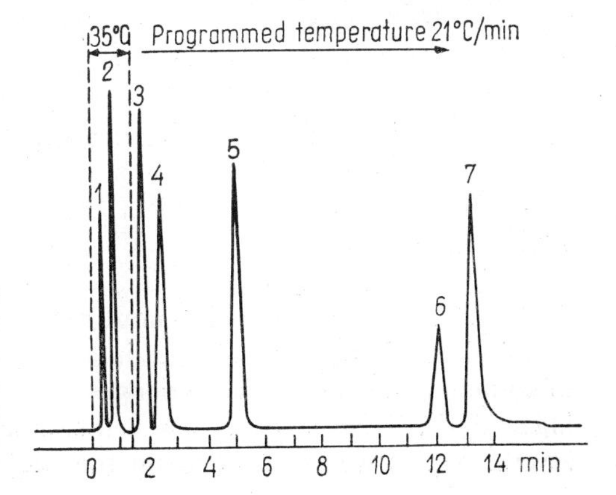

Fig. 7.2. Gas analysis on a programmed temperature molecular sieve column. 1. Neon; 2. argon; 3. nitrogen; 4. methane; 5. carbon monoxide; 6. nitrogen peroxide; 7. carbon dioxide

heavy contaminants.[34] If necessary the contaminants can be concentrated in the column by repeated sample injection and then eluted by raising the column temperature. In this way ppb quantities can be detected with the flame ionization detector.

In the case of heat sensitive samples it is a great advantage that the evaporation of the sample can be omitted and thus thermal degradation during injection eliminated.

Programmed temperature may also greatly raise the efficiency of preparative gas chromatography.[35] At higher temperature the peaks will be narrower thereby facilitating separation on the one hand, and on the other hand the recovery of the component by raising its concentration in the carrier gas (see Chapter 9).

Replacement of analytical distillation by programmed temperature gas chromatography is another important new application (see Chapter 11).

Thermochromatography. This is another way of raising column temperature during analysis and has been in use in the Soviet Union [36–38] for GSC since the beginning of the 1950's. In thermochromatography the column is surrounded by a mobile furnace which proceeds at an adjustable rate in the direction of the gas flow. On the column section heated by the furnace a definite temperature gradient develops which travels along the column with the movement of the furnace. With the progression of the temperature zone the component zones are compressed and even the heavy components will produce sharp peaks within a shorter analysis time. Some detailed papers on the theory and technique of thermochromatography have been published.[39–41] Reports have appeared on the modification of thermochromatography.[42, 43] However, due to the inherent difficulties of moving the furnace and of constant operating conditions the method has failed to gain wide-spread acceptance.

7.2 Flow programming

Method. To accelerate the elution of heavy components in the analysis of wide boiling range samples in addition to the gradual increase of column temperature, the carrier gas flow rate can be increased during analysis. Such a method[44] was first described in 1959 and in 1962 several papers were published on the subject,[45, 46] though the first comprehensive reviews[47–49] on the implementation and possibilities of flow programming appeared only in 1964.

In flow programming the velocity of the carrier gas is gradually increased by raising the inlet pressure. With an exponential increase

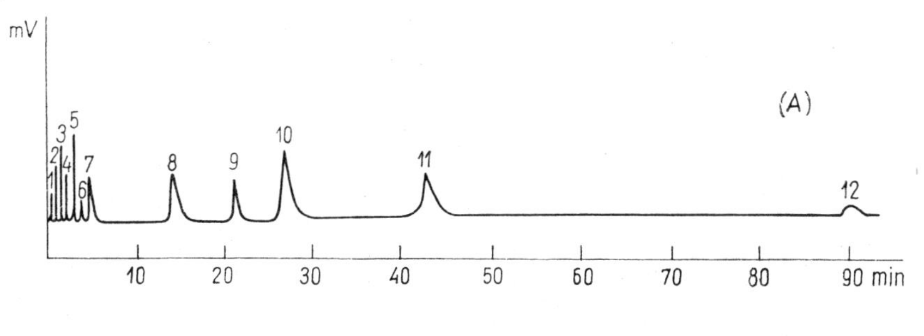

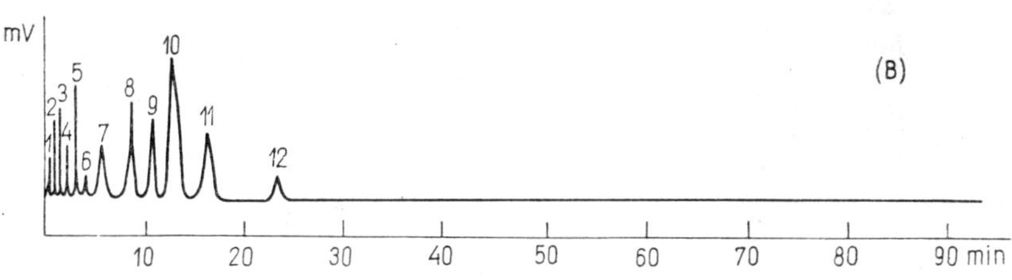

Fig. 7.3. Effect of programmed flow on the chromatogram. (A) Constant velocity; (B) flow program (starting in the 7th min). 1. Air; 2. acetone; 3. benzene; 4. toluene; 5. m-, p-xylene 6. o-xylene; 7. anisole; 8. nitrobenzene; 9. methyl salicylate; 10. o-nitro-ethylbenzene; 11. p-nitro-ethylbenzene; 12. methyl phthalate

of the pressure the velocity of the carrier gas increases almost linearly as a function of analysis time.[48] At higher carrier gas rates the heavier components are eluted within a shorter period of time and will give symmetric peaks as illustrated in Fig. 7.3. The theoretical problems and application possibilities of linear flow programming are discussed in detail by Mázor and Takács.[50, 51]

Flow programming opens the way to analysis at low temperatures, which extends on the one hand the number of possible liquid phases, while on the other hand by reducing the evaporation of the liquid phase it raises the stability of the base line and column life. Low temperature is particularly favourable in the analysis of heat sensitive substances. Low temperature has the other advantage that relative retention, α, generally increases with decreasing temperature, that is, more efficient separation can be achieved.

In the analysis of wide boiling range samples flow programming results in a considerable shortening of analysis time with the simultaneous production of symmetrical sharp peaks which are even appropriate to trace analysis. Compared to programmed temperature, flow programming has the advantage of a reduced base line drift. At programmed temperature the vapour pressure of the liquid phase increases exponentially with the temperature, while in flow programming the evaporation rate will change linearly with the flow rate and will be negligible because of the low temperatures. Flow programming has the further advantage that the effect of pressure change is compensated within a very short time, and there is no "cooling" period after analysis.

Apparatus construction. Flow programming makes no particular demands on apparatus, a simple gradual flow program can be effected in fact with most commercial equipment by manual adjustment of the inlet pressure. The most important requirement is an adequate inlet pressure which in the case of longer packed or open tubular columns may reach values of 10 to 12 atm. Flow programming may however be used efficiently even with a very slight increase of the pressure.[52] In recent years special flow programming units have been developed which can be acquired as ancillary devices and fixed to most commercial equipment. Flow programming can be effected with both packed and open tubular columns and with various types of detectors. With concentration sensitive detectors (thermal conductivity cell) the carrier gas is split into two streams after the flow programmer and one stream flows through a parallel second column to the reference branch of the detector. Lately Halász and Deininger[53] have described a simple highly efficient device for the continuous programming of the inlet pressure which can be connected to existing equipment and will provide a pressure program with better than 1% reproducibility.

Column efficiency. According to the van Deemter equation column efficiency decreases with increasing carrier gas rate (H increases). This effect is not significant in the case of columns, e.g. with small particle size packings, for which the H *vs.* $\bar{u}$ function will be a flat curve. For the investigation of column efficiency the average linear gas velocity, $\bar{u}$, or the corrected average gas rate, v_o, has to be known.

Both can be calculated by means of the pressure correction factor, j (see Chapter 2).

Several correlations have been suggested for calculating retention time or retention volume depending on the method of programming.[48, 52] Similar problems arise here in qualitative evaluations as with programmed temperature, consequently identification with the help of the injection of pure components or by other ancillary methods are the most reliable.

With mass sensitive detectors (e.g. with the flame ionization detector) quantitative evaluation will present no problems as the method involves no change in the overall mass velocity of the component. With concentration sensitive thermal conductivity cells every given flow program requires separate quantitative calibration.

Applications. Essentially flow programming can be applied to all the tasks for which programmed temperature is recommended and may often satisfactorily replace the latter. Hence it is primarily suited to the analysis of wide boiling range samples, of gas mixtures on adsorbent packings (e.g. molecular sieves) and in addition also to analyses at temperatures lower than needed in constant rate (isorheic) methods. Occasionally in some highly complex analyses the best separation can be achieved by a combination of flow programming and programmed temperature.[48]

Flow programming may be advantageously used in preparative gas chromatography.[54] By raising the carrier gas rate the asymmetry of the larger peaks can be reduced, while separation can be performed at a lower temperature. The temperature plugs and retardations which may occur with programmed temperature can be eliminated by programming the flow (see Chapter 9).

7.3 Pyrolysis gas chromatography (PGC)

Method. The pyrolysis of certain substances results in characteristic decomposition products which after separation on the gas chromatographic column can be qualitatively and quantitatively determined and from these results the initial substance identified or valuable data on its chemical structure obtained. The method is essentially similar to that of a mass spectrometer which affords means of analysis based on the nature and quantitative distribution of molecular fragments obtained through the bombarding of the test substance with high speed electrons. The pyrolytic chromatogram, also called the pyrogram, clearly indicates in relatively simple cases, e.g. of linear polymers, the presence of a small number of well identifiable components. In the case of most substances however it will be difficult to identify accurately the decomposition products because of their large number. Such pyrograms are primarily suited to comparisons as "fingerprints", or to the detection of the presence of certain components or contaminants.

Pyrolysis gas chromatography was first applied to the analysis of non-volatile solid substances which could otherwise not be analyzed by gas chromatography. Davison *et al.*[55] subjected various polymers to pyrolysis and collected the decomposition products in cooled traps, from which they injected them on to the gas chromatographic column and identified the initial substance from the chromatograms of its products.

The method was improved by connecting the pyrolysis reactor to the inlet end of the gas chromatograph so that the pyrolysis products were directly and rapidly transferred to the separating column. Today the in-line method is used with the advantage that all the volatile products reach the column without loss, while the possibility of secondary reactions is reduced by the rapid sweeping of the reactor. This method has the third important advantage that due to the high sensitivity of the detectors very low sample quantities (of the order of micrograms) are sufficient for analysis. Pyrolysis gas chromatography has gained wide acceptance in the analysis of non-volatile substances and has risen to the rank of an indispensable method in the investigation of polymers, drugs and other important natural and synthetic substances. The method was later adapted to the analysis of volatile components. Keulemans *et al.*[56, 57] demonstrated the identification of various hydrocarbons from their decomposition products when the quality and structure of the initial substance could be ascertained in a way similar to mass spectrometric methods.

In recent years the importance of pyrolysis gas chromatography has risen not only in direct analysis but also in other fields. One of its important applications is the qualitative identification of the components which had been separated on the gas chromatographic column. In this method an arbitrarily chosen component emerging from the separating column is led by means of a selector valve into the pyrolysis reactor and the pyrolysis products are then analyzed after separation on a second column.[58]

Pyrolysis gas chromatography offers a new way to study the structure of organic compounds. In the study of the structure of pure substances the pyrolysis reactor is usually mounted before the column. Provided pyrolysis is conducted under adequate conditions, the nature and quantitative distribution of the decomposition products will provide valuable information on the chemical structure of the substance.[59]

The most important requirements which pyrolysis gas chromatography has to meet are: 1. accurately controlled reproducible conditions of pyrolysis, 2. the decomposition products must rapidly leave the high temperature zone to prevent secondary reactions and 3. for efficient separation the products must reach the separating column rapidly and in a plug.

Construction of the pyrolysis reactor. Several pyrolysing units have been described in the literature and some excellent reviews have been published recently on the comparison of the generally used types and on the general aspects of their construction.[60, 61]

The pyrolysis reactor has to meet the following requirements: the sample should be heated to the decomposition temperature within the shortest possible time, decomposition temperature should be accurately controlled and reproducible and the material of the reactor should have no catalytic effect on the substance under investigation.

From the aspect of heat input the operation of pyrolysis reactors can be classified into two groups: 1. pulse reactors, 2. tube reactors. Both the temperature profile and the residence (contact) times are rather different in the two types of reactors, consequently there will be a greater difference between the results obtained with different types of reactors than with similar types.

Pulse reactors

The common characteristic of reactors of this type is that heating begins after the sample has been introduced, usually in the form of a solid particle or thin liquid film directly on to the heating element. Depending on the method of heat input the reactors may be provided either with a heated filament, induction heating, electrical discharge device or arc lamp.

Pyrolysis unit with a heated filament. This oldest and still popular pyrolysis device uses a platinum or chromium–nickel spiral.[62, 63] Figure 7.4 is a schematic diagram of a simple pyrolysis unit. The sample under investigation is placed in the form of solid granules or

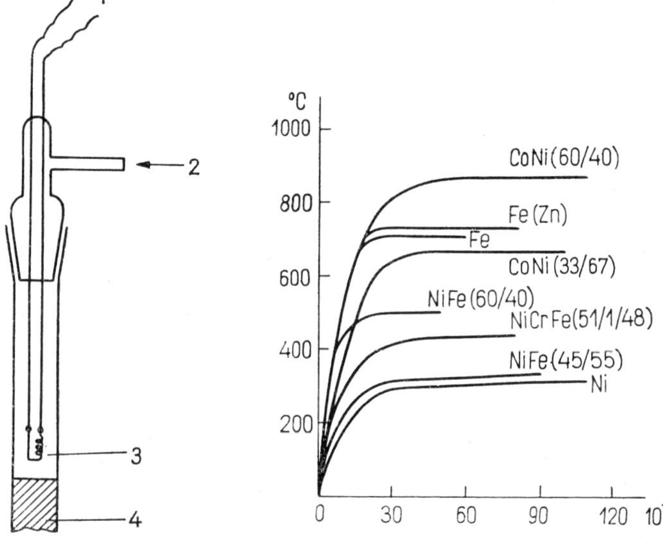

Fig. 7.4. Schematic diagram of a heated filament pyrolysis unit. 1. Electric terminals; 2. carrier gas inlet: 3. Pt spiral; 4. column packing

Fig. 7.5. Heating curves of some ferromagnetic conductors[65]

dissolved in an appropriate solvent on the spiral. The resistance wire is heated by means of an electric timer for 1–30 sec when the wire temperature is varied by the adjustment of the current intensity. Instead of the spiral a metal disk is occasionally used which facilitates the introduction of the sample and has a longer life.[64]

The operation of a heated filament pyrolysis unit involves several deficiencies and interfering factors of which the most important is measurement of the filament temperature which is cumbersome and inaccurate and is usually performed with an optical pyrometer. In the case of solid samples a significant temperature gradient develops in the sample itself which completely impedes an even approximate estimation of the decomposition temperature. The filament may have a catalytic effect on the decomposition of the sample, though with very small quantities (20–30 μg) identical product distributions were obtained with different resistance wires.[62] The resistance wire has a short life, often not more than 20 or 50 cycles and its characteristics change with age. In spite of all these drawbacks the heated filament pyrolysis unit is still the most widely used in industrial practice, mainly for its simplicity and easy handling. It was proved by several authors that for the analysis of polymers and other non-volatile substances reproducibility is quite satisfactory and thus the device is adequate for routine investigations.

A pyrolysis unit on similar lines has been evolved in recent years for several commercial instruments (Perkin–Elmer, Pye, Hewlett–Packard, Carlo Erba, Virus, etc.) which can be mounted directly before the sample injector or in place of the latter. This construction reduces the dead volume to a minimum and the pyrolysis products are rapidly swept into the column. Filament temperature *vs.* current intensity is determined by preliminary calibration.

High frequency pyrolysis unit. A new type of pyrolysis device was evolved by Simon and Giacobbo[65, 66] who used ferromagnetic conductors. The method is based on the fact that when certain metal alloys are induction heated their ferromagnetic characteristics will drop suddenly at a given temperature (the so-called Curie point) so that they cannot be heated to higher temperatures. With high frequency induction heating the required temperature is reached in a very short time, within 20–30 msec, and temperature can be maintained accurately at the required level. Temperature can be adjusted to different values by the application of several interchangeable ferromagnetic conductors of different compositions. Figure 7.5 shows the characteristic heating curves of some substances.[65] A schematic diagram of the pyrolysis unit is shown in Fig. 7.6.[67] This pyrolyser approximates best the ideal conditions; it has a very small dead volume (5–10 μl), pyrolysis is rapid and can be accurately controlled with a high degree of reproducibility. No detrimental catalytic effect was observed with the various ferromagnetic conductors. The pyrolysis unit can be joined directly to an open tubular column so that the decomposition products can be analyzed with very high resolution. So far this technique has been applied mainly to the investigation of the structure of organic substances but its use will probably spread. Recently a pyrolysis unit of this type has been marketed commercially by Philips Co. (now Pye Unicam).

Reactors using decomposition by electrical discharge[68] or in the

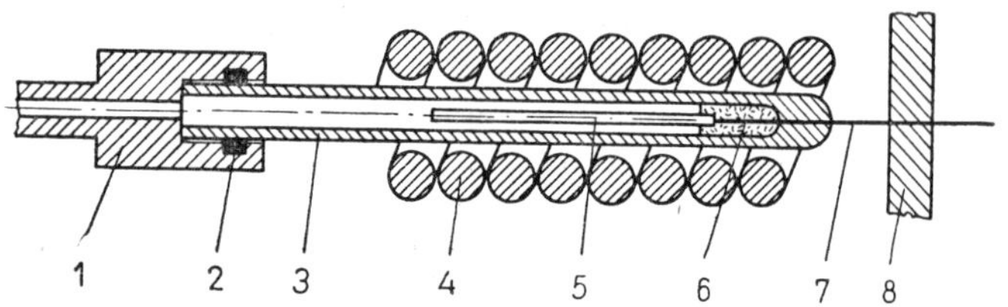

Fig. 7.6. High frequency heated pyrolysis unit.[67] *1. Carrier gas connection; 2. O-rings; 3. glass capillary; 4. high frequency induction coil (copper); 5. ferromagnetic conductor (0.5 × 20 mm wire); 6. sintered glass beads; 7. Pt Ir capillary (0.3 mm internal diameter); 8. rubber septum of the sample injector*

electric arc also belong to the category of pulse reactors, but have no special practical significance.

Tube reactors

The common feature of this second group of pyrolysis reactors is continuous heat input with sample introduction into the reactor heated to constant temperature which is then maintained during the pyrolysis process.

The pyrolysis unit contains a quartz or metal tube provided with electric heating or placed into a heated metal block. The construction and operation of the tube reactors depend on the state of the sample under investigation. Ettre and Várady[69] developed a quartz tube reactor for the investigation of solid substances with a small quartz boat for the sample. This boat is placed outside the heated zone of the reactor in a separate tube from which it is pushed once heat equilibrium has been reached in the reactor by an appropriate device into the heated zone for pyrolysis. The method has the advantages that at high temperatures the samples come into contact with quartz only and that the quantity of the sample can be accurately determined both before and after pyrolysis. The sample may be introduced also by other methods and in addition packed pyrolysis reactors have also been used.[70, 71]

The tube reactor has the apparent advantage that it affords accurate adjustment and measuring of the temperature. In fact considerable discrepancies may however occur between the temperatures of the tube wall and of the sample and in the case of solid samples a temperature gradient may develop even within the particle. The other difficulty is caused by the temperature differences along the tube so that only a small zone can be considered isothermic. As the sample is heated the production of decomposition products is accelerated and the process cannot be assigned to a certain temperature. A further problem arises from the fact that the decomposition products travel through the heated zone of the reactor where they may be subject to further decomposition or to secondary reactions. The temperature distribution along the tube reactor depends on the operating conditions, primarily on the carrier gas rate changes which may affect the distribution of the products.

The general difficulties encountered with tube reactors are less prominent in the pyrolysis of liquid and gas samples where introduction is much simpler and may be performed e.g. with a syringe. Keulemans *et al.*[56, 57] have done considerable work on the development of tube reactors for the pyrolysis of hydrocarbons. The schematic diagram of a simple tube reactor is shown in Fig. 7.7. In this type the reactor tube is a stainless steel or a noble metal (gold, silver or platinum) capillary which provides for uniform, satisfactorily controlled temperature and for alteration of the residence time by changes in the length of the capillary. The construction of the tube reactor has to meet the following important requirements: 1. uniform and rapid heat transfer; 2. possibility of the exact measurement of the temperature and of the residence time. This last can be realized by means of sensing elements (thermistors) mounted at the inlet and outlet ends.

For the pyrolysis of gaseous samples, e.g. for the qualitative identification of gas chromatographically separated components the pulse reactor is unsuitable and only tube reactors can be used. There is a difference of opinion in the literature concerning the advantages and disadvantages of the two types of reactors. It is however certain that some improvement in pyrolysis reactors in general is needed for the reliable control of the process and for accurately reproducible quantitative results.

From the point of view of reactor technique the tube reactor is not a continuous apparatus, as sample introduction and thus pyrolysis are intermittent and the reaction proceeds under non-stationary conditions. Continuous operation of the tube reactor can be realized by continuously feeding the sample into a reactor in a side branch,[57]

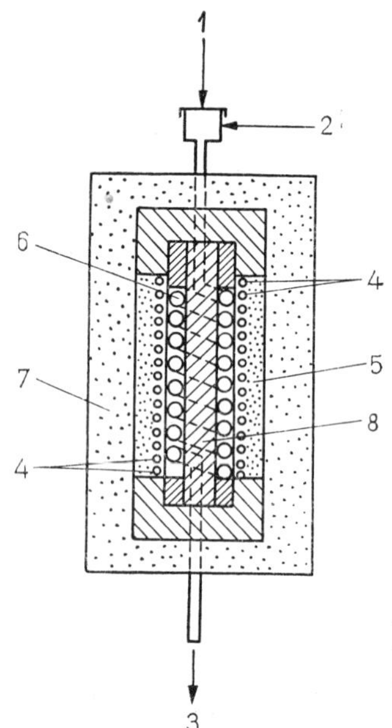

Fig. 7.7. Schematic diagram of a pyrolysis tube reactor. 1. Sample injection; 2. carrier gas inlet; 3. carrier gas + pyrolysis products effluent; 4. heated filament terminal; 5. heating coil; 6. tube reactor; 7. insulation (diatomite)

when the main reactor will operate under stationary conditions and the samples taken at intervals from the effluent can be led to the separating column of the gas chromatograph. This method obviously raises different demands on the quantity of the sample and on the method of sample injection.

Effect of operating conditions. The mechanism of pyrolysis reactions is in general extremely complex and only little understood. The nature and quantitative distribution of the products depend on a great number of factors, some of which were discussed in the section on reactor construction. Temperature and residence time are the most important operation conditions which together will determine the severity of the pyrolysis. The thermal decomposition of compounds of various types may proceed under very different conditions. Some compounds, for example, quaternary ammonium salts, and dialkyl phosphates, decompose even at 200–300 °C, while the decomposition of other compounds may require much higher temperatures, e.g. the decomposition of hydrocarbons or of porphyrins. The temperature range of pyrolysis reactors is usually 200–1000 °C, most often between 500 and 800 °C. Depending on the temperature, the nature and amount of decomposition products may greatly differ; at high temperatures the formation of small fragments (e.g. in the case of hydrocarbons of H_2 and of CH_4) is favoured and these are less characteristic of the initial substance than the products obtained at lower temperatures.

As well as the maximum temperature, the temperature distribution of the reactor and the residence time also have a significant influence on the products. When not rapidly removed from the reaction zone, the primary decomposition products may undergo secondary reactions and furnish secondary products.

The measure of the secondary reactions, and thus changes in product distribution greatly depend on the quantity of the sample which may be considered one of the most important factors. It was proved by several authors that in the case of very small (microgram) samples the secondary reactions are negligible and product distribution will be only slightly sensitive to fluctuations in the operating conditions.

The influence of pressure and carrier gas rate is not significant in the case of pulse reactors, at least not when short, 1–2 sec heating periods are applied. In the case of tube reactors these factors will however influence the temperature distribution of the reactor on the one hand and the measure of the secondary reactions on the other.

Analysis of the pyrolysis products. In general a great number of components of different types and with considerably different boiling points will be formed by pyrolysis and the separation and identification of these components is indeed a difficult task. In certain cases the heavy components may be retained on a pre-column and the analysis of the light components may then be sufficient.[72, 73] Another possibility is analysis by the combination of two or more columns.[63] Another possibility of analyzing wide boiling range components is the application of programmed temperature. Quite often for the satisfactory separation of the great number of components a high efficiency open tubular column is indispensable.[74, 75]

Applications. The most important and most widely accepted application field of pyrolysis gas chromatography is the investigation of polymers. For identification and routine analytical tasks the distribution of the decomposition products gives a characteristic "fingerprint" of the sample.[68–70, 76–79] In this respect Groten's work[79] is quite outstanding as it includes the study of 150 different substances, such as polymers, elastomers, resins, natural substances, cellulose esters, polymer mixtures and copolymers, products with and without

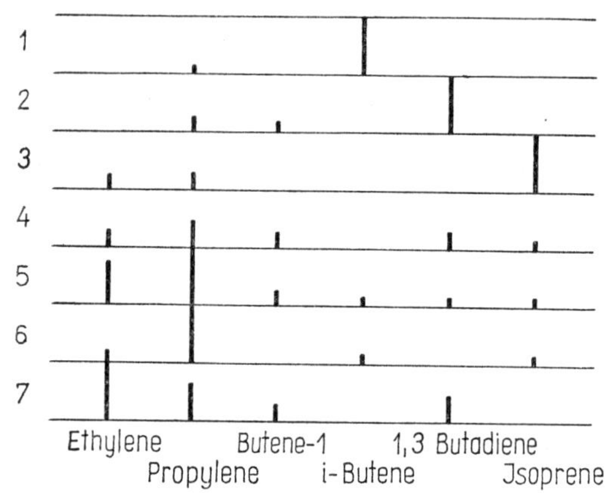

Fig. 7.8. Identification of polyolefins from their pyrogram.[60] *1. Butyl rubber; 2. polybutadiene; 3. natural rubber; 4. hypalon; 5. ethylene–propylene elastomer; 6. polypropylene; 7. polyethylene*

fillers, etc. A product spectrum of polyolefins[60] is shown in Fig. 7.8. The method is also applicable to the determination of small quantities of contaminations and of the composition of copolymers.[79, 80] Pyrolysis gas chromatography furnishes valuable information on the microstructure of polymers.[76] The decomposition mechanism of the various polymers is affected by their molecular weight, molecular weight distribution, branching, cross linking steric configuration, etc. Thus a different product distribution is obtained for ethylene–propylene copolymers than for the mixtures of these polymers, and atactic, isotactic and mixed polypropylenes can also be distinguished by this method. The third field in the investigation of polymers is the study of thermal stability and decomposition.[81] The decomposition of polymers made from monomers containing labelled atoms offers a possibility of investigating their formation mechanism.

Pyrolysis gas chromatography is also applied to the identification of various drugs and other organic compounds, such as barbiturates and atropine derivatives,[82] proteins and amino acids,[83] phenothiazines,[84] purines and pyrimidines,[85] alkaloids and other substances. Its application is spreading in the paint and varnish industry for the analysis of resins, varnishes and paints.[86, 87] It is beginning to be used in the petroleum industry for the analysis of heavy distillates, as a qualitative test of crude oils, distillation residues, tars, etc.[88] The pyrolysis of rock samples indicates the presence of organic substances which is an important factor in oil prospecting[88] and tests of this type are part of the space research programs for the investigation of the surface of the Moon and of the planets.[89]

In addition to direct analytical applications, pyrolysis gas chromatography is also used in the identification of the peaks obtained by gas chromatographic separation.[58] Tests have been made not only with hydrocarbons of different types but also with alcohols, mercaptans, fatty acid esters and terpene acetates. Pyrolysis gas chromatography has opened new possibilities for the study of the kinetics and mechanism of thermal decomposition processes.[57] The advantages of the method are its low sample requirement and its rapidity giving a large number of measurements within a relatively short time.

An interesting application of pyrolysis gas chromatography has been established in forensic medicine for the identification primarily of drugs, poisons and natural organic substances, but also for industrial products, such as plastics and paints.[90] The substance is not only identified but its origin can also be determined from the contaminants. The identification of micro-organisms from their pyrolysis products appears to be a highly important new application.[91] Pyrolysis gas chromatography by ensuring rapid, simple and reliable identification may soon replace the lengthy morphological, serological and biochemical methods, now in use in microbiology. Promising results have been achieved recently also in the identification and characterization of antibiotics.[92]

In addition to the investigation of polymer structures pyrolysis gas chromatography has been gaining in importance in the study of the structure of other, mainly high molecular, organic compounds.[59, 65–67] The method was applied to the study of the structure of various aromatic compounds, steroids, amino acids, proteins, etc. and was found to be particularly valuable in cases unsuitable for mass spectrometry. Wolf and Rosie[93] have investigated thoroughly the influence of temperature on thermal degradation of approximately 90 organic compounds in connection with the determination of functional groups. Using compounds with labelled atoms a combination of pyrolysis with radio-gas chromatography can provide valuable information on structure and decomposition processes.

7.4 Reaction gas chromatography

Method. Under reaction gas chromatography those methods are generally understood which involve the direct gas chromatographic separation of the products of some chemical reaction, that is when chemical reaction and gas chromatographic separation proceed continuously in a closed system with units connected in series. The first experiments of this type were performed by Kokes, Tobin and Emmett[94] who combined a microreactor packed with a cracking catalyst and a gas chromatograph. The term "reaction gas chromatography" was coined by Drawert *et al.*[95–97] and has been generally accepted in recent years.

The reaction may take place at different points in the system; the microreactor can be mounted before the sample injector, in the sample injector itself or on the chromatograph column or connected to the outlet end of the column. In the last case a secondary column or columns will provide for the separation of the reaction products. Methods which involve the reaction of the substance under investigation in a closed system and the injection of an aliquot part of the products into the gas chromatograph are not considered reaction gas chromatographic processes. If however the reaction takes place in a closed system joined directly to the gas chromatograph and after the reaction the product is led rapidly and directly into the separation column then this might be considered a reaction gas chromatographic method. Neither are methods involving the identification of the chromatographed products by colour reactions included into the methods to be discussed here; these were already described in Chapter 6.

Sometimes the so-called subtractive procedures are classified among reaction gas chromatographic methods; these consist in the selective removal of certain compounds or group of compounds prior to the injection of the sample into the gas chromatograph.[98] This view is however highly debatable and in this book the subtractive methods were discussed in connection with the pre-columns. The subtractive technique aims in general at the selective retention of certain components and not at the transformation and analysis of the products. In addition, certain compound types are removed by means of physical adsorption (e.g. the adsorption of n-hydrocarbons on a 5A molecular sieve) which can by no means be regarded as a reaction gas chromatographic process. All methods however which involve the chemical transformation of a certain component on the pre-column (e.g. the reaction of water and calcium carbide to give acetylene) belong to the group of reaction gas chromatographic processes.

The method has been applied in conjunction with diverse homogeneous and heterogeneous reactions and an extension of their field of application is expected. A comprehensive survey of the various methods is presented in the review by Beroza and Coad.[98]

By definition, pyrolysis gas chromatography would also belong to this group, but because of its special technique and wide application it was nevertheless discussed separately.

Elementary analysis is one of the special fields of reaction gas chromatography which involves chemical reactions leading finally to well defined decomposition products formed from the atoms constituting the compound under investigation, so that from the analysis of these decomposition products the elementary composition of the compound can be determined. Elementary analysis will be described in greater detail in the chapter on special applications of gas chromatography.

Apparatus and methods. An attempt to give more than an outline of the apparatus and methods used in reaction gas chromatography would surpass the scope of this book, so that we shall restrict ourselves to the most important trends in this field only. One group of methods is based on the reaction proceeding in the gas chromatographic column itself. After the injection of the sample a compound is injected which will form with certain components of the sample a volatile derivative. In this way e.g. alcohols can be converted into trimethylsilyl ethers,[99] alkaloids and steroids to acetates and propionates[100] and analyzed as such.

Certain reactions may take place in the pre-column before the separating column, thus for instance amino acids are converted on a pre-column containing ninhydrin into aldehydes[101], or HCl may be determined from the analysis of CO_2, evolved on a $NaHCO_3$ packed pre-column.[102]

In other instances the reaction occurs in a closed capillary joined to the sample injector or in the injector head proper under the action of some reagent added to the sample.

Most of the reaction gas chromatographic methods use flow-through microreactors. There are two types of operation techniques with the microreactor, namely intermittent "plug-like" sample injection (pulse reactors), or continuous flow. The continuous method involves the intermittent injection of single doses into the chromatographic column from the constant stream of material travelling through the reactor in a by-pass by turning a valve or by automatic change over. The pulse method is the most frequently used as it requires very small sample quantities. The microreactor is usually a tube packed with some catalyst which can be inserted between the sample injector and the column or after the column. It is of fundamental importance from the aspect of an efficient separation of the products that the reaction should proceed rapidly and the products should be swept quickly from the reactor. To accelerate the reaction usually a high operation temperature (200–400 °C) is chosen and the reactor is provided with separate heating. The dimensions of the reactor depend on the reaction involved and mainly on the quantity of the sample which is again a function of detector sensitivity. Generally 5–10 cm long and 2–8 cm diameter tubes are used. The reactor is conveniently made of quartz, but the use of stainless steel tubes is also quite wide-spread. For reproducible results the temperature of the reactor must be accurately measurable and controlled. From the aspect of product analysis it is highly important to have a minimum dead volume in the reactor and at the connections and to ensure a uniform flow through the reactor.

Sample injection. Liquid samples are usually injected with a microsyringe directly on to the catalyst bed. Solids can be dissolved in a solvent for injection. To eliminate interference by the solvent a direct solid sample injector has been devised.[103]

Different reactor designs have been described in the literature[95, 97, 103, 104] and recently microreactors and reactor units which can be mounted on commercial gas chromatographs have appeared on the market.

Applications. Reaction gas chromatography may be applied to the selective detection of certain compounds or groups of compounds, to the identification of gas chromatographically separated components, to the study of the structure of pure substances, to the study of the kinetics and mechanism of various reactions as well as to the testing of catalysts.

Various non-catalytic and catalytic reactions have been used for the identification and study of the structure of compounds, which through producing derivatives of certain compound types, by degrading the compounds, or by the quantitative determination of certain functional groups, provide information on the nature and structure of the starting substance. The analyses of alcohols,[95, 99] of amino acids,[101] aldehydes and ketones,[105] acids and esters[96] and amino groups[106] etc. have been performed through preparation of their derivatives. One of the most important applications of this method is the blood alcohol test. The quantity of ethylene obtained by the dehydration of alcohol is determined by gas chromatographic analysis. The method is very rapid (5–10 min), and can be carried out from 0.1 ml of blood with an average error less than 0.01 % absolute.[96, 97]

Catalytic reactions in the gas phase are preferred for the conversion of the components, primarily hydrogenation, dehydrogenation and hydrocracking. The simplest form of hydrogenation reactions is hydrogenation of the unsaturated bonds in the molecule. From the chromatograms recorded before and after hydrogenation the unsaturated compounds can be determined. In addition to hydrocarbons[107] this method was also successfully applied to the methyl esters of unsaturated fatty acids.[108] Naphthenes were determined by conversion into aromatic compounds by means of a dehydrogenating catalyst.[107, 109]

Hydrocracking is the most important group of the hydrogen reactions. Thompson *et al.*[110–114] worked out a method for the identification of compounds with heteroatoms and for the study of their structure by combining the microhydrogenation reactor and gas chromatography. The method was adopted for the identification of gas chromatographically separated components, to the study of the structure of synthetic organic compounds and of the components of petroleum fractions. The reaction proceeds in a microreactor packed with a catalyst at a temperature of about 200 °C in a stream of hydrogen, the reaction products are collected in a cooled trap and taken up in a solvent or injected directly on the column. A commercial grade catalyst consisting of 0.5 % Pd on alumina was used. The method gives satisfactory results in the analysis of compounds containing sulphur,[110] oxygen,[111] halogens[112] and nitrogen[113] atoms. For nitrogen compounds

Table 7.1

Products of "carbon skeleton" chromatography by the hydrogenolysis of compounds of various types

	Type of compound	*Product*
Paraffin hydrocarbon		Unchanged
Unsaturated hydrocarbon	$RCH=CHR'$	RCH_2CH_2R'
	$RC\equiv CR'$	RCH_2CH_2R'
Halide	$R\downarrow X$	RH
	$R\downarrow X_2$	RH_2
Mercaptan	$R\downarrow SH$	RH
Thiophene	$R\downarrow S$	RH_2
Sulphide	$R\downarrow SS\downarrow R'$	RH, $R'H$
Alcohol, primary	$R\downarrow CH_2\downarrow OH$	RH, RCH_3
secondary	$RR'CH\downarrow OH$	$RR'CH_2$
tertiary	$R'(R)C(R'')\downarrow OH$	$R'(R)C(R'')H$
Aldehyde	$R\downarrow CH\downarrow O$	RH, RCH_3
Ketone	$RC(=O)R'$	RCH_2R'
Ether	$R\downarrow CH_2\downarrow O\downarrow CH_2\downarrow R$	RCH_3,RH, $R'CH_3$, $R'H$
Acid	$R\downarrow COOH$	RH
Ester	$R\downarrow CH_2\downarrow OC(=O)\uparrow R'$	RCH_3, RH, $R'H$
Anhydride	$R\uparrow C(=O)-O-C(=O)\uparrow R'$	RH, $R'H$
Amine, primary	$R\downarrow CH_2\downarrow NH_2$	RCH_3, RH
secondary	$RR'CH\downarrow NH_2$	$RR'CH_2$
tertiary	$R'(R)C(R'')\downarrow NH_2$	$R'(R)C(R'')H$
Amide	$R\downarrow C(=O)\downarrow N\uparrow R'R''$	RH, RCH_3, $R'H$, $R''H$

porous glass powder coated with 5% Pt was found to be a more efficient catalyst. The reaction involves hydrocracking at the heteroatom and depending on the structure of the initial compound one or more paraffins, naphthenes or aromatic hydrocarbons will be the main products. These hydrocarbons are then gas chromatographically identified whereby valuable data are obtained on the nature of the initial substance. The method is further applicable to the removal of other heteroatoms, such as phosphorus, silicon and metals.[114] Quantities of 0.01–0.1 μl are sufficient for the analysis.

Beroza *et al.*[115–118] improved the method of hydrogenolysis by working out a microreactor unit which can be directly joined to the sample injector of the gas chromatograph. They used as catalyst 1% Pd on the gas chromatographic support and heated the reactor to 300 °C. In the reaction the unsaturated bonds will become saturated and the molecule cleaved at the heteroatoms whereby hydrocarbons with the same number and/or with one less carbon atom are formed. The method is called "carbon skeleton chromatography" and has been applied up to carbon numbers C_{30} in the most diverse types of compounds (Table 7.1).

By raising the temperature to 360 °C dehydrogenation may be performed in the same reactor, e.g. naphthenes may be converted into aromatics.[117] In addition to the applications mentioned reaction gas chromatography may also be used in conjunction with many other reactions, or as a combination of several different reactions, e.g. pyrolysis and hydrogenation in series.[119]

Another important and extensive field of research is the investigation of the reaction kinetics and mechanisms by means of reaction gas chromatography[120–123] and the study of the operation and activity of catalysts.[124–126] These latter may be performed with semi-automatic and automatic apparatus suitable for serial tests.[127]

References

1. W. E. Harris and H. W. Habgood, *J. Gas Chromatog.*, **4**, 144 (1966)
2. W. E. Harris and H. W. Habgood, *J. Gas Chromatog.*, **4**, 168 (1966)
3. W. E. Harris and H. W. Habgood, *J. Gas Chromatog.*, **4**, 217 (1966)
4. W. E. Harris, "*Programmed Temperature Gas Chromatography*", J. Wiley, New York (1966)
5. J. Griffiths, D. H. James and C. S.G. Phillips, *Analyst*, **77**, 897 (1952)

6. S. Dal Nogare and C. E. Bennett, *Anal. Chem.*, **30**, 1157 (1958)
7. G. F. Harrison, P. Knight, R. R. Kelly and M. T. Heath, Chapter 1, ref. 60, p. 216
8. A. J. P. Martin, C. E. Bennett and F. W. Martinez, Chapter 1, ref. 64, p. 363
9. S. Dal Nogare and J. C. Harden, *Anal. Chem.*, **31**, 1829 (1959)
10. R. Teraniski, C. C. Nimmo and J. Corte, *Anal. Chem.*, **32**, 1384 (1960)
11. M. J. Golay, L. S. Ettre and S. R. Norem, Chapter 1, ref. 66, p. 139
12. L. S. Ettre and F. I. Kabot, *Anal. Chem.*, **34**, 1431 (1962)
13. G. Deininger and I. Halász, *Z. anal. Chem.*, **228**, 81 (1967)
14. R. M. Bethea and F. S. Adams Jr., *J. Chromatog.*, **8**, 532 (1962)
15. F. Baumann, R. F. Klaver and J. F. Johnson, Chapter 1, ref. 66, p. 152
16. C. Merritt Jr. and J. T. Walsh, *Anal. Chem.*, **35**, 110 (1963)
17. H. G. Baettger, Chapter 1, ref. 37, p. 133
18. E. M. Emery and W. E. Koerner, *Anal. Chem.*, **34**, 1196 (1962)
19. H. W. Habgood and W. E. Harris, *Anal. Chem.*, **32**, 450 (1960)
20. J. C. Giddings, *J. Chromatog.*, **4**, 11 (1960)
21. J. C. Giddings, Chapter 1, ref. 67, p. 57
22. A. S. Said, Chapter 1, ref. 67, p. 78
23. H. van den Dool and P. D. J. Kratz, *J. Chromatog.*, **11**, 463 (1963)
24. G. Guiochon, *Anal. Chem.*, **36**, 661 (1964)
25. J. A. Schmit and R. B. Wynne, *J. Gas Chromatog.*, **4**, 325 (1966)
26. C. Merritt Jr. *et al.*, *Anal. Chem.*, **36**, 1502 (1964)
27. J. D. Mold *et al.*, *Biochem.*, **3**, 1293 (1964)
28. J. G. Nikelly, *Anal. Chem.*, **36**, 2244 (1964)
29. R. V. Blundell *et al.*, Chapter 1, ref. 63, p. 360
30. H. W. Habgood and W. E. Harris, *Anal. Chem.*, **34**, 882 (1962)
31. H. Schulz and H. O. Reitemeyer, *Chromatographia*, **1**, 315 (1968)
32. H. Schulz and H. O. Reitemeyer, *Chromatographia*, **1**, 364 (1968)
33. P. Solbrig, W. Saffert and H. Schubert, *Chem. Techn.*, **16**, 745 (1964)
34. J. J. Kirkland, *Anal. Chem.*, **34**, 428 (1962)
35. M. Verzele, *J. Chromatog.*, **9**, 116 (1962)
36. A. A. Zhukhovitskii, O. V. Solotaryeva, V. A. Sokolov and N. M. Turkeltaub, *Doklady Akad. Nauk SSSR*, **77**, 435 (1951)
37. A. A. Zhukhovitskii, N. M. Turkeltaub and T. V. Georgievskaya, *Doklady Akad. Nauk SSSR*, **92**, 987 (1953)
38. A. A. Zhukhovitskii, E. V. Vagin and S. S. Petruchov, *Doklady Akad. Nauk SSSR*, **102**, 771 (1955)
39. A. A. Zhukhovitskii, N. M. Turkeltaub and V. A. Sokolov, *Doklady Akad. Nauk SSSR*, **88**, 859 (1953)
40. A. A. Zhukhovitskii and N. M. Turkeltaub, *Doklady Akad. Nauk SSSR*, **94**, 77 (1954)
41. A. A. Dachkevich, A. A. Zhukhovitskii and N. M. Turkeltaub, *Ind. Laboratory*, **25**, 222 (1959)
42. A. G. Nerheim, *Anal. Chem.*, **32**, 436 (1960)
43. R. W. Ohline and D. D. De Ford, *Anal. Chem.*, **35**, 227 (1963)
44. S. R. Lipsky, R. A. Landowne and J. E. Lovelock, *Anal. Chem.*, **31**, 852 (1959)

45. J. M. VERGNEAUD, *Bull. Soc. chim. France*, **19,** 14 (1962)
46. S. VALUSSI and G. COFLERI, *Boll. Lab. Chim. Prov.*, **13,** 1 (1962)
47. C. COSTA NETO, J. T. KOFFER and J. W. DE ALENCAR, *J. Chromatog.*, **15,** 310 (1964)
48. A. ZLATKIS *et al.*, *J. Gas Chromatog.*, **3,** 75 (1965)
49. L. MÁZOR, F. BALLA and J. TAKÁCS, *J. Chromatog.*, **20,** 221 (1965)
50. L. MÁZOR and J. TAKÁCS, *J. Chromatog.*, **29,** 24 (1967)
51. L. MÁZOR and J. TAKÁCS, *J. Gas Chromatog.*, **6,** 58 (1968)
52. L. MÁZOR and J. TAKÁCS, *J. Gas Chromatog.*, **4,** 322 (1966)
53. I. HALÁSZ and G. DEININGER, *Z. anal. Chem.*, **228,** 321 (1967) and **229,** 14 (1967)
54. R. P. W. SCOTT, Chapter 1, ref. 70, p. 25
55. W. H. DAVISON, S. SLANEY and A. L. WRAGG, *Chem. and Ind.*, 1356 (1954)
56. A. I. M. KEULEMANS and S. G. PERRY, Chapter 1, ref. 66, p. 356
57. C. A. M. CRAMERS and A. I. M. KEULEMANS, *J. Gas Chromatog.*, **5,** 58 (1967)
58. E. J. LEVY and D. G. PAUL, *J. Gas Chromatog.*, **5,** 136 (1967)
59. W. SIMON *et al.*, *J. Gas Chromatog.*, **5,** 53 (1967)
60. S. G. PERRY, *J. Gas Chromatog.*, **5,** 77 (1967)
61. R. L. LEVY, *J. Gas Chromatog.*, **5,** 107 (1967)
62. C. E. R. JONES and A. F. MOYLES, *Nature*, **189,** 222 (1961)
63. S. E. MARTIN and R. W. RAMSTAD, *Anal. Chem.*, **33,** 982 (1961)
64. D. F. NELSON and P. L. KIRK, *Anal. Chem.*, **34,** 899 (1962)
65. W. SIMON and H. GIACOBBO, *Chem. Ing. Techn.*, **37,** 709 (1965)
66. W. SIMON and H. GIACOBBO, *Angew. Chem. int. Ed.*, **4,** 938 (1965)
67. J. A. VOELLMIN *et al.*, *Microchem. J.*, **11,** 73 (1966)
68. A. BARLOW, R. S. LEHRLE and J. C. ROBB, *Polymer*, **2,** 27 (1961)
69. K. ETTRE and P. F. VÁRADI, *Anal. Chem.*, **35,** 69 (1963)
70. B. C. COX and B. ELLIS, *Anal. Chem.*, **36,** 90 (1964)
71. J. H. DHONT, *Analyst*, **69,** 71 (1964)
72. G. C. HEWITT and B. T. WHITMAN, *Analyst*, **86,** 643 (1961)
73. D. DEUR-SIFTAR, *J. Gas Chromatog.*, **5,** 72 (1967)
74. E. W. CIEPLINSKI *et al.*, *Z. anal. Chem.*, **205,** 357 (1964)
75. B. KOLB *et al.*, *Z. anal. Chem.*, **209,** 302 (1965)
76. C. E. R. JONES and G. E. J. REYNOLDS, *J. Gas Chromatog.*, **5,** 25 (1967)
77 C. KARR *et al.*, *Anal. Chem.*, **35,** 1441 (1963)
78. E. W. NEUMANN and H. G. NADEAU, *Anal. Chem.*, **35,** 1454 (1963)
79. B. GROTEN, *Anal. Chem.*, **36,** 1206 (1964)
80. J. STRASSBURGER *et al.*, *Anal. Chem.*, **32,** 454 (1960)
81. R. S. LEHRLE and J. C. ROBB, *J. Gas Chromatog.*, **5,** 89 (1967)
82. J. JANÁK, Chapter 1, ref. 63, p. 387
83. C. MERRITT JR. and D. H. ROBERTSON, *J. Gas Chromatog.*, **5,** 96 (1967)
84. C. R. FONTAN, N. C. JAIN and P. L. KIRK, *Mikrochim. Acta*, **1964,** 326 (1964)
85. E. C. JENNINGS JR. and K. P. DIMICK, *Anal. Chem.*, **34,** 1543 (1962)
86. D. BRAUN and G. VORENDOHRE, *Farbe u. Lack*, **69,** 820 (1963)
87. F. SADOWSKI and E. KULM, *Farbe u. Lack*, **69,** 267 (1963)
88. P. LEPLAT, *J. Gas Chromatog.*, **5,** 128 (1967)
89. W. F. WILHITE and M. R. BURNELL, Chapter 1, ref. 68, p. 243

90. P. L. KIRK, *J. Gas Chromatog.*, **5**, 11 (1967)
91. E. REINER, *J. Gas Chromatog.*, **5**, 65 (1967)
92. T. F. BRODASKY, *J. Gas Chromatog.*, **5**, 311 (1967)
93. T. WOLF and D. M. ROSIE, *Anal. Chem.*, **39**, (7) 725 (1967)
94. R. J. KOKES, H. TOBIN JR. and P. H. EMMETT, *J. Amer. Chem. Soc.*, **77**, 5860 (1955)
95. F. DRAWERT, R. FELGENHAUER and G. KUPFER, *Angew. Chem.*, **72**, 555 (1960)
96. F. DRAWERT, Chapter 1, ref. 66, p. 347
97. F. DRAWERT, Chapter 1, ref. 69, p. 339
98. M. BEROZA and R. A. COAD, *J. Gas Chromatog.*, **4**, 199 (1966)
99. S. H. LENGER and P. PANTAGES, *Nature*, **191**, 141 (1961)
100. M. W. ANDERS and G. J. MANNERING, *Anal. Chem.*, **34**, 730 (1962)
101. A. ZLATKIS, J. F. ORO and A. P. KIMBALL, *Anal. Chem.*, **32**, 162 (1960)
102. G. F. HARRISON, Chapter 1, ref. 58, p. 332
103. C. J. THOMPSON *et al.*, *J. Gas Chromatog.*, **5**, 1 (1967)
104. M. BEROZA and R. SARMIENTO, *Anal. Chem.*, **38**, 1042 (1966)
105. L. A. JONES and R. I. MONROE, *Anal. Chem.*, **37**, 935 (1965)
106. E. R. HOFFMANN and I. LYSYJ, *Microchem. J.*, **6**, 45 (1962)
107. R. ROWAN JR., *Anal. Chem.*, **33**, 658 (1961)
108. T. L. MOUNTS and H. G. DUTTON, *Anal. Chem.*, **37**, 641 (1965)
109. A. I. M. KEULEMANS and H. H. VOGE, *J. Phys. Chem.*, **63**, 476 (1959)
110. C. J. THOMPSON, H. J. COLEMAN, C. C. WARD and H. T. RALL, *Anal. Chem.*, **32**, 424 (1960)
111. C. J. THOMPSON *et al.*, *Anal. Chem.*, **32**, 424 (1960)
112. C. J. THOMPSON *et al.*, *Anal. Chem.*, **34**, 154 (1962)
113. C. J. THOMPSON *et al.*, *Anal. Chem.*, **34**, 151 (1962)
114. C. J. THOMPSON *et al.*, *Anal. Chem.*, **37**, 1042 (1965)
115. M. BEROZA, *Anal. Chem.*, **34**, 1801 (1962)
116. M. BEROZA and R. SARMIENTO, *Anal. Chem.*, **35**, 1353 (1963)
117. M. BEROZA and R. SARMIENTO, *Anal. Chem.*, **36**, 1744 (1964)
118. M. BEROZA and R. SARMIENTO, *Anal. Chem.*, **37**, 1040 (1965)
119. B. KOLB and K. H. KAISER, *J. Gas Chromatog.*, **2**, 233 (1964)
120. J. NAKANISHI and K. TAMARU, *Trans. Farad. Soc.*, **59**, 1470 (1963)
121. J. E. STAUFFER and W. L. KRANICH, *Ind. Eng. Chem. Fundamentals*, **1**, 107 (1962)
122. S. OGASAWARA and R. J. CVETANOVIC, *J. Catalysis*, **2**, 45 (1963)
123. J. G. LARSON *et al.*, *J. Amer. Chem. Soc.*, **87**, 1880 (1965)
124. L. S. ETTRE and N. BRENNER, *J. Chromatog.*, **3**, 524 (1960)
125. W. K. HALL, P. H. EMMETT, *J. Amer. Chem. Soc.*, **79**, 2091 (1957)
126. C. J. NORTON, *Chem. and Ind.*, 258 (1962)
127. D. P. HARRISON, J. W. HALL and H. F. RASE, *Ind. Chem.*, **57**, 18 (1965)

8. Analytical Applications of Gas Chromatography

Even compared to the general rapid development in other branches of instrumental analysis, the speed at which gas chromatographic analytical methods developed and spread is quite unique. Gas chromatography can be applied to the analysis of highly diverse gases and liquids in the boiling range between -269 and $+450\,°C$, from rare gases to heavy petroleum fractions. Volatile solids can be injected in the form of their melts, dissolved in an appropriate solvent or directly on the column. Non-volatile solids and heat sensitive liquids are either converted first into a volatile derivative or their pyrolysis products are analyzed. Gas chromatography has opened a way to the investigation of multicomponent, complex mixtures whose analysis hitherto required the combination of several methods, or were impossible to analyze. By means of the very great number of adsorbents and liquid phases with diverse properties it has become possible to investigate and analyze inorganic and organic compounds of all types.

Gas chromatographic analytical methods are used in every branch of the chemical industry, primarily in the petroleum, petrochemical and pharmaceutical industries, in the manufacture of synthetic organic substances, where it is applied to routine analyses, in research and process control. The importance and utilization of gas chromatography are rapidly rising also in the other industries and sciences, such as metallurgy, plastics, food and fermentation industries, in biochemistry, medical chemistry and forensics.

There is almost no type of compound left today whose analysis by gas chromatographic methods has not been described in the literature.

By the end of 1968 the number of gas chromatographic publications was over 17,000, so a detailed discussion of the applications is well beyond the scope of the present book. Instead of some randomly chosen examples we prefer to present two tables containing some references on the analysis of the most important groups of compounds. Table 8.1 summarizes the analyses of gases, Table 8.2 that of liquids.

Table 8.1

Gas analysis

Type	*References*
Rare gases	1, 2, 3, 4, 5
Permanent gases	2, 3, 6, 7, 8, 9, 10, 11, 179
Gases with CO and CO_2 content	8, 9, 10, 11, 12, 13, 14, 179, 180
Gaseous nitrogen compounds	8, 9, 15, 16, 17, 18, 19, 179, 180
Gaseous sulphur compounds	8, 9, 20, 21, 22, 23, 24, 179
Halogens and halogen compounds	8, 9, 25, 26, 27, 28, 179, 181
Other inorganic gases and vapours	8, 9, 29, 30, 31, 32, 179
Separation of isotopes	33, 34, 35, 36, 37, 179
Hydrocarbon gases	8, 9, 38, 39, 40, 41, 42, 179
Gases dissolved in liquids	43, 44, 45, 46
Gases in solids (metals, glass, etc.)	47, 48, 49, 50, 51
Trace analysis	
Atmospheric contaminants	52, 53, 54, 55
Exhaust gases	56, 57, 58, 59
Gas purity tests	60, 61, 62, 182, 183

In both tables some literature on the application of gas chromatography to trace analysis is quoted, as this is one of the important applications of this method. Solids are analyzed in the form of their melt or solutions by methods similar to the analysis of liquids or by one of the special techniques (see Chapter 7).

No attempt at completeness was made in this compilation, either with regard to the type of compounds or to the references. The tables cannot give more than some information on the possibilities of gas chromatography and some references into the various fields. The references in the tables were selected in general from the literature of the past few years, with the advantage of including the most up-to-date methods and at the same time offering the reader more literature references, as most of the later communications contain references of earlier published works in the field. In the selection of the references it was borne in mind to quote popular and easily accessible periodicals.

However should the reader require further and more detailed information on some question, we refer him to the general literature in the appendix of Chapter 1. The volumes of Gas Chromatography Abstracts[201–210] contain a subject index according to types of compounds. The weekly published punched card abstracts of the Preston Technical Abstracts Co., 909 Pitner Ave., Evanston, Ill., 60202 USA

Table 8.2

Liquid analyses

Type	*References*
Hydrocarbons	
Low boiling (up to n-C_6)	11, 174, 175, 179
Medium boiling (up to C_{10})	64, 71, 176, 177, 184, 185
High boiling (above n-C_{10})	72, 77, 178
Aliphatics	11, 63, 64, 65, 66, 67, 184
Alicyclics	64, 68, 69, 70, 184, 185
Monocyclic aromatics	71, 72, 73, 74, 185, 186
Polycyclic aromatics	75, 76, 77, 78
Petroleum fractions	79, 80, 81, 82
Oxygen compounds	
Alcohols	83, 84, 85, 187, 188
Aldehydes	86, 87, 88, 189
Ketones	89, 90, 91
Ethers, lactones	92, 93, 94, 187
Acids	95, 96, 97, 190, 191
Esters	98, 99, 100
Phenols and their derivatives	101, 102, 103, 192
Heterocyclic compounds	104, 105, 193
Carbohydrates	106, 107, 108
Volatile oils	109, 110, 111
Steroids	112, 113, 114
Nitrogen compounds	
Aliphatic and aromatic amines	115, 116, 117, 194
Amino acids and their derivatives	118, 119, 120, 195
Nitriles	121, 122
Nitro compounds	123, 124, 125
Heterocyclic N compounds	126, 127, 128, 193
Alkaloids	129, 130, 131, 141
Sulphur compounds	
Mercaptans, sulphides	132, 133, 134, 138
Sulphonic acids, sulphones	135, 136, 196
Heterocyclic S compounds	137, 139
Others	140, 142, 197
Halogen compounds	
Chlorine compounds	143, 144, 147, 148, 150
Bromine compounds	151, 153, 154
Fluorine compounds	145, 146, 149, 152, 165

Table 8.2 (continued)

Type	*References*
Metal organic and inorganic compounds	
Boron compounds	163, 169
Phosphorus compounds	149, 161, 164, 166
Silicon compounds	156, 167, 168, 198
Metals	157, 158, 159, 199
Others	155, 160, 162, 198, 200
Trace analysis, purity tests	170, 171, 172, 173

offer perhaps an even greater help in literature search. [211–214] There are also some other useful compilations.[215–220]

The December issue of the *Journal of Gas Chromatography* publishes a detailed bibliography of the gas chromatographic publications of the past year with a subject index which helps to find the publications pertaining to the most important types of compounds.[221–224]

References

1. M. Krejci, K. Tesarik and J. Janák, Chapter 1, ref. 64, p. 255
2. R. Berry, Chapter 1, ref. 66, p. 321
3. A. Purer and C. A. Seitz, *Anal. Chem.*, **36,** 1694 (1964)
4. R. Aubeau, L. Champeil and J. Reiss, *J. Chromatog.*, **6,** 209 (1961)
5. J. A. J. Walker, *Nature*, **209,** 197 (1966)
6. R. F. Putman and H. W. Myers, *Anal. Chem.*, **34,** 486 (1962)
7. D. P. Manka, *Anal. Chem.*, **36,** 480 (1964)
8. O. L. Hollis, *Anal. Chem.*, **38,** 307 (1966)
9. O. L. Hollis and W. V. Hayes, Chapter 1, ref. 73, p. 57
10. L. R. Roberts and J. J. McKetta, *J. Gas Chromatog.*, **1,** (3) 14 (1963)
11. J. O. Terry and J. H. Futrell, *Anal. Chem.*, **37,** 1165 (1965)
12. I. Lysyj and P. R. Newton, *J. Chromatog.*, **11,** 173 (1963)
13. T. Doran and J. P. Cross, *J. Gas Chromatog.*, **4,** 260 (1966)
14. G. J. Cvejanovich, *Anal. Chem.*, **34,** 654 (1962)
15. R. P. De Grazio, *J. Gas Chromatog.*, **3,** 204 (1965)
16. P. J. Kipping and P. G. Jeffery, *Nature*, **200,** 1314 (1963)
17. J. M. Travell, *Anal. Chem.*, **37,** 1152 (1965)
18. B. J. Gudzinovicz and W. R. Smith, *Anal. Chem.*, **35,** 465 (1963)
19. R. E. Isbell, *Anal. Chem.*, **35,** 255 (1963)
20. H. Nestler and W. Berger, *Chem. Techn.*, **15,** 616 (1963)
21. C. Bighi and G. Saglietto, *J. Chromatog.*, **17,** 13 (1965)
22. L. A. Robbins, R. M. Bethea and T. L. Wheelock, *J. Chromatog.*, **13,** 361 (1964)

23. C. T. HODGES and R. F. MATSON, *Anal. Chem.*, **37,** 1065 (1965)
24. H. L. HALL, *Anal. Chem.*, **34,** 61 (1962)
25. A. G. HAMLIN, G. IVESON and T. R. PHILIPS, *Anal. Chem.*, **35,** 2037 (1963)
26. R. A. LAUTHEAUME, *Anal. Chem.*, **36,** 486 (1964)
27. A. KARMEN, *Anal. Chem.*, **36,** 1416 (1964)
28. J. JANÁK and V. SVOJANOVSKY, Chapter 1, ref. 73, p. 166
29. T. DUMAS, *J. Agr. Food Chem.*, **12,** 257 (1964)
30. C. S. G. PHILLIPS and P. TIMMS, *Anal. Chem.*, **35,** 505 (1963)
31. B. LENGYEL, G. GARZÓ and T. SZÉKELY, *Acta Chim. Hung.*, **37,** 37 (1963)
32. J. TADMOR, *J. Gas Chromatog.*, **2,** 385 (1964)
33. M. MOHNKE and W. SAFFERT, Chapter 1, ref. 66, p. 218
34. E. H. CARTER JR. and H. A. SMITH, *J. Phys. Chem.*, **67,** 1515 (1963)
35. M. E. GERSH, *Anal. Chem.*, **37,** 1786 (1965)
36. F. BRUNNER, G. P. CARTONI and A. LIBERTI, *Anal. Chem.*, **38,** 298 (1966)
37. I. LYSYJ, Chapter 1, ref. 68, p. 149
38. ANON., *Erdöl u. Kohle*, **16,** 319 (1963)
39. ASTM Standard: D 1945—64
40. R. L. CHURCHWELL and A. ZLATKIS, *J. Gas Chromatog.*, **2,** 275 (1964)
41. P. D. KOONS and J. S. WALKER, *Petrol. Refiner*, **42,** (4) 133 (1963)
42. M. G. BLOCH, Chapter 1, ref. 64, p. 133
43. J. P. PAGLIS, Chapter 1, ref. 64, p. 351
44. K. E. GUBBINS, S. N. CARDON and R. D. WALKER, *J. Gas Chromatog.*, **3,** 98 (1965)
45. R. H. WILSON, *J. Gas Chromatog.*, **2,** 365 (1964)
46. A. A. KILNER and G. A. RATCLIFF, *Anal. Chem.*, **36,** 1615 (1964)
47. R. J. HYNEK and J. A. NELCU, *Anal. Chem.*, **35,** 1655 (1963)
48. J. NICKEL, Chapter 1, ref. 67, p. 431
49. R. K. WINGE and V. A. FASSEL, *Anal. Chem.*, **37,** 67 (1965)
50. F. R. BRYON and J. C. NEERMAN, *Anal. Chem.*, **34,** 278 (1962)
51. P. D. GARN, *Talanta*, **11,** 1417 (1964)
52. A. P. ALTSHULLER, *J. Gas Chromatog.*, **1,** (7) 6 (1963)
53. T. A. BELLAR, M. F. BROWN and J. E. SIGSBY, *Anal. Chem.*, **35,** 1924 (1963)
54. J. NOVAK, V. VASÁK and J. JANÁK, *Anal. Chem.*, **37,** 660 (1965)
55. E. C. TABOR, *Trans. N. Y. Acad. Sci.*, **28,** II. (5) 569 (1966)
56. D. J. MCEWEN, *J. Chromatog.*, **9,** 266 (1962)
57. D. J. MCEWEN, *Anal. Chem.*, **38,** 1047 (1966)
58. W. B. INNES and W. E. BAMBRICK, *J. Gas Chromatog.*, **2,** 309 (1964)
59. I. M. WILLIAMS, *Anal. Chem.*, **37,** 1723 (1965)
60. A. ZLATKIS and H. R. KAUFMANN, *J. Gas Chromatog.*, **4,** 240 (1966)
61. F. STEINBACH, *J. Chromatog.*, **15,** 432 (1964)
62. C. G. SCOTT and C. S. G. PHILLIPS, *Nature*, **199,** 66 (1963)
63. C. R. FERRIN, J. O. CHASE and R. W. HURN, Chapter 1, ref. 67, p. 423
64. A. G. POLGÁR *et al.*, *Anal. Chem.*, **34,** 1226 (1962)
65. R. L. MARTIN, *Anal. Chem.*, **32,** 336 (1960)
66. L. E. GREEN, D. K. ALBERT and H. H. BARBER, *J. Gas Chromatog.*, **4,** 319 (1966)
67. G. SCHOMBURG, *J. Chromatog.*, **23,** 1 (1966)
68. G. SCHOMBURG, *J. Chromatog.*, **23,** 18 (1966)

69. G. L. K. Hunter and W. B. Brogden, *J. Food Sci.*, **30**, 383 (1965)
70. W. R. Roth, *Annalen*, **671**, 10 (1964)
71. L. C. Case, *J. Chromatog.*, **6**, 381 (1961)
72. F. Baumann, A. E. Straus and J. F. Johnson, *J. Chromatog.*, **20**, 1 (1965)
73. E. W. Cieplinski, *Anal. Chem.*, **37**, 1160 (1965)
74. G. M. Mamedaliev *et al.*, *Neftekhimiya*, **4**, 618 (1964)
75. C. J. Norton and T. E. Moss, *Ind. Eng. Chem., Proc. Design and Dev.*, **3**, 23 (1964)
76. F. J. Pinchin and E. Pritchard, *Chem. and Ind.*, **40**, 1753 (1962)
77. M. I. Gerber *et al.*, *Neftekhimiya*, **5**, 776 (1965)
78. H. D. Sauerland and M. Zander, *Erdöl u. Kohle*, **19**, 502 (1966)
79. R. L. Martin, Chapter 1, ref. 68, p. 127
80. P. I. Sidorov, *Khim. i Tekhnol. Topliv i Masel*, **10**, 20 (1965)
81. T. C. Davies, J. C. Petersen and W. F. Haines, *Anal. Chem.*, **38**, 241 (1966)
82. G. Kemmer, E. Kolb and H. Pauschmann, Bodenseewerk Perkin–Elmer Co. GmbH. Appl. GC No 2E
83. D. G. H. Daniels, *J. Gas Chromatog.*, **4**, 305 (1966)
84. B. Drews, H. Specht and G. Offer, *Z. anal. Chem.*, **189**, 325 (1962)
85. W. J. A. Van den Heuvel, W. L. Gardiner and E. C. Horning, *J. Chromatog.*, **19**, 263 (1965)
86. G. Mizuno, E. McMeans and J. R. Chipault, *Anal. Chem.*, **37**, 151 (1965)
87. M. P. Stevens, *Anal. Chem.*, **37**, 167 (1965)
88. E. Fedeli and M. Cirimele, *J. Chromatog.*, **15**, 435 (1964)
89. J. K. Haken and P. Souter, *J. Gas Chromatog.*, **3**, 348 (1965)
90. I. R. Hunter and M. K. Walden, *J. Gas Chromatog.*, **4**, 246 (1966)
91. R. I. Morrison and W. Bick, *Chem. and Ind.*, 596 (1966)
92. V. Schölkopf and W. Pitteroff, *Chem. Ber.*, **97**, 636 (1964)
93. M. S. Northington and G. Owens, *Anal. Biochem.*, **9**, 48 (1964)
94. E. Bayer and R. Widder, *Annalen*, **686**, 181 (1965)
95. R. W. McKinney, *J. Gas Chromatog.*, **2**, 108 (1964)
96. R. B. Jackson, *J. Chromatog.*, **16**, 306 (1964)
97. M. I. Rogozinski, *J. Gas Chromatog.*, **2**, 136 (1964)
98. N. Ruseva-Atanasova and J. Janák, *J. Chromatog.*, **21**, 207 (1966)
99. W. Ebing and H. G. Henkel, *J. Gas Chromatog.*, **2**, 207 (1964)
100. R. L. Glass, R. Jenness and H. A. Troolin, *J. Dairy Sci.*, **48**, 1106 (1965)
101. D. W. Grant and G. A. Vaughan, Chapter 1, ref. 66, p. 305
102. R. W. Freedman and G. O. Charlier, *Anal. Chem.*, **36**, 1880 (1964)
103. J. R. Smith, R. O. Norman and G. U. Radda, *J. Gas Chromatog.*, **2**, 146 (1964)
104. M. A. Gianturco, P. Friedel and V. Flanagan, *Tetrahedron Letters*, No. 23, 1847 (1965)
105. A. F. Thomas and M. Stoll, *Chem. and Ind.*, 1491 (1963)
106. J. S. Sawardeker and J. H. Sloncker, *Anal. Chem.*, **37**, 945 (1965)
107. C. C. Sweeley and B. Walker, *Anal. Chem.*, **36**, 1461 (1964)
108. C. C. Sweeley *et al.*, *J. Amer. Chem. Soc.*, **85**, 2497 (1953)
109. R. W. Wolford and J. A. Attaway, Chapter 1, ref. 67, p. 457
110. K. H. Miltenberger and G. Keicher, *Farbe u. Lack*, **69**, 677 (1963)
111. W. D. McLeod and N. M. Buigues, *J. Food Sci.*, **29**, (1964)
112. B. A. Knights, *J. Gas Chromatog.*, **2**, 160 (1964)

113. R. W. JELLIFFE and D. H. BLANKENHORN, *J. Chromatog.*, **12**, 268 (1963)
114. B. A. KNIGHTS, *J. Gas Chromatog.*, **4**, 329 (1966)
115. E. W. CIEPLINSKI, *Anal. Chem.*, **38**, 928 (1966)
116. I. K. LEWIS, G. B. RUSSELL, R. D. TOPSOM and J. VAUGHAN, *J. Org. Chem.*, **29**, 1160 (1964)
117. C. LANDAULT and G. GUIOCHON, *J. Chromatog.*, **13**, 327 (1964)
118. S. MAKISUMI and H. A. SAROFF, *J. Gas Chromatog.*, **3**, 21 (1965)
119. B. POTTEAU, *Bull. Soc. chim. France*, 3747 (1965)
120. P. A. CRUICKSHANK and J. C. SEEHAN, *Anal. Chem.*, **36**, 1191 (1964)
121. M. TARAMASSO and A. GUERRA, *J. Gas Chromatog.*, **3**, 138 (1965)
122. R. M. MORIARTY and M. RAHMAN, *Tetrahedron*, **21**, 2877 (1965)
123. G. PARASKEVOPOULOS and R. J. CVETANOVIC, *J. Chromatog.*, **25**, (1966)
124. R. KETCHAM, R. CAVESTRI and D. JAMBOTKAR, *J. Org. Chem.*, **28**, 2139 (1963)
125. J. C. COURTIER, L. ÉTIENNE, J. TRANCHANT and S. VERTALIER, *Bull. Soc. chim. France*, 3367 (1965)
126. G. ALBERINI, V. CANTUTI and G. P. CARTONI, Chapter 1, ref. 73, p. 258
127. G. P. BEAN, *Anal. Chem.*, **37**, 756 (1965)
128. K. GROB, *J. Gas Chromatog.*, **2**, 80 (1964)
129. E. BROCHMANN-HANSSEN and C. R. FONTAN, *J. Chromatog.*, **20**, 394 (1965)
130. E. BROCHMANN-HANSSEN and C. R. FONTAN, *J. Chromatog.*, **20**, 296 (1965)
131. E. BROCHMANN-HANSSEN and A. B. SVENDSEN, *J. Pharm. Sci.*, **51**, 1095 (1962)
132. H. J. COLEMAN, C. J. THOMPSON, R. L. HOPKINS and H. T. RALL, *J. Chromatog.*, **25**, 34 (1966)
133. V. T. BRAND and D. A. KAYWORTH, *Anal. Chem.*, **37**, 1424 (1965)
134. D. M. OAKS, H. HARTMANN and K. P. DIMICK, *Anal. Chem.*, **36**, 1560 (1964,
135. E. BENDEL, B. FELL, A. COMMICHAU, H. HÜBNER and W. MELTZOW, *J. Chromatog.*, **19**, 277 (1965)
136. V. E. CATES, C. E. MELOAN, *J. Chromatog.*, **11**, 472 (1963)
137. H. J. COLEMAN, C. J. THOMPSON, R. L. HOPKINS and H. T. RALL, *J. Chromatog.*, **20**, 240 (1965)
138. R. BLOEMBERGEN and C. VERMAAK, *Erdöl u. Kohle*, **18**, (3) 185 (1965)
139. R. L. MARTIN and J. A. GRANT, *Anal. Chem.*, **37**, 649 (1965)
140. E. MÜLLER and E. W. SCHMIDT, *Chem. Ber.*, **97**, 2622 (1964)
141. H. FRAUENDORF and H. VOGEL, *Z. anal. Chem.*, **205**, 460 (1964)
142. T. J. WALLACE and J. J. MAHON, *Nature*, **201**, 4921 (1964)
143. B. FELL and L. H. KUNG, *Chem. Ber.*, **98**, 2871 (1965)
144. S. H. LANGER and J. H. PURNELL, *J. Phys. Chem.*, **67**, (2) 263 (1963)
145. C. A. CLEMONS and A. P. ALTSHULLER, *Anal. Chem.*, **38**, 133 (1966)
146. H. GERSHON and J. A. A. RENWICK, *J. Chromatog.*, **20**, 134 (1965)
147. L. D. HINSHAW, *J. Gas Chromatog.*, **4**, 300 (1966)
148. R. L. JOHNSON and D. J. BUTON, *J. Chromatog.*, **20**, 138 (1965)
149. A. KARMEN, *J. Gas Chromatog.*, **3**, 336 (1965)
150. J. PETRANEK, M. KOLLINSKY and D. LIM, *Nature*, **207**, 1290 (1965)
151. M. ROGOZINSKI and L. M. SHORR, *J. Org. Chem.*, **29**, 948 (1964)
152. J. C. MAILEN, T. M. REED and J. A. YOUNG, *Anal. Chem.*, **36**, 1883 (1964)
153. W. R. MOORE, W. R. MOSER and J. E. LA PRADE, *J. Org. Chem.*, **28**, 2200 (1963)
154. J. VAUGHAN, J. WELCH and G. J. WRIGHT, *Tetrahedron*, **21**, 1665 (1965)

155. R. S. JUVET and R. L. FESHER, *Anal. Chem.*, **38**, 1860 (1966)
156. C. R. THRASH, *J. Gas Chromatog.*, **2**, 390 (1964)
157. R. S. JUVET and R. P. DURBIN, *Anal. Chem.*, **38**, 565 (1966)
158. W. D. ROSS, R. E. SIEVERS and G. WHEELER, *Anal. Chem.*, **37**, 598 (1965)
159. J. E. DENNISON and H. FREUND, *Anal. Chem.*, **37**, 1770 (1965)
160. C. S. EVANS and C. M. JOHNSON, *J. Chromatog.*, **21**, 202 (1966)
161. C. A. BACHE and D. J. LISK, *Anal. Chem.*, **37**, 1477 (1965)
162. J. A. SEMLYEN and C. S. G. PHILLIPS, *J. Chromatog.*, **18**, 1 (1965)
163. P. A. MCCUSKER, F. M. ROSSI, J. H. BRIGHT and G. F. HENNION, *J. Org. Chem.*, **28**, 2889 (1963)
164. W. STEINDORF, E. JUST and H. W. ARDELT, *Z. Chem.*, **5**, 388 (1965)
165. A. ENGELBRECHT, E. NACHBAUR and E. MAYER, *J. Chromatog.*, **15**, 228 (1964)
166. K. D. BERLIN *et al.*, *J. Gas Chromatog.*, **3**, 256 (1965)
167. D. VRANDI-PISKOU and G. PARISSAKIS, *J. Chromatog.*, **22**, 449 (1966)
168. I. M. T. DAVIDSON, C. EABORN and M. N. LILLY, *J. Chem. Soc.*, 2624 (1964)
169. R. KÖSTER, W. LARBIG and G. W. ROTERMUND, *Annalen*, **862**, 21 (1965)
170. R. E. LEONARD and J. E. KIEFER, *J. Gas Chromatog.*, **4**, 142 (1966)
171. D. W. GRANT and G. A. VAUGHAN, *Nature*, **208**, 75 (1965)
172. W. H. MCCURDY JR. and R. W. REISER, *Anal. Chem.*, **38**, 795 (1966)
173. K. ABEL, *J. Chromatog.*, **13**, 22 (1964)
174. D. M. OTTMERS, G. R. SAY and A. F. RASE, *Anal. Chem.*, **38**, 148 (1966)
175. R. W. CARR, *J. Phys. Chem.*, **70**, 1970 (1966)
176. I. HALÁSZ and E. HEINE, *Nature*, **194**, 971 (1962)
177. W. HEINEMANN, *Erdöl u. Kohle*, **13**, 828 (1960)
178. O. T. CHORTYK, W. S. SCHLOTZHAUER and R. L. STEDMAN, *J. Gas. Chromatog.*, **3**, 394 (1965)
179. R. J. LEIBRAND, *J. Gas Chromatog.*, **5**, 518 (1967)
180. C. G. CRAWFORTH and D. J. WADDINGTON, *J. Gas Chromatog.*, **6**, 103 (1968)
181. L. G. SPEARS and N. HACKERMAN, *J. Gas Chromatog.*, **6**, 392 (1968)
182. F. ZOCCHI, *J. Gas Chromatog.*, **6**, 100 (1968)
183. G. M. SASSU, F. ZILIO-GRANDI and A. CONTE, *J. Chromatog.*, **34**, 394 (1968)
184. H. WIDMER, *J. Gas Chromatog.*, **5**, 506 (1967)
185. R. D. SCHWARTZ, R. G. MATHEWS and D. J. BRASSEAUX, *J. Gas Chromatog.*, **5**, 251 (1967)
186. P. L. GUPTA and P. KUMAR, *Anal. Chem.*, **40**, 992 (1968)
187. J. R. LINDSAY SMITH and D. J. WADDINGTON, *J. Chromatog.*, **36**, 145 (1968)
188. R. WOOD, *J. Gas Chromatog.*, **6**, 94 (1968)
189. K. JONES, *J. Gas Chromatog.*, **5**, 432 (1967)
190. J. R. P. CLARKE and K. M. FREDRICKS, *J. Gas Chromatog.*, **5**, 99 (1967)
191. J. HRIVNAK and V. PALO, *J. Gas Chromatog.*, **5**, 325 (1967)
192. J. T. CLARK, *J. Gas Chromatog.*, **6**, 53 (1968)
193. G. A. KALUZA, F. MARTIN, *J. Gas Chromatog.*, **5**, 562 (1967)
194. J. R. LINDSAY SMITH and D. J. WADDINGTON, *Anal. Chem.*, **40**, 522 (1968)
195. F. SHAHROKHI and W. GEHRKE, *J. Chromatog.*, **36**, 31, (1968)
196. J. S. PARSONS, *J. Gas Chromatog.* **5**, 254 (1967)
197. A. GIRAUD and M. A. BESTOUGEFF, *J. Gas Chromatog.*, **5**, 464 (1967)
198. G. GARZÓ, J. FEKETE and M. BLAZSÓ, *Acta Chim. Acad. Sci. Hung.*, **51**, (4) 359 (1967)

199. R. E. SIEVERS, J. W. CONNOLLY and W. D. ROSS, *J. Gas Chromatog.*, **5,** 214 (1967)
200. N. L. SOULAGES, *J. Gas Chromatog.*, **6,** 356 (1968)
201. C. E. H. KNAPMAN (Ed.), *Gas Chromatography Abstracts*, 1958, Butterworths, London (1960)
202. C. E. H. KNAPMAN (Ed.), *Gas Chromatography Abstracts*, 1959, Butterworths, London (1960)
203. C. E. H. KNAPMAN (Ed.), *Gas Chromatography Abstracts*, 1960, Butterworths, London (1961)
204. C. E. H. KNAPMAN (Ed.), *Gas Chromatography Abstracts*, 1961, Butterworths, London (1962)
205. C. E. H. KNAPMAN (Ed.), *Gas Chromatography Abstracts*, 1962, Butterworths, London (1963)
206. C. E. H. KNAPMAN (Ed.), *Gas Chromatography Abstracts*, 1963, Butterworths, London (1964)
207. C. E. H. KNAPMAN (Ed.), *Gas Chromatography Abstracts*, 1964, Butterworths, London (1965)
208. C. E. H. KNAPMAN (Ed.), *Gas Chromatography Abstracts*, 1965, Butterworths, London (1966)
209. C. E. H. KNAPMAN (Ed.), *Gas Chromatography Abstracts*, 1966, Institute of Petroleum, London (1967)
210. C. E. H. KNAPMAN (Ed.), *Gas Chromatography Abstracts*, 1966, Institute of Petroleum, London (1968)
211. Card Abstracts, Abstracting Service, Preston Technical Abstracts Co., Evanston, Ill., USA
212. Termatrex Index — Gas Chromatography Literature, Preston Technical Abstracts Co., Evanston, Ill., USA
213. Microfilmed Gas Chromatography Abstracts, Preston Technical Abstracts Co., Evanston, Ill., USA
214. S. T. PRESTON and G. HYDER, *A Comprehensive Bibliography and Index to the Literature on Gas Chromatography*, Preston Technical Abstracts Co., Evanston, Ill., USA (1964)
215. A. V. SIGNEUR, *Guide to Gas Chromatography Literature*, Plenum Press, New York (1964)
216. A. V. SIGNEUR, *Guide to Gas Chromatography Literature*, Supplement No. 1, Plenum Press, New York (1966)
217. A. V. SIGNEUR, *Guide to Gas Chromatography Literature*, Vol. 2, Plenum Press, New York (1967)
218. J. S. LEWIS, *Compilation of Gas Chromatography Data*, ASTM Publ. No. 343, Philadelphia (1963)
219. O. E. SCHUPP and J. S. LEWIS, *Compilation of Gas Chromatographic Data DS 25 A*, ASTM Publ. Philadelphia (1968)
220. W. O. MCREYNOLDS, *Gas Chromatographic Retention Data*, Preston Technical Abstracts Co., Evanston, Ill., USA (1966)
221. S. T. PRESTON, G. HYDER and M. GILL, *J. Gas Chromatog.*, **2,** 391 (1964)
222. M. GILL and S. T. PRESTON, *J. Gas Chromatog.*, **2,** 391 (1964)
223. S. T. PRESTON and M. GILL, *J. Gas Chromatog.*, **3,** 399 (1965)
224. S. T. PRESTON and M. GILL, *J. Gas Chromatog.*, **4,** 435 (1966)

9. Preparative Gas Chromatography

9.1 Development and applications

The outstanding separation power of gas chromatography offers unique possibilities of preparing high purity substances. This possibility was soon recognized and publications on the preparative application of gas chromatography appeared already in the 1950's[1–3] and by the end of that decade detailed preparative gas chromatographic methods and appropriate apparatus began to emerge. Today highly efficient, high capacity preparative set-ups are available.

Theoretically every analytical application of gas chromatography can be utilized for preparative purposes. The objectives and applications of preparative gas chromatography can be classified as follows:[4]

1. Preparation of small (milligram) quantities of pure substances for identification purposes, e.g. spectrometric tests, with laboratory analytical gas chromatographs.
2. Preparation of pure substances in quantities ranging from 1 to 1000 g for organic syntheses, reaction kinetic studies, measurement of physical properties with special preparative apparatus.
3. Preparation of pure substances in quantities ranging from 1 to 1000 kg for marketing or for further processing with production type equipment.

Like other separation and purification methods (distillation, crystallization, etc.) preparative gas chromatography can be applied to the removal of contaminations, the preparation of high purity substances, to the separation of the contaminants for identification, and to the separation of multicomponent mixtures into narrow fractions or individual components.

The great assortment of gas chromatographic column packings makes possible the separation of substances with very small differences in boiling points or with similar chemical properties, thus for instance azeotrope formation encountered in distillation processes has no effect on the separation. Compared to distillation, preparative gas chromatography has the added advantage of being applicable to the separation of the components which might polymerize or decompose during a lengthy distillation process.

There are considerable differences in the actual methods of preparative gas chromatography depending on the objective of the procedure as set out above. The preparation of milligram quantities (paragraph 1) can essentially be solved with an analytical gas chromatograph by repeated sample injection and the collection of the separated components. To comply with requirements of this type the gas chromatograph is provided with ancillary devices (injector, collector) for the production of small quantities of substances.

The objective of the other branch of preparative gas chromatography is the production of larger quantities which raises special demands on the equipment. The failure of the first trials in this field were mainly due to attempts to solve this task with the analytical chromatograph. The maximum sample size which can still be processed by an analytical column increases roughly with the fourth power of the column diameter, but in the case of samples above 1 g the operation of the other parts of the analytical equipment is no longer satisfactory, which has necessitated the elaboration of special preparative gas chromatographs. The third branch of preparative gas chromatography, namely production scale chromatography, will be dealt with at the end of this chapter.

We shall now survey the problems of construction and efficient operation of preparative gas chromatographs. The preparative use of analytical apparatus presents no problem with an appropriate device for collecting the fractions and perhaps a special sample injector head.

To increase the capacity of preparative gas chromatography, i.e. the maximum permissible sample size and the column performance, various methods have been suggested, such as: 1. application of large diameter columns (5–20 cm); 2. several smaller diameter columns (15–20 mm) in a parallel arrangement; 3. long small diameter (8–12 mm) columns; 4. automatic sample injection and fraction collection; 5. special rotating columns and moving bed equipments.

Automatic sample injection and fraction collection may be combined by any of the methods in paragraphs 1–3, and the latest preparative chromatographs are indeed automatically operated. Special techniques will be dealt with in another section.

9.2 Columns

Column construction

In the section on packed columns the effect of column diameter on efficiency has already been discussed. By increasing the column diameter the homogeneity of the packing is reduced, flow rate will fluctuate and the zone occupied by a component will undergo a distortion in its cross-section, leading to higher *H* values. Besides non-uniform flow conditions the deterioration in column efficiency is also related to the following factors: a) non-uniform liquid coating which influences mass transfer velocity; b) radial temperature gradient caused by the dissolution and desorption heats or local temperature changes due to the low heat conductivity of the packing; c) in spiral columns the path (i.e. the rate) of the gas stream will be different on the outer and inner part of the bend; d) end effects at the column outlet due to sudden constrictions or dead spaces. The effect of these factors will be the greater the greater the column diameter is. With the first 2–3 cm diameter preparative columns a considerable decrease in efficiency was observed, but later work proved that this was mainly due to inadequate packing. The thorough investigations of Huyten *et al.*,[5] and of Bayer *et al.*[6–7] have proved that provided packing is adequate the deteriorated efficiency due to larger column diameters is not a decisive factor and *H* values of 2–3 mm can be achieved with 10–20 cm diameter columns. Columns of this size can be operated with samples of the order of 100 g which with automatic operation provides for the processing of several kg per day.

It was pointed out in connection with the hydrodynamics of packed beds (Chapter 2) that flow rate will have a maximum in the vicinity of the wall. To eliminate this effect Frisone[8] placed in the column at 30 cm intervals filter paper rings soaked with the liquid phase. Wright[9] divided the packing into smaller sections by inserting radial metal baffles all along the column mainly to eliminate temperature fluctuations. The porous metal disks used by Carel and Perkins[10] were found adequate for the homogenization of the flow and are also applied in commercial equipment.[4] By placing these baffle disks at every 50 cm in a 10 cm diameter column *H* values less than 2 mm may be achieved, while with the same method the value of *H* will be less than 1 mm in 15–20 mm diameter columns.

Another way to reduce the difficulties raised by greater column diameters is the construction of special column geometry. Recently columns with four leaf clover cross-section were prepared which in fact corresponds to five parallel but radially communicating columns.[11] Another special construction is the ring shaped cross-section column of Nester–Faust Ltd. which can also be provided with internal heating.[12] These devices are applicable with programmed temperature in the case of high capacity columns.

Not only the dimension and the internal construction of the column but also the shape of the column as a whole is important. In coiled, spiral columns the homogeneity of the packing is lower in the bends, so that the column should be prepared from straight tubes and these should be joined with small diameter tubes, through conical joints and wire mesh plugs.

The problems arising with larger column diameters may also be eliminated by using several parallel columns of smaller (15 mm) diameter.[13] This method however necessitates exactly identical flow conditions in every column, so that the components shall emerge from all columns at the same time, a requirement which is difficult to fulfil from the point of view of both packing and sample injection. There are some commercial preparative set-ups with parallel columns, but the method has failed to gain wide acceptance.

The third possibility of increasing the sample size is the use of small diameter long columns. The quantity of the sample can be increased roughly linearly with the length of the column, but for the same resolution analysis time will increase proportionally.[14] Retention time may be reduced by using large particle size (10–20 mesh) solid supports, by diminishing the quantity of the liquid phase, by applying less compact packing or by raising the flow rate. Under these conditions several millilitres of the sample can be injected into a 9 mm diameter column and e.g. 10 ml samples of a mixture of aromatic hydrocarbons can be separated on a 75 m long column.[14] In addition to their high separation power the long columns have the further significant advantage that they require less packing and their carrier gas consumption is also lower than that of columns with larger internal diameters. As a consequence of the large sample size the concentration of the components in the carrier gas will be higher which results in a better efficiency of collection. Programmed temperature for instance cannot be used efficiently with large diameter columns.

The most up-to-date trend in the development of preparative gas chromatographs provides for the use of both short, large diameter and long, small diameter columns. It will depend on the separation task in hand which one of the two possibilities will be chosen. For the preparation of large quantities of a certain substance or for the collection of contaminants large diameter columns are needed. An appropriately long, small diameter column will give the best results when complicated separation problems have to be solved or when smaller quantities of the product are required.

Automatic sample injection and fraction collection which was mentioned as the fourth way of raising column capacity can be combined with any column type and is the fundamental condition for effective preparative work.

Column packing

There is a general agreement in the literature with respect to the requirements the packing of preparative columns have to meet, but this agreement does not hold for the method by which the column should be packed. Both description and evaluation of the different methods are rather ambiguous, so that the problem of preparing a reproducible, effective packing is not yet completely solved. The greatest problem encountered in packing large diameter columns is the radial separation of the particles, with the larger particles tending towards the tube wall and the smaller remaining in the middle considerable deviations in packing density and thus in flow rate distribution may result.[15]

Such a separation occurs with the "mountain packing" method formerly used, when a tube connected to the feeding funnel extends into the column so that the outlet end should always be somewhat above the packing level.[15] In the so-called "bulk packing" method the diameter of the inlet tube is only slightly smaller than that of the column and the full quantity of the packing is poured practically at once into the column when the required packing density is achieved by tapping the tube wall.[16] Higher packing densities may be obtained by the so-called "snow-packing" method[16] which involves the introduction of separate particles from a small opening of the vibrator-operated feeding funnel into the bottom of the column and then tapping the column to obtain the required packing density.

Several methods have been applied to achieve the packing density needed, by vibrating the column during or after filling, tapping the tube wall and tapping the tube vertically. Bayer *et al.*[7] found that vibrating in one direction only is unsatisfactory and a compact packing needs both vertical and horizontal vibration on a vibrating table. Other authors claim that the best results can be achieved by vertically vibrating the column and packing it with small portions at a time.[17] The density of the packing is not a decisive factor with long, small diameter columns, moreover it may be expedient to use loose packing, whereby pressure drop will be reduced and separation accelerated.[14]

A further problem connected with the preparation of a uniformly dense packing is introduced by the fact that to achieve optimum results with supports of different qualities and mesh sizes different methods are required.[17]

Column efficiency and effect of sample size

The efficiency of a preparative column is usually characterized by the number of theoretical plates or by the H value. Relative peak broadening is known to depend on a number of factors among which the quantity of the sample has the greatest influence. For the characterization of the efficiency of analytical columns the N or H values extrapolated to zero sample quantity should be used. In general for the investigation of preparative column efficiency again very small samples are used, when the values obtained will satisfactorily approximate the data of analytical columns.[16] Preparative columns however must work with large samples often in an overloaded condition. The most important characteristics of the preparative column are the maximum permissible sample size and the dependence of column efficiency on the latter.

Study of the efficiency of preparative columns revealed that into the van Deemter equation for analytical columns a further term has to be introduced representing the contribution of the non-uniform flow distribution in the larger diameter[7] which can be expressed as:

$$H_{\mathrm{p}} = E\frac{d^{0.58}}{u^{1.806}} \qquad 9.1$$

where H_{p} is the preparative H term in the van Deemter equation,

E is the correlation factor, d is the column diameter, and u the linear gas velocity. Factors affecting zone broadening and the difference between analytical and preparative columns have also been studied.[18] Sie and Rijnders investigated in detail the effect of particle size, column diameter and carrier gas rate on the development of the flow profile and on column efficiency.[19]

In the case of preparative columns however column efficiency cannot be evaluated from the effect on zone broadening alone. The purpose of preparative work is the production of a maximum quantity of the pure substance within the shortest possible time. In the majority of cases the column may be expediently overloaded and the pure components collected after the discarding of the intermediate fractions corresponding to overlapping peaks. The shape of the peak is not important in preparative gas chromatography, as larger samples in practice will always give distorted peaks. From the aspect of yield it would be the most satisfactory if the component would produce a square signal and not a peak, as yields are always related to the concentration of the component in the carrier gas.

The work of Verzele *et al.*[20–22] includes some findings of practical importance concerning column efficiency and sample quantity. With increasing quantities of the sample the peaks broaden and consequently the number of theoretical plates diminishes. Figure 9.1 shows the changes in the number of theoretical plates measured on two different packings *vs.* the quantity of the sample. This number of preparative "plates" is characteristic of the given separation and for its unequivocal definition the relevant sample size must also be known. According to Eq. 3.42 the number of theoretical plates necessary for a given separation is:

$$N_{\mathrm{ne}} = 16R_{\mathrm{s}}^2 \left(\frac{\alpha}{\alpha - 1}\right)^2 \left(\frac{1 + k'}{k'}\right)^2 \qquad 3.42$$

As already discussed in Chapter 3 when $k' > 5$, as is always the case with preparative columns, the factor containing k' can be neglected and the necessary number of theoretical plates is essentially defined by the relative retention, α. A high value of α thus represents a high number of theoretical plates and accordingly a larger sample can be injected. With the help of Eq. 3.42 the maximum permissible sample size for a given preparative column can be determined. Provided the value of α is known (which can be determined quite simply in an analytical column) the value of N can be calculated and from the correlation plotted in Fig. 9.1 the maximum permissible sample size can be determined. Thus for any given column the maximum permissible sample size varies considerably depending on the components

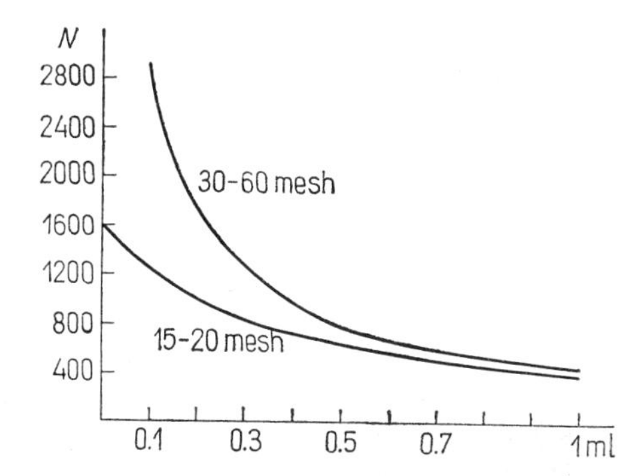

Fig. 9.1. Number of theoretical plates vs. sample size on two supports with different mesh sizes (Chromosorb W)[21]

to be separated. The shape of the curves in Fig. 9.1 will depend on the column dimensions, on the packing, etc., but another correlation is valid for short, large diameter and for long, small diameter columns.

As a general approximation for the maximum permissible sample size it may be said that this: a) increases proportionally to the column diameter; and b) in the case of constant column diameter increases proportionally to the length of the column. In addition, the permissible sample size may also be increased between certain limits almost proportionally to the quantity of the liquid phase, though this effect is significant in the case of small sample quantities only.[20] Columns with different diameters can be compared only from the results obtained with the injection of identical sample sizes and converted to unit cross-section.

Recently Rose *et al.* have carried out detailed investigations of the development of the flow profile in the column and of the effect of the size and method of introduction of the sample and of other operation conditions.[23–25]

Besides the maximum permissible sample size, references to the so-called critical sample size may also be found in the literature, meaning that samples greater than the critical will cause permanent damage to the packing. For a 9 mm column this value is claimed to be 1 ml, for a 10 cm column the critical sample is 161 ml.[4] This, however, had not been confirmed by other authors.[14, 22]

Effect of the nature of the packing

The solid support. In the choice of the solid support mechanical firmness is the most important aspect, to avoid attrition. Thus for instance Celite 545 and Chromosorb W are less suitable from this point of view. The support must not catalyze the decomposition of the liquid phase whose products may contaminate the separated components. Problems of this type may arise with Chromosorb A and P. Chromosorb G and other corresponding products were found to be the most satisfactory. The particle size of the support for preparative work has to meet different demands from those for analytical columns; from the point of view of efficiency the small particle size, narrow mesh fraction (60–80, 80–100 mesh) support is the most suitable for preparative columns too, but as the price of the packing enters as a considerable factor into the cost of the method, wider mesh fractions (10–60 mesh) are used in practice for which the price is only one-tenth of the price of the narrow mesh fractions. For long, small diameter columns the coarser grades (10–20, 20–30 mesh sizes) are quite satisfactory[21, 22] and have the advantage of producing a smaller pressure drop and reducing analysis time. For the separation of substances liable to decomposition (steroids, pesticides, etc.) an inert support is needed, such as glass beads.[26]

In the preparation of the support the removal of fines is very important and can be carried out most efficiently by wet flotation.

The liquid phase. In the choice of the liquid phase the volatility and heat stability of the liquid are highly important aspects to exclude all possibilities of contamination of the separated components by evaporation or by decomposition products of the liquid phase. From the aspect of separation it may be desirable to use a liquid phase providing for the highest value of α, this would however necessitate the preparation of a number of different packings. In practice relatively few liquid phases with general character are used and higher efficiency is provided by other means. The most generally used liquid phases are: Apiezon L, SE-30, Carbowax, Carbowax 20 M; recently polyaromatic resins without liquid coating have appeared.

With respect to the quantity of the liquid phase opinions differ, generally 20–25 %, in some cases 10–15 % of liquid is applied. Higher liquid phase loadings are associated with higher capacities with the simultaneous greater time required for separation. These two points will have to be considered when deciding on the quantity of the liquid to be used. To increase efficiency, lately, columns with gradient packing have been used in which the quantity of the liquid phase decreases towards the end of the column.[27]

The support is usually coated by mixing with a dilute solution of the liquid phase and subsequent evaporation of the solvent. Attrition of the support is less significant in the frontal technique when a 10–15% solution of the liquid phase is passed through the column, the liquid phase displaced by a gas stream and the packing settled.[7] Before use the finished packing should be carefully conditioned at the maximum operation temperature by passing a carrier gas stream through it at least overnight.

Effect of operation conditions. In preparative work the effect of operating conditions is less critical and H_2, He or N_2 may be equally well used. Large diameter columns require considerably more carrier gas which is again an important cost factor. In general, flow rate is proportional to the cross-section, but in preparative work the effect of gas velocity is not so important.[28] Should carrier gas consumption be exceedingly high, the effluent gas may be recirculated after removal of the contaminants by adsorption. Separation efficiency may be raised especially in the case of wide boiling range samples by flow programming which has been recently introduced in preparative column techniques.

The influence of temperature is similar to that in analytical columns, at higher temperatures less time is required for separation, but this is associated with reduced column capacity.

Temperature programming has an important role in preparative gas chromatography, as it not only reduces separation time, but considerably improves the injection of larger samples and yield by raising the concentration of the component in the effluent through narrowing the component fronts. In large diameter columns considerable temperature differences between the column wall and centre may however arise as a consequence of the poor heat conductivity of the packing. The resulting fluctuations in flow rate may cause distortions of the component zones and thus reduced efficiency.[29] Consequently large diameter (> 15–20 mm) columns cannot be used very well with programmed temperature, though favourable results were obtained with internal metal baffles[30] and certain special designs[11, 12] of large diameter columns. Programmed temperature is particularly advantageous in combination with small diameter long columns.[20]

9.3 Construction and operation

Though the construction and packing of the column is the main problem of preparative gas chromatography, the construction and operation of the other parts of the apparatus are also of decisive importance for effective preparative work. Figure 9.2 shows a general

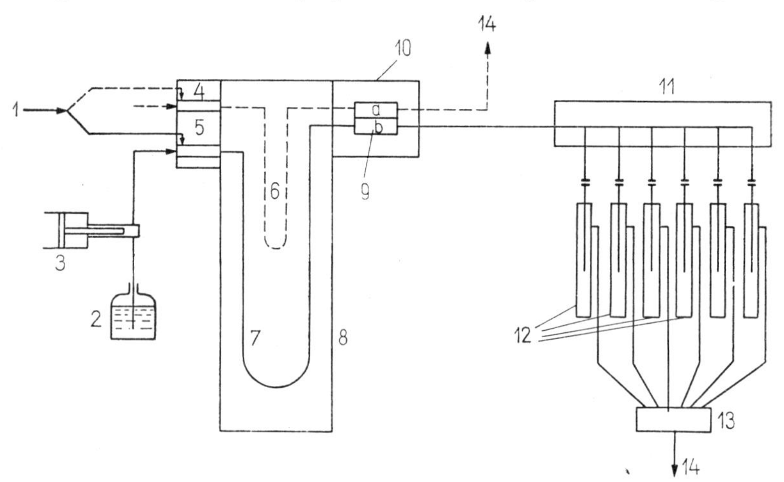

Fig. 9.2. Schematic diagram of the preparative gas chromatograph. Full line: preparative section; dashed line: analytical section. 1. Carrier gas inlet; 2. liquid tank; 3. dosing pump; 4. analytical column, injection port; 5. preparative column, injection port; 6. analytical column; 7. preparative column; 8. column thermostat; 9. thermal conductivity cell; a reference branch; b measuring branch; 10. detector thermostat; 11. manifold; 12. traps; 13. collector valve; 14. carrier gas outlet

schematic diagram of a preparative apparatus. The construction of some parts of the preparative apparatus, notably the sample injector, the operation of the detectors, the collection of the separated components and the automation of the apparatus significantly differ from those used with analytical columns and will therefore be discussed in greater detail.

Sample injection

Larger sample sizes raise problems of injection and the injection method has a significant influence on separation efficiency. For the evaporation of 1 g of a liquid sample a quantity of heat of the order of 100 cal is needed and with still larger samples the rapid transfer of the necessary heat presents a serious problem. Slow evaporation of the sample considerably reduces separation efficiency. This effect can be diminished by four methods:

a) Greatly overheated, high heat capacity injector and rapid injection. Some very efficient evaporators have been evolved for the latest types of apparatus which are capable of evaporating 150 ml of the sample within less than 50 sec with only a minimum increase in the temperature.[4] This injection method results however in an exponential distribution in the column.

b) Evaporation of the sample in a separate chamber and injection in vapour form. Injection should be performed preferably at constant vapour pressure near a saturation value corresponding to column temperature.[16] With this method a distribution corresponding approximately to a quadratic response may be obtained.

c) Slow liquid injection at column temperature with intermittent reduction of carrier gas rate.

d) On-column liquid injection at low initial temperature and programmed temperature operation.

Methods c) and d) are of special importance in the separation of heat sensitive substances. As already mentioned, programmed temperature can be applied only to smaller column diameters. Most commercial equipment uses method a). The liquid is injected manually by means of a syringe or with an automatic injector. For automatic injection piston-type feeders or liquid injectors under gas pressure (timed-pressure mode) are used. At each stroke the piston injects on the average 0.5–15 ml of liquid. The feeding tank may have any size from 100 to 1000 ml. In the timed-pressure injectors the liquid tank is under the pressure of some inert gas and injection takes place when under the effect of a pressure pulse on the tank the liquid passes through a capillary choke directly into the evaporater. The injected quantity depends on the pressure and the choke. With this method larger quantities can be satisfactorily injected.

Detectors

Preparative gas chromatographs generally use thermal conductivity cells or flame ionization detectors. The entire gas quantity may pass through an appropriately constructed detector but recently some thermal conductivity cells with higher sensitivity are used in by-pass arrangement. The flame ionization detector must operate in a by-pass when only a fraction of the effluent (0.1–1 %) is led through the detector by means of an appropriate splitter. To ensure stable operation the detector is provided with a separate gas system. It has been established as a general principle that detectors with low response time and wide linear dynamic range are the most appropriate here.

Collection of the components

The most important requirement which the sample collecting system has to meet is the transfer of the separated components without contamination and condensation into their appropriate receivers or traps where the components should condense. The sample collecting system has three parts: a) gas outlet pipe and collecting manifold; b) collecting traps with cooling bath and; c) automation of the collecting system.

The gas to the traps is led through a heated collecting manifold, generally in descending arrangement, and connected to trap holders. To avoid back-mixing and contamination, as well as dead volumes, the gas should travel over a short path. The manifold is heated to 250–350 °C to avoid condensation. The gas to the traps may be led by two essentially different methods of which several modifications are known.

In the first method the traps are placed at the circumference of a *turntable* which is provided with a gas outlet. The table may hold 6 to 12 traps and is turned by a motor through the required angle. By the manual or automatic operation of the turntable the traps come in succession under the gas outlet. The gas outlet ends in a pneumatically operated needle which will perforate the rubber septum of the traps just under the outlet end, so that the gas will flow through the trap.[31, 32] In another arrangement the traps are provided with an open inlet and the gas outlet is joined at the required moment through a special ground glass joint or through an O-ring seal to the trap.

The other type of sample collecting system has no rotating part and each of the traps in the row has a fixed inlet which is branched off a common tube. The trap is connected to the gas stream by means of *automatically operated electropneumatic* or *magnetic valves.*[6, 33, 34] Earlier the valves were included in the heated zone before entering the trap, but more recently the collecting system is constructed in such a way that the valves are incorporated after the traps as this

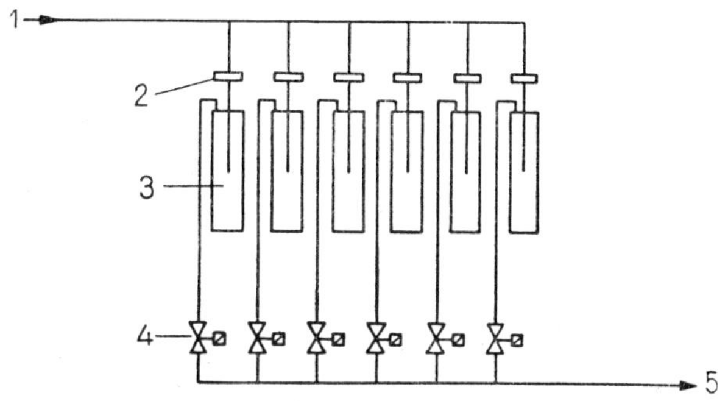

Fig. 9.3. Schematic diagram of the collecting system for preparative equipment.[34] *1. Gas inlet; 2. diffusion seal (porous metal disk); 3. trap; 4. magnetic valve; 5. carrier gas outlet*

arrangement ensures safer operation. A schematic diagram[34] of such a collecting system is shown in Fig. 9.3, where the traps are placed in such a way that in addition to the traps for the pure fractions there are one or more common collecting traps for the intermediary fractions or for the collection of components which have not been intended for separation.

The *construction of the traps* is highly important in preparative gas chromatography, as incomplete condensation may cause considerable losses. The efficiency of collection depends on the vapour pressure of the component at the temperature of the trap, on the component concentration in the carrier gas, the total quantity of the component, the available heat transfer surface and on the conditions of heat transfer (flow rate, heat conduction properties, etc.). With certain components fog formation may present a special problem, as this will have a detrimental effect on the yield; 60–99% yields are quoted by various manufacturers for different components; in the latest apparatus 90–95% yields are quite common. The gas chromatograph manufacturing companies have devoted much work to the development and comparison of the collecting systems which resulted in a considerable improvement in collection efficiency.[35]

Usually the traps are made of glass, but with high capacity equipment occasionally metal traps are also used.[4] Figure 9.4 shows several

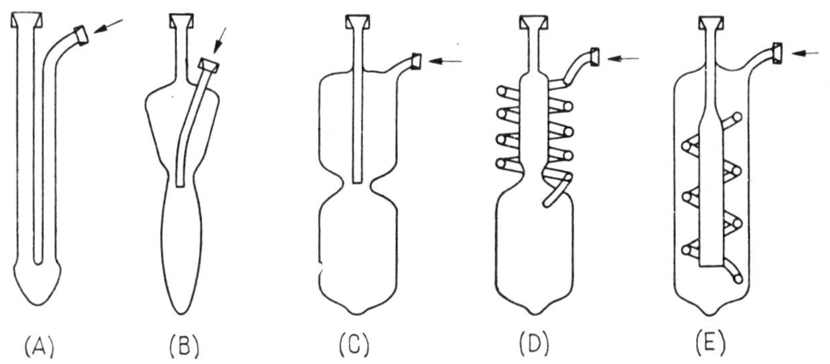

Fig. 9.4. Schematic diagram of the various types of traps

characteristic trap types. The optimum construction of the trap will depend on the nature and quantity of the component to be collected. The upper part of the trap is appropriately filled with glass wool to accelerate condensation. Traps provided with external cooling coils are the most effective for lengthening the residence time of larger samples. The efficiency of these traps can be further increased by the incorporation of a porous disk after the spiral to decompose the formed aerosols. The construction must be modified for the collection of solids to avoid blocking. Various suggestions have been put forward to increase collection efficiency and to eliminate fog formation, such as multistep cooling, intermittent cooling and heating[36] electrostatic fog precipitation,[37] filling of the trap with some gas chromatographic packing or adsorbent.[35] With an adequately constructed packed trap the yield may reach almost 100%, but packing has the drawback of raising difficulties to the recovery of pure samples. Special methods were described for the separation of the liquid by centrifuging.

The liquid can be removed from the trap by means of an injection syringe or from larger traps with a vacuum pump.

The receivers are immersed in Dewar vessels filled with some cooling medium; the turntable system uses a common Dewar vessel for all traps, while the valve system has a special cold bath for each trap.

For cooling liquid air (−180 °C), dry ice–diethylether (−100 °C), dry ice–acetone (−86 °C), salt–ice (−21 °C) mixtures are used depending on the boiling point of the component.

The traps may be changed manually as the peaks appear on the recorded chromatogram, or by an automatic device operated by the recorder pen or on the basis of a fixed time program.

Automatic operation

The performance of preparative apparatus rises considerably by automatic operation; with small diameter columns 100–1000 g of products, with larger diameter columns 1–10 kg daily can be prepared. Automatic preparative columns were first reported at the end of the 1950's[38–40] and soon high performance automatic preparative chromatographs appeared on the market.

Automatic operation involves automated sample injection and fraction collection and in some models automatic programming of the temperature. An automatic apparatus repeats the set cycle of operations automatically, so that it can be operated continuously for a long period of time up to several days.

Automation of sample injection can be easily solved by the pneumatic operation of the piston or gas pressure type injector systems. After adjustment of the sample size to be injected and of the cycle period a timer controls injection. The injection cycle can be accelerated when separation time is relatively long (with long columns) and the mixture to be separated contains only a few components, so that the separation of several sample doses proceeds simultaneously in the column. Instead of a preset time cycle the sample injector can be operated from the appearance of the last or of a randomly chosen peak. The problems involved in automatic sample introduction are discussed in a paper by Jentzsch.[41]

One way of *automated fraction collection* is to change the traps by means of a time programmer on the basis of predetermined retention times.[38] In spite of its simplicity the method has the disadvantage that retention times may change depending on the column parameters or operation conditions. In a recently evolved method the traps are changed according to the actual chromatogram, corresponding to the deflections of the recorder. The components to be collected are first selected by preliminary analysis and a selector is marked with plugs or press-buttons to show which of the components (e.g. the 5th, 7th, 9th, 10th) should actually be collected. These apparatuses are usually suited to the processing of mixtures containing anything up to 16–18 components of which they can actually collect 5–8.

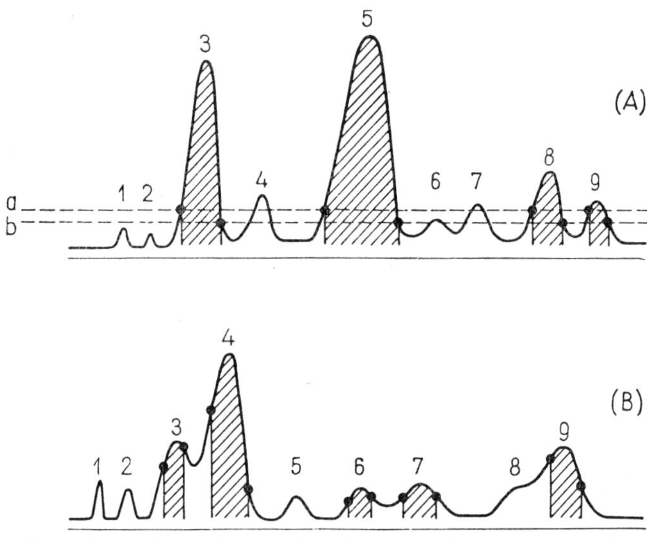

Fig. 9.5. Methods of controlling fraction collection. (A) Collection at constant response level: a–opening: 10% of the scale; b-closing: 6% of the scale; (B) collection at variably adjusted response level (peak height discriminator)

The simplest way of fraction collection is to change the trap at a preset, but always identical signal level of the recorder. This is usually 5–10% of the whole scale. The opening and closing position of the trap can be adjusted to different values, but the values are the same for every peak. However, this method cannot be applied to poorly separated, overlapping peaks, or to components present in very different concentrations. Overlapping peaks often occur when the column is overloaded to bring about a considerable increase of its capacity. In these cases a peak height discriminator should be used with which the level for the connection and disconnection of the traps can be randomly adjusted on the ascending and descending branches of the chosen component peaks.[34]

This method ensures the most appropriate change-over for any given task, independently of the shape of the chromatogram. The two methods are compared in Fig. 9.5.

Fraction collection on the basis of the detector response is operated by means of an electromechanical, photocell or electronic system. In the latest systems the change-over is operated after appropriate amplification by the output signal of an a.c. bridge formed by connecting a potentiometer parallel to the measuring potentiometer of the recorder and adjusting potentiometers corresponding to the single change-over points.

The third automated circuit serves the *programming of the temperature*. The programmer is operated by the timers and ensures the automatic repetition of the various isothermal and heating periods, as well as the cooling of the column at the end of the cycle and of the temperature stabilization period.

With respect to the demands which the preparative apparatus has to meet the other parts of the apparatus must also be mentioned.

In the construction of the thermostat, difficulties are encountered in providing a sufficiently large air space which by necessity is considerably larger than in the case of analytical columns and which must ensure uniform and constant temperature.

The preparative apparatus contains three or four separately controlled heating circuits, namely that for the sample injector, for the thermostat, for the detector and for the manifold of the collecting system.

As carrier gas consumption rises considerably when larger diameter columns are used, the carrier gas is often recirculated. This necessitates careful purification of the carrier gas, i.e. the removal of the component vapours by low temperature condensers or by adsorbers.

High capacity, large preparative equipment often contains a parallel special analytical column for the preliminary determination of the separation conditions.

9.4 Special techniques and industrial application

Recirculated column

By the alternative arrangement of three short columns Golay *et al.*[42] first separated from the components to be collected the heavier and lighter fractions and then recirculated the selected components till the desired degree of separation was achieved. This affords highly efficient separation and the method seems applicable to larger quantities.

Multicolumn rotating unit

An interesting method for increasing the capacity of preparative gas chromatography is the Roto–Prep apparatus evolved in the SNAM Laboratory.[43–45] This consists of 100 small diameter (6–8 mm) straight columns placed on the circumference of a metal cylinder which revolves at constant speed. The evaporated sample is injected from a fixed sample injector into the column as it comes round, while all the other columns are flushed with carrier gas which elutes previously injected samples. During rotation the outlet end of a given column comes into contact with the subsequent traps of the fixed collecting system. The flow rate of the carrier gas and the rotation speed of the cylinder are adjusted in such a way that the components separated on the different columns should be collected by the same trap. Both sample injection and collection are automatic and the continuously operating apparatus can process 200–1000 ml of liquid per hour. No detector or recorder are included with the apparatus which operates according to a program previously determined by a laboratory chromatograph.

Circular chromatographic columns[46–48]

In recent years reports have been published on the application of circular columns whose operation is shown diagrammatically in Fig. 9.6. The apparatus consists of a static and a rotating part with an

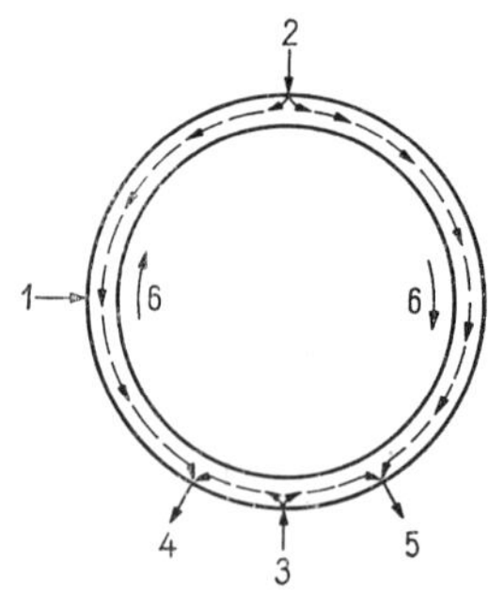

Fig. 9.6. Flow sheet of a circular column. 1. Sample injection; 2. carrier gas inlet; 3. purge gas inlet; 4. light product outlet; 5. heavy product outlet; 6. rotation direction

appropriate driving motor. The packing is in the ring shaped column of square cross-section which is wound round the circumference of a disk rotating with 1–10 rev/h. The sample is injected through an injector mounted on the static part which also carries the carrier gas inlet and the sample collector. The more readily sorbed components are adsorbed by the packing and travel along the direction of the rotation, while the lighter components flow in a counter-current to the packing and emerge from the column at one of the sample collecting heads together with the purge gas which is used to ensure complete separation. The adsorbed components are desorbed by the introduction of the carrier gas and by heating. In this form the apparatus is suitable for the separation of two components, or fractions. The liquid sample is injected in the vapour phase after preliminary evaporation. Both sample injection and the collection of the two products are continuous. This method differs essentially from the hitherto described procedures by being a continuous counter-current chromatographic process. The greatest problem of this apparatus is the proper sealing of the continuous gas inlet and outlet. The method was designed to eliminate the problems arising in moving bed continuous counter-current methods (see below), mainly to reduce attrition of the packing and efficiency deterioration, but is as yet in an experimental stage.

Moving bed continuous counter-current gas chromatography

Moving bed gas chromatography was first applied to various gas separation problems on activated charcoal adsorbent by Benedek and Szepesy [49–52] as an improvement of Berg's hypersorption method.[53] The theoretical problems involved in this type of separation were also thoroughly studied. Pichler and Schulz[54] and later several other authors [55–57] performed moving bed separation with liquid coated packings and succeeded in the production of a good yield of high purity substances. The principle of operation of moving bed continuous gas chromatography is shown in Fig. 9.7. The packing moves slowly downwards by gravity, in a vertical tube, the velocity of the packing will depend on the discharge rate at the bottom of the column. The evaporated sample is fed continuously together with the carrier gas to the middle section of the column. Packing velocity and the carrier gas velocity are adjusted in such a way that the less readily sorbed component or fraction flows upwards in the column and can be withdrawn at the top. The heavier sorbed component moves downwards and is eluted by heating the lower section of the column with a purge gas. This fraction is discharged at the bottom.

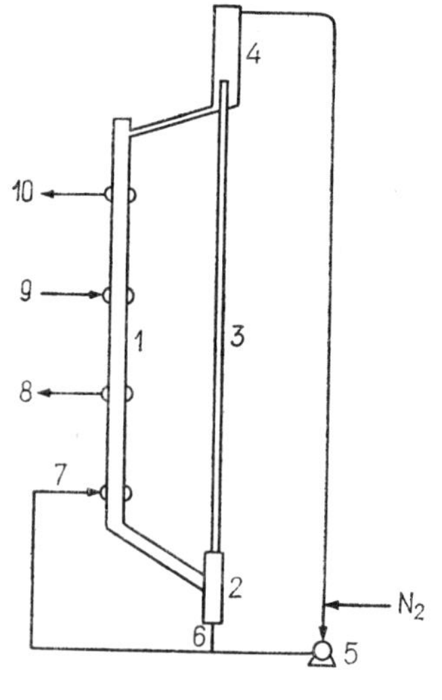

Fig. 9.7. Flow sheet of moving bed continuous gas chromatography. 1. Column; 2. lift feeder; 3. lift line; 4. discharge port; 5. blower; 6. lift gas inlet; 7. carrier (purge) gas inlet; 8. bottom product discharge; 9. feed; 10. overhead product discharge

The packing discharged at the bottom of the column is continuously conveyed to the column top by an appropriate device (cup conveyor, gas lift) completing the cycle. By some constructional modifications the column can be made suitable for the recovery of three or four products.

With the moving bed method very high capacities, even commercial scale production can be realized. The drawback of the method appears in the loosening of the packing, due to its movement which results in non-uniform density and consequently greatly reduced efficiency. Another problem is the attrition of the packing during the cycle, which excludes the use of all but the mechanically strong packings.

Scale-up and industrial application of preparative gas chromatography

With the latest high capacity preparative gas chromatographs (e.g. 10 cm diameter columns) daily quantities up to 1–10 kg may be processed. Their importance is continuously increasing in the commercial production of standard substances, analytical grade chemicals, and highly valuable pharmaceutical raw materials.[58] Operations which formerly required cumbersome laboratory distillations can now be solved far more efficiently and economically with preparative gas chromatography.

Considerable progress has been made in recent years with respect to scaling-up columns and the development of production scale units. In the USA Abcor Inc. has successfully solved the problems of scaling up, including the assembly of 30 cm diameter, 4.5 m long columns, and even greater units (120 cm diameter) are being planned.[59, 60] The increase in column diameter is achieved by a special internal baffle system without significantly impairing efficiency. Some serious engineering problems had to be solved with respect to sample injection, detection, fraction collection and the recirculation of the carrier gas. The mathematical model of the system has proved to be of great help in the adjustment of the optimum operation conditions.

Plant scale preparative gas chromatography will soon compete with commercial distillation and other separation processes, especially when substances of closely similar properties ($\alpha < 1.3$) have to be prepared in high purity ($> 99.5\%$). In the separation of heat sensitive substances gas chromatography has the great advantage over distillation that the residence time of the substance in the column is quite short. According to preliminary calculations a 30 cm diameter column is suitable for the processing of 120 kg per day, the 120 cm diameter column for 1400 kg per day with a production cost of £ 0.28 or $ 0.67 and £ 0.16 or $ 0.45 per kg, respectively. Units with even higher capacities are being planned. In spite of the many sceptic forecasts uttered in connection with the industrial application of gas chromatography, production scale gas chromatography is expected to be realized within a very short time. In addition to the economic production of pharmaceutical products (alkaloids, steroids, barbiturates, amino acid derivatives) and analytical grade reagents by these high capacity units, the latter will provide possibilities of commercially producing certain important intermediaries, essential oils, high purity hydrocarbons, solvents and other petrochemical products.

References

1. D. E. EWANS and J. C. TATLOW, *J. Chem. Soc.*, 1184 (1955)
2. D. E. EWANS and J. C. TATLOW, Chapter 1, ref. 58, p. 256
3. B. T. WHITHAM, Chapter 1, ref. 58, p. 194
4. F. J. DEBBRECHT *et al.*, Hewlett–Packard Technical Paper No. 33., F. & M. Scientific Corporation, Avondale, Pa. (1965)
5. F. HUYTEN, W. VAN BEERSUM and G. W. A. RIJNDERS, Chapter 1, ref. 63, p. 224
6. E. BAYER, K. P. HUPE and H. G. WITSCH, *Angew. Chem.*, **73,** 525 (1961)
7. E. BAYER, K. P. HUPE and H. MACK, *Anal. Chem.*, **35,** 492 (1963)
8. G. J. FRISONE, *J. Chromatog.*, **6,** 97 (1961)
9. J. L. WRIGHT, *J. Gas Chromatog.*, **1,** (1) 10 (1963)
10. A. B. CAREL, G. PERKINS JR., *Anal. Chim. Acta*, **34,** 83 (1966)
11. Short Notes 2—66 Carlo Erba, Milan (1966)
12. B. M. MITZNER and W. V. JONES, *J. Gas Chromatog.*, **3,** 294 (1965)
13. T. JOHNS, M. R. BURNELL and D. W. CARLE, Chapter 1, ref. 64, p. 207
14. M. VERZELE and M. VERSTAPPE, *J. Chromatog.*, **19,** 504 (1965)
15. J. C. GIDDINGS and E. N. FULLER, *J. Chromatog.*, **7,** 255 (1962)
16. G. M. C. HIGGINS and J. F. SMITH, Chapter 1, ref. 70, p. 94
17. S. F. SPENCER and P. KUCHARSKI, Technical Paper No. 37. Hewlett–Packard F. & M. Scientific Division, Avondale, Pa. (1966)
18. J. C. GIDDINGS and G. E. JENSEN, *J. Gas Chromatog.*, **2,** 290 (1964)
19. S. T. SIE and W. A. RIJNDERS, *Anal. Chim. Acta*, **38,** 3 (1967)
20. M. VERZELE *et al.*, *J. Chromatog.*, **18,** 253 (1965)
21. M. VERZELE, *J. Gas Chromatog.*, **3,** 186 (1965)
22. M. VERZELE, *J. Gas Chromatog.*, **4,** 180 (1966)
23. A. ROSE, D. J. ROYER and R. S. HENLY, *Separ. Sci.*, **2,** 211 (1967)
24. A. ROSE, D. J. ROYER and R. S. HENLY, *Separ. Sci.*, **2,** 229 (1967)
25. A. ROSE, D. J. ROYER and R. S. HENLY, *Separ. Sci.*, **2,** 257 (1967)
26. M. VERZELE and M. VERSTAPPE, *J. Chromatog.*, **26,** 485 (1967)
27. R. C. DUTY, *J. Gas Chromatog.*, **6,** 193 (1968)
28. M. VERZELE, *J. Chromatog.*, **15,** 482 (1964)
29. K. P. HUPE and E. BAYER, Chapter 1, ref. 70, p. 62
30. R. W. REISER, *J. Gas Chromatog.*, **4,** 390 (1966)
31. G. KRONMÜLLER, Chapter 1, ref. 64, p. 199
32. Autoprep A-700, Aerograph, Wilkens Instr. Inc. (1964)
33. H. WIEGLEB and H. PRINZLER, *Chem. Tech.*, **15,** 98 (1963)
34. K. P. HUPE, V. BUSCH and W. KUHN, *J. Gas Chromatog.*, **3,** 92 (1965)
35. R. J. HUNT, *Column*, Pye Gas Chromatography Bulletin, **2,** (1) 6 (1967)
36. R. TERANISHI *et al.*, *J. Chromatog.*, **9,** 244 (1962)
37. W. P. ROSS, J. F. MOON and R. L. EVERS, *J. Gas Chromatog.*, **2,** 341 (1964)
38. D. AMBROSE and R. R. COLLERSON, *Nature*, **177,** 84 (1956)
39. E. HEILBRONNER, E. KOVÁTS and W. SIMON, *Helv. Chim. Acta*, **40,** 2410 (1957) and **41,** 275 (1958)
40. A. P. ATKINSON and G. A. TUEY, Chapter 1, ref. 60, p. 270
41. D. JENTZSCH, *J. Gas Chromatog.*, **5,** 226 (1967)

42. W. J. M. Golay, H. I. Hill and S. D. Norem, *Anal. Chem.*, **35**, 488 (1963)
43. D. Dinelli, S. Polezzo and M. Taramasso, *J. Chromatog.*, **7**, 477 (1962)
44. S. Polezzo and M. Taramasso, *J. Chromatog.*, **11**, 19 (1963)
45. M. Taramasso and D. Dinelli, *J. Gas Chromatog.*, **2**, 150 (1964)
46. P. E. Barker and D. H. Huntington, *J. Gas Chromatog.*, **4**, 59 (1966)
47. D. Glasser, Chapter 1, ref. 73, p. 119
48. P. E. Barker and D. H. Huntington, Chapter 1, ref. 73, p. 135
49. P. Benedek and L. Szepesy, *Erdöl u. Kohle*, **9**, 593 (1956)
50. M. Freund, P. Benedek and L. Szepesy, Chapter 1, ref. 58, p. 359
51. P. Benedek, L. Szepesy and I. Szépe, Chapter 1, ref. 59, p. 225
52. P. Benedek, L. Szepesy and I. Szépe, *Acta Chim. Hung.*, **14**, 339, 353 and 359 (1958)
53. C. Berg, *Chem. Eng. Progr.*, **47**, 585 (1951)
54. H. Pichler and H. Schulz, *Brennstoff. Chem.*, **39**, 149 (1958)
55. R. P. W. Scott, Chapter 1, ref. 60, p. 287
56. P. E. Barker and D. Critcher, *Chem. Eng. Sci.*, **13**, 82 (1960)
57. H. Schulz, Chapter 1, ref. 70, p. 225
58. Anon., *Chem. Eng. News*, **43**, (7) 46 (1965)
59. Anon., *Chem. Eng. News*, **43**, (26) 46 (1965)
60. Anon., *Chem. Eng. News*, **44**, (21) 52 (1966)

10. Process Gas Chromatographs

10.1 Development and characteristics

The high efficiency and wide scope of gas chromatographic analytical methods soon turned attention to its possible application in process control. The fundamental condition for the economic operation of new, highly complex technological processes, that is, of the high capacity plants employing them is the appropriate instrumentation for automatic monitoring and process control. Analytical instruments are gaining in importance in process monitoring as witnessed by the increasing numbers of refractometers, photometers, infrared and ultraviolet analyzers, mass spectrometers, etc. being incorporated in the production lines. Owing to the high efficiency of separation, the sensitivity of the detectors and the accuracy of its results, gas chromatography promises new possibilities of process monitoring, especially in the monitoring of multicomponent, complex process streams, problems which were insoluble or only partly soluble by other methods.

The first process gas chromatographs appeared in the factories in 1956–57. Based on experience with these first trials the production and application of gas chromatographs designed for process control rapidly expanded at the beginning of the 1960's.

Comparison with laboratory gas chromatographs

The first process gas chromatographs were modifications of the laboratory apparatus with automatic sample injection. The more rigorous requirements which the process instrument had to meet necessitated however entirely new constructions. The most important of these requirements are:[1, 2]

1. Automatic and stable operation without supervision from sample injection to the processing of the analytical results for a relatively long (3–6 months) period.

2. Construction conforming to the demands of plant conditions and labour safety.

3. Analysis time should correspond to the time constant of the investigated process and should be as short as possible.

4. Absolute constancy of operation conditions to ensure accurate and reproducible results.

5. The apparatus should require only simple and rapid maintenance and repair work.

Besides the higher stability and reliability required from the process chromatograph the latter differs from the laboratory apparatus also in that it is designed to analyze one or more mass streams of constant quality and that analysis conditions cannot be varied as in laboratory apparatus. A laboratory apparatus consists generally of several units with ancillaries which may be used with different columns in a combination most appropriate to the task in hand. With the process chromatograph the operating conditions have to be adjusted to a given task and their choice is limited by the demands of stability and minimum analysis time. From the aspect of the analytical method this involves the use of multicolumn arrangements and different column switchings as already discussed in the chapter on columns.

Comparison with other process analytical methods[3, 4]

Gas chromatography is by its nature an intermittent process in which the analytical cycles are periodically repeated. This is a certain disadvantage, compared to other in-line analyzers which produce continuous signals. With the first apparatuses analysis time varied between 15 and 40 min which was unsatisfactory for monitoring, but mainly for process control. This drawback was practically eliminated in the later gas chromatographs for which analysis time is considerably shorter, on the average not more than 1–10 minutes which is suitable for process control. With some special equipment the analysis time of e.g. C_4–C_5 hydrocarbons is not more than 20–50 sec.

The greatest advantage of process chromatographs over other in-line analytical instruments is that they first measure the concentration not of a single component only but of multicomponent mixtures and, therefore, they provide more detailed information on the process under investigation. Owing to the high separation power of the gas chromatograph and to the possibilities of various column arrangements, components having slightly different boiling points or of a similar chemical nature may also be analyzed. This is of particular importance in the monitoring of many petroleum and petrochemical processes where other analytical instruments fail. With appropriate detectors gas chromatography can be applied to trace analysis and to the detection of impurities in ppm concentrations. In the analysis of liquid mixtures the gas chromatograph has opened new possibilities for with other analytical instruments only binary mixtures or certain components can be analyzed. The gas chromatograph can be simply incorporated into the process line, the results can be read off or converted into the desired signal.

The accuracy and reproducibility of the gas chromatograph is similar to that of other instruments, moreover it produces quite often more favourable results. The accuracy of the apparatus depends on the accuracy of the calibration method and on the constancy of operation conditions. With the latest gas chromatographs reproducibility is about $\pm 2\%$.

The reliability experienced with process chromatographs has been very favourable; the latest apparatuses operate quite smoothly for three to six months. Most of the difficulties arise from the unsatisfactory construction and operation of the sampling system.

A gas chromatograph has considerably more complicated parts than simple process analyzers, thus they are more liable to breakdowns and their maintenance and repair require highly skilled operators. Against this stands the fact that by monitoring simultaneously several components the gas chromatograph is equivalent to several simple process analyzers.

Quite often the gas chromatograph can be applied to the monitoring of processes, i.e. material streams which escape analysis by other instruments. In cases when the task in hand appears to be soluble by other, simpler analyzers (e.g. the analysis of non-hydrocarbon gases, trace analysis in gases, etc.) the most appropriate method must be selected after thorough consideration.

Economic aspects of process chromatographs

The application of process gas chromatographs is decided by economic and safety considerations.[5] It is not possible to decide directly the economy of the apparatus chosen for safety reasons, though its price is rapidly amortized by its preventing work accidents. The most important economic effects are manifest in higher production, improved quality of the products and lower costs of analyses and operation.

The price of process chromatographs is two to three times higher than that of laboratory apparatus, generally around £ 3150–£ 5000 or $ 9000–$ 12,000, plus 15–20% for installation and starting operation, while the price of the sampling system may amount to 30–50% of the price of the chromatograph, resulting in a total of £ 5000–£ 8000 or $ 12,000–$ 20,000 for the whole apparatus. Without the price of the sampling system this is about two to four times the price of simpler (IR, UV) analyzers.

Process chromatographs are rapidly amortized. According to the calculations of Philips Co., which are often referred to in the literature, the total investment cost of a gas chromatograph built in for closed-loop control of a distillation column which separates n-butane from isobutane is amortized: a) in a fortnight by higher production; b) in six weeks by the improved quality of the product (higher purity); c) in six months by the diminished maintenance costs.[6] Some other data[7] claim that the amortization of process gas chromatographs takes three to six months, at the best one month in the petroleum, natural gas and petrochemical industries. These data are clear indications of the economic advantages of process gas chromatographs, nevertheless, in Europe the application of gas chromatographs to process control is not yet generally accepted.

The detailed data provided by process chromatographs are the basic conditions for computer monitored and controlled operations and processes. These are rapidly gaining ground so that e.g. in a petrochemical plant of Monsanto Chemical Co. the composition of the mass streams is fed into four computers from 36 gas chromatographs, while Union Carbide Olefins Company has 13 gas chromatographs to provide the data of 269 peak areas for computer processing and evaluation.[8]

10.2 Construction and application of process gas chromatographs

General construction of the apparatus

In its simplest form the process chromatograph consists of three units:

analyzer;
programming unit (timer);
recorder.

These basic units may be complemented by a data processing unit, units recording each component separately and units for storing and converting the responses for control purposes. The system for sample handling may be considered a separate unit which in fact is not a part of the gas chromatograph but an important ancillary device of every process instrument.

The analyzing unit is constructed so as to be explosion proof and is most expediently located in the vicinity of the actual place of sampling in the plant. The other parts of the apparatus are in the instrument or control room generally mounted on the instrument panel. The distance between the analyzing and control units may be 150–300 m.

The first process gas chromatographs were constructed in a way that was supposed to ensure their versatile wide applicability. It was however soon clear that this would increase not only the price but also the complexity of the apparatus with considerably higher probabilities of errors. Another trend of apparatus development aimed at simple construction, suitable for certain fixed tasks. The latest trend is the modular type which by the combination of various module units (e.g. for the analysis of one or several streams, with gas or liquid injection and different detectors) offers possibilities of constructing apparatus suitable for widely diverse analytical tasks (e.g. Pye Modular Plant Chromatograph).

Sampling and preparation of the sample

The analyzer unit of the process gas chromatograph is connected to the process stream to be investigated by the sampling system through which a branch of the main stream continuously flows into the sampling device of the analyzer. The sampling device introduces at definite intervals constant quantities of the sample from this continuous stream. Preliminary purification and preparation of the sample is one of the basic conditions for the proper operation of the chromatograph. Sampling systems consisting of many units, such as micro-filters, adsorbers, scrubbers, pressure and flow regulators, switching valves, heat exchangers and evaporators, have been constructed. When sampling is carried out from a system at atmospheric or reduced pressure, a membrane compressor has to be incorporated into the sampling system.

Several concomitant sample streams are analyzed by using a separate preparation system for each stream when an automatic switch directs the various streams into the sampling device. These units can be assembled on a common panel or into a box which can be heated if necessary and is conveniently placed before the analyzer unit. The construction of the sampling devices varies with the task in hand; several special designs are known from the literature.[9]

Analyzer

The analyzer unit contains the sampling device, the columns, the switch valves, the detector and the temperature and flow regulators. It is usually cylindrical or box-shaped with an explosion-proof jacket quite often with an air purge to increase safety. The analyzer consists usually of two parts, one being the thermostat to ensure the appropriate temperature of the parts in contact with the sample.

The *thermostat* is constructed in such a way that changes in the ambient temperature (−25 to +40 °C) shall have no effect on the operation of the analyzer. Metal block[10] or air[10, 11] thermostats are used, the metal block thermostat is usually cylindrical with a heated metal block on which the spiral column is wound. Some commercial apparatuses are designed for the analysis of gases or volatile liquids and can be thermostated at temperatures between 50 and 100 °C. The more recent chromatographs may operate up to 220–250 °C. Temperature control has to meet even stricter requirements than in the case of laboratory apparatus, with a tolerance of ±0.1–0.2 °C, down to 0.01 °C in certain types. Some of the latest apparatuses use programmed temperature.

Carrier gas system. The carrier gas may be hydrogen, helium or nitrogen. Regulation of the flow rate is extremely important from the point of view of the stable operation of the apparatus, so that both pressure and flow regulation is used with thermostated regulators. In the latest types flow rate can be regulated with ± 0.1% accuracy. To accelerate certain analyses some chromatographs work with flow programming.[12]

Sample introduction. One of the most critical points of the process gas chromatographs is the proper construction of the sampling device. The first process chromatographs used the automated variation of the rotating laboratory sampling valve which was however far from satisfactory.[1]

There are several methods for the introduction of gas samples. The size of the gas sample varies usually between 1 and 5 ml, in certain

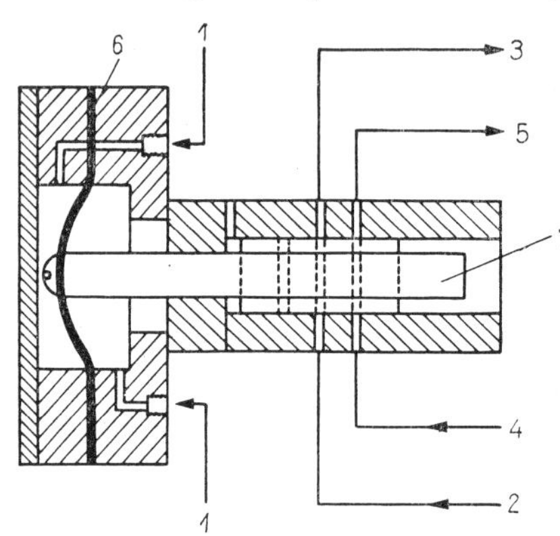

Fig. 10.1. Schematic diagram of a piston type sample injector. 1. Air inlet; 2. sample stream inlet; 3. sample stream outlet; 4. carrier gas inlet; 5. carrier gas + sample to column; 6. membrane; 7. moving part

cases, especially for trace analysis, larger samples are introduced. In the first apparatus of this type the liquid samples were introduced in vapour form by evaporation of the liquid. Reliable sampling devices for the direct introduction of very small quantities of liquid samples (1–10 μl) appeared only at the beginning of the 1960's. Plant experience and development work led to several types of sampling devices which in general may be used for both gas and liquid introduction.

Piston type sampling valves for gas introduction have been known for some time; these can be simply automated by a pneumatically operated membrane. The valve in Fig. 10.1 is an improved version for the introduction of very small quantities of liquids or gases. The moving part is usually Teflon, the holes drilled in it form the sample volume ($< 1\ \mu$l) and are alternately purged or used for introduction.[13] For the introduction of larger gas or vapour samples the sampling device may have, instead of four, six ports with an external sample loop.

The other moving sampling device type is the so-called *flat sliding valve* where a Teflon plate with appropriate grooves moves back and forth over a finely machined stainless steel surface when the grooves will connect different in- and outlet openings.

Rotating valves operating on the same principle have also been designed. Yet another modification of the rotating valve rotates not between two opposed positions but performs continuous slow rotations.[14]

All these variations can be produced for very small sample volumes and adapted to liquid introduction. In some types the port with the

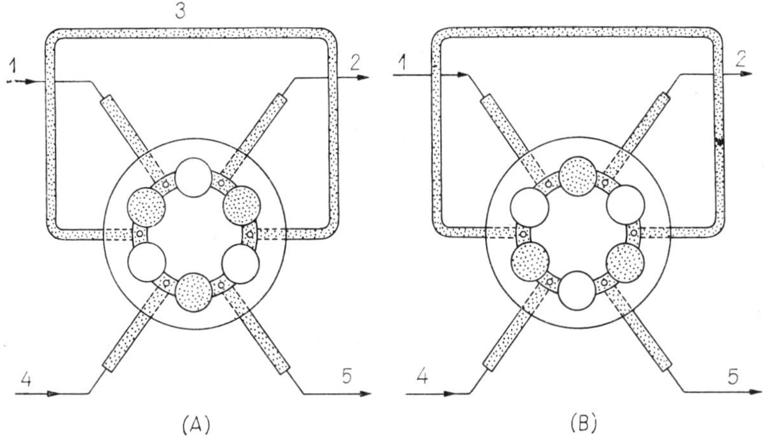

Fig. 10.2. Flow sheet of the membrane gas sampling valve. (A) Sample flush; (B) sample introduction. 1. Sample stream inlet; 2. sample stream outlet; 3. sample volume; 4. carrier gas inlet; 5. carrier gas to column

liquid sample leads into a heated chamber where the liquid evaporates, while in other variations a pre-evaporator is used and the vapour is introduced.[15]

Constructions with moving Teflon parts can be used generally up to 150 °C, and when very carefully designed up to 220 °C.

Pneumatically operated membrane valves are extensively used for the introduction of gases. Figure 10.2 shows the operation principle of such a valve.[16] Under the action of air pressure on the membrane three of the six valves are open and six closed. By changing the direction of the pressure on the membrane the open valves are closed and *vice versa.* The sample is introduced as shown in diagrams (A) and (B). The membrane may be a thin plastic film (Teflon, Neoprene, polyethylene).

The new sampling devices are highly reliable, they are capable of operating with ±1% accuracy. The sampling device can be operated by pneumatic, hydraulic or mechanical transmission controlled by a timer.

Valves similar to the sampling valve can be used to effect the various column switchings. For this 4, 6, 8 — in the case of rotating valves even 16 — ports are used. When used as switch valves the inlets and outlets can be short circuited in different ways to form the required gas paths.

Column. Because of their low space requirement spiral, flat spiral or U tube columns are used in process gas chromatographs, while to shorten analysis time small diameter (2–3 mm) columns with low liquid loaded packings are preferred. Efficient separation is provided by the appropriate arrangement of columns with different packings. The demands on column stability are high, thus the choice of the stationary phase is of great importance.[17, 18] The column has to operate with the required stability for at least six months, as changes in packing characteristics will result in a change in the retention of the components and in the peak heights.

In the case of adsorbent packings column stability is affected by the heavy impurities in the sample, that is, by the irreversible adsorption of certain components. This effect can be eliminated partly by the adequate purification of the sample stream and partly by pre-columns.

When the packing is liquid loaded, the evaporation of this liquid will interfere with the operation of the detector and result in a change in packing activity after a certain time. The liquid phase should be chosen in such way that it shall not affect the activity of the packing even after operation for several months and that no drift of the base line shall occur. Consequently not all the liquid phases used in laboratory columns are suitable for process chromatography and packings must be operated at considerably lower than the maximum permissible temperature.

To shorten analysis time liquid coated tubular columns have been tried, but proved to be inadequately stable.[18] In this field new possibilities were opened by the latest column types, e.g. by the solid coated open tubular columns or packed capillaries.

Column arrangements. Process gas chromatographs are very often used in various arrangements or as multicolumn systems to shorten analysis time, to improve the efficiency of separation and the recovery of the main components or simply to save the packing from contamination by heavy impurities.[2, 10, 17–19] These arrangements involve the back-flush of the column, or series, alternative or parallel arrangement of several columns.

Provided analysis of the light components only is required, on a single column, *back-flush* will considerably reduce analysis time. The light components travel rapidly through the column, while the heavy components are retained. With reversed gas flow the heavy components may be eluted in the opposite direction and are sensed by the

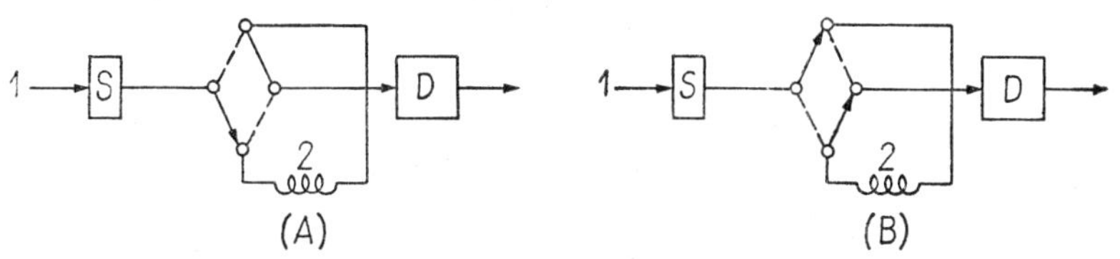

Fig. 10.3. Column back-flush arrangement. (A) Pre-flush; (B) back-flush; S—Sample injector; D—detector; 1. carrier gas inlet; 2. column

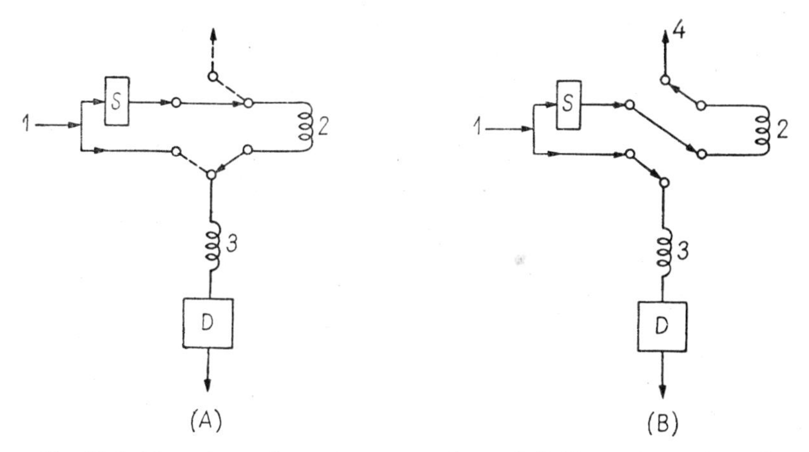

Fig. 10.4. Flow sheet of a stripper pre-column. (A) Two columns in series; (B) stripper column back-flush. S—sample introduction; D—detector; 1. carrier gas inlet; 2. stripper pre-column; 3. analytical column

detector as a single peak (Fig. 10.3). This operation is called flushing. If the sensing of the heavy components is unnecessary they are back-purged by a separate carrier gas stream into the open, so that the column will be ready for the injection of the next sample.

A frequently applied dual column arrangement is the application of the so-called *stripper pre-column* (Fig. 10.4). The injected sample passes first through the stripper column which retains the heavy components and only the components to be analyzed reach the separating column. Simultaneously with the elution of the separating column, the carrier gas stream elutes the stripper column by back-flushing. This method is suitable for the retention of the contaminants on the one hand, and for the rapid analysis of certain components on the other. Thus from a mixture of C_1–C_4 saturated and unsaturated

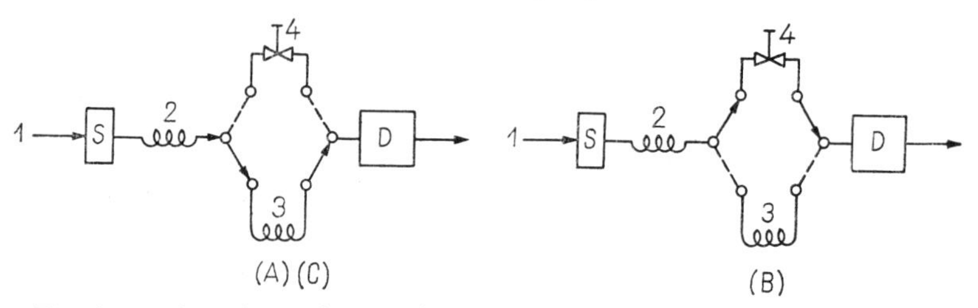

Fig. 10.5. Flow sheet of two columns in series. (A) and (C) two columns in series; (B) second column by-passed. S—sample introduction; D—detector; 1. carrier gas inlet; 2. first column; 3. second column; 4. choke

hydrocarbons ethane and ethylene can be analyzed within 1–2 min with a silica gel packed column provided the heavier hydrocarbons are retained by a pre-column with liquid loaded packing.

Analysis of wide boiling range complex samples requires usually more than one or one extremely long single column. With the *serial arrangement and alternative operation of two columns* with different packings, such analyses can be efficiently performed (Fig. 10.5). When the light components, having travelled through the first column, reach the second column (A) this will be by-passed by the carrier gas stream by the turning of a valve and the heavy components eluted from the first column (B) are fed directly to the detector. The second column is again switched into the carrier gas stream after the detection of the heavy components and the light components are then separated (C). Such a procedure is necessary for example for the analysis of a mixture of permanent gases + CO_2 when the second column is packed with a molecular sieve for the separation of O_2 from N_2 and the irreversible adsorption of CO_2.

In another dual column arrangement the first column is by-passed and then switched into the carrier gas stream after the elution of the light component. These methods are known by the name of "*column storing*", as certain components are temporarily stored in one of the columns.

With dual column and back-flush arrangements small quantities of contaminants can be concentrated or separated from greater quantities of other components of similar chemical nature. Thus, e.g. small quantities of ethane in ethylene may be concentrated in a short pre-

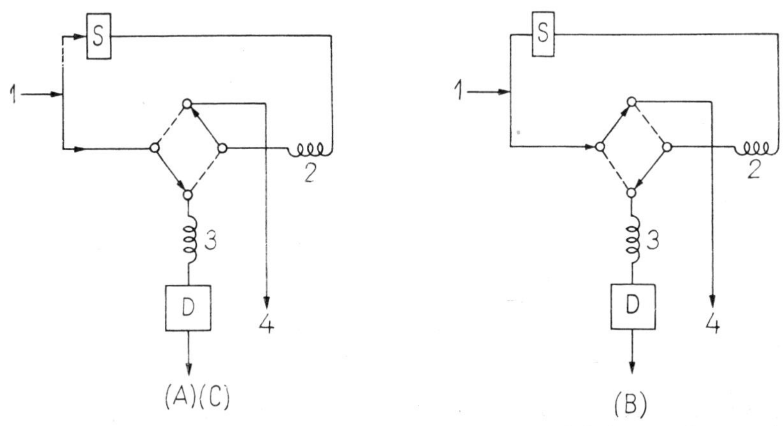

Fig. 10.6. Flow sheet of the heart cut method. (A) and (C) pre-column and analytical column in parallel arrangement; (B) pre-column and analytical column in series. S–Sample introduction; D–detector; 1. carrier gas inlet; 2. pre-column; 3. analytical column; 4. vent

column with silica gel packing which can then be back-flushed and the fraction rich in ethane led to the separating column.

The so-called "*heart cut*" method is a recent variation of dual column arrangement which provides means for the examination of any randomly chosen narrow range on the chromatogram, thus for the analysis of the component present in small quantity which is eluted after the main component (Fig. 10.6). Essentially the two columns are in parallel arrangement, from the sample injected into the first column the unwanted components are purged into the open and only a narrow fraction of the components is led into the second column for analysis.[2]

For the analysis of complex samples *three or more different columns* may be needed. Theoretically any number of columns might be used, but for practical reasons of space three, or at most four, columns are combined. With three or four columns different arrangements of back-flush, temporary storing, etc. can be realized.

A special arrangement of three columns is the "*flip-flop*" *method* for the re-grouping of the components.[2, 18, 20] This is used when instead of separation into individual components certain component groups, e.g. C_1–C_3, C_4–C_6, C_7–C_9 are to be analyzed. The principle of the

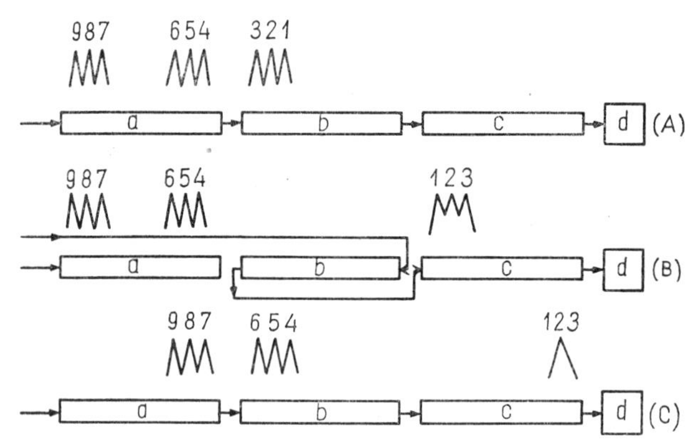

Fig. 10.7. Flow sheet of the three column flip-flop arrangements. (A) Three columns in series, first component group into the storing column; (B) first column by-passed, second column back-flushed; (C) three columns in series, first component group giving a single peak; a–separating column; b–storing column; c–re-grouping column; d–detector; 1–9–components

method is shown in Fig. 10.7. The sample is separated by the first "splitter" column from where the components reach the second, storing column (A). After the elution of the first chosen group of components, the first column is by-passed and from the second column the components are back-flushed into the third column where re-grouping takes place (B). From this column the selected group will appear as a single peak on the recorder (C). This process is repeated for different groups of components. The method is applied mainly in process control.[2]

The satisfactory operation of multicolumn methods is based on properly constructed and reliably operating switch valves. Though the various column arrangements may considerably raise the performance of the process gas chromatographs by their higher separation efficiency, and considerably reduced analysis time, certain drawbacks are involved in their application. The apparatus will be more complicated, requiring very accurate adjustment and strict observation of the time programs. Switchings may cause drift of the base line which in some cases has to be corrected by zero setting after each switching.

Detector. Most of the process gas chromatographs use *thermal conductivity detectors* which are quite adequate because of their low price, simplicity and reliable operation; it is applicable to the detection of 0–50 ppm concentration in four channel bridge circuits and to 0–1% concentrations with dual channel bridge circuits.[18] The detector is constructed with minimum gas volume to reduce the time constant and a separate carrier gas stream flows through the reference branch.

For trace analyses *flame ionization or argon detectors* are used to sense concentrations in the ppm range. Thus for instance C_2H_2 traces of 0–1 ppm concentration in oxygen can be detected with the ionization detector.[5] Lately, flame ionization detectors are preferred because of their simplicity and reliability.[18] Some favourable results have been obtained with gas density balances. Provided only moderate sensitivity is required, the gas density balance may have great advantages because of its stable operation and minimum maintenance needs. Gas chromatographic detectors in general have an extremely wide linear dynamic range which enables the measurement of components in the 0–0.1% and 0–100% concentration ranges with equal accuracy by means of adequate attenuation.

Programmer unit (timer)

Control and automatic repetition of the entire analysis is provided by the timer which controls and checks the various functions according to a pre-adjusted time program.

The timer controls the following functions:

1. in the case of several sample streams switching from one sample to the other;
2. sample injection and its automatic repetition;
3. switching on and switching off the columns;
4. transmission of the response of the selected component for recording;
5. variable attenuation according to the component;
6. change in the polarity of the detector response (in certain cases only);

7. operation of the recording chart;
8. transmission of the electric or pneumatic signal for process control;
9. automatic zero correction after each analysis.

The various time switches operate according to the relative times for each sample injection.

From the point of view of construction, the timer may: a) include disks with cams whose position can be altered and which are rotated by a synchronous motor;[21] b) operate on the basis of charging and discharging capacitors;[22] or c) by the optical follow-up of signals on a transparent film tape;[10] d) control the various functions from signals on a magnetic tape.[23]

In general the analysis cycle may be varied between 2 and 40 min and six to twelve components can be recorded on the chromatogram. The operation of the timer can be followed from the signal lamps corresponding to the various functions. The automatic timer can be by-passed and the process controlled by the manual operation of press-buttons.

The programming unit is assembled on a panel which usually carries the control elements and also the electrical circuit and electronic parts of the detector.

Recording

The analytical results are usually plotted by a recorder in the range between 0–1 or 0–10 mV, by the methods illustrated in Fig. 10.8. When recording, the chart stops between the peaks so that a series of successive peaks appear (A). In recording the bar graph the chart progresses at equal intervals producing vertical lines at the places of the peaks. In the case of multicomponent mixtures a pattern is produced which is impossible to survey, so that only the responses of the selected components are recorded (B). Another possibility is to transmit the signal of the important components through a signal store to a multicolour recorder, which however requires very expensive equipment and is therefore not widely used.

In trend recording, the response of the selected key component is continuously recorded up to the appearance of the response of the next analysis by the interpolation of a pneumatic or electronic memory unit. The step-wise function obtained in this way is a good approximation of the continuous trend function (C). This method is applied mainly in the analysis of a single component or in rapid analysis, though with multicolour recording the responses of several components may also be recorded.

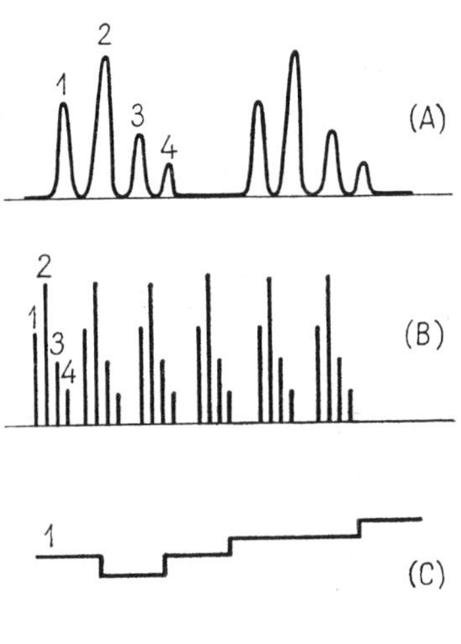

Fig. 10.8. Methods of recording. (A) Peak recording; (B) bar graph; (C) trend recording; 1–4–components

Evaluation of the results

Quantitative evaluation is usually based on measuring the peak heights, which provides under the satisfactorily controlled constant operation conditions of process gas chromatography the required accuracy. Peak areas are determined by means of the integrator joined to the gas chromatograph.

Inspection of the peak heights provides a direct guide source for the control of production processes, while the signal is at the same time a danger signal from safety aspects. For process evaluation, e.g. estimation of the material balance, or for computer control a percentage value of the components is required which can be achieved by the incorporation of a digital calculator.[24] Evaluation is based either on the peak height or the peak area. The results are printed in digital form or the responses of the individual components are recorded or transmitted for further processing by means of a digital-analogue signal transformer. The output signal might be a current intensity, voltage, resistance or pressure signal which can be fed into the computer for process control. It has been estimated that one in four process gas chromatographs produces a continuous output signal.

The accuracy of the results depends primarily on the accuracy of calibration which is performed by the injection of pure components or reference mixtures. The calibration of chromatographs designed for trace analysis presents a specially complicated problem. The apparatus has to be checked during operation at certain intervals (weekly or even daily) by the injection of standard mixtures of known compositions.

10.3 Process control by means of gas chromatography

On the basis of the analytical data furnished by the process gas chromatograph the operator should in fact carry out the regulation of the process. The next obvious step was the application of the signal of the gas chromatograph to the automatic control of the process. Any obstacle in the way of this application was eliminated by the development of high speed gas chromatographs. Analysis times between one and five minutes are in most cases quite satisfactory for process control.

The first task in process control is the transformation of the intermittent signals of recurring cycles into a continuous electrical or pneumatic signal. This can be performed in the way described for trend recorders by means of the interpolation of electronic or pneumatic memory units and signal transformers. In general the process is controlled on the basis of the concentration changes of one or two selected key components, by using the peak height values. The signal transformer provides continuous electrical or pneumatic signals which it corrects after each cycle, resulting in a step-like outgoing signal corresponding to the concentration changes of the component.

When *process control* is based on the concentration changes of a single component, then the signal of the component is compared to a reference signal and the difference between the two is transmitted to the process control unit.

Another possibility is to use the signal corresponding to the concentration difference or ratio between two components.

Figure 10.9 shows the principle of a gas chromatograph used for automatic process control. The output signal is suitable for both cascade and direct process control. In the first case this signal adjusts or corrects the set point of some control loop (e.g. temperature).[6, 25, 26] In the second case the output signal of the gas chromatograph operates the controlling unit, usually a pneumatic valve.[27, 28] Schematic diagrams of the two types of control are shown in Fig. 10.10.

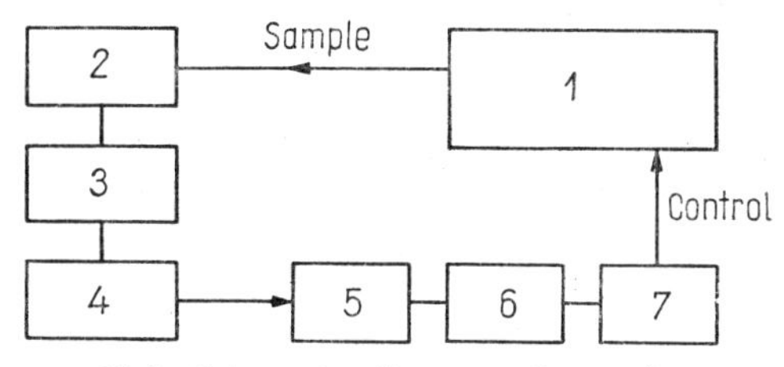

Fig 10.9. Schematic diagram of gas chromatographic process control. 1. Process; 2. analyzer; 3. programmer; 4. recorder and/or signal transformer; 5. memory unit; 6. signal transmitter; 7. control unit

In *computer process control* usually the data of several components are processed which a digital calculator unit transmits on the basis of the peak heights or peak areas. The peak height is determined with an analogue digital signal transformer, peak area with a voltage-frequency transformer and pulse counter. The digital calculator stores the values obtained till the arrival of the next cycle and transmits a proportional signal to the control system.[29, 30]

One of the most important applications of process control by gas chromatography is the *control of distillation columns*. Regulation of product streams, reflux recirculation or heating of the reboiler are usually based on the temperature, measured at a certain point in the column. Temperature changes are however rather insensitive to changes in composition, while in multicomponent systems the temrerature–composition correlation is unreliable, so that the duality of the product may vary between wide limits. Separation efficiency and the purity of the product may be greatly improved by gas chromatographic control.[6, 22, 31, 32] Selection of the place of sampling and the quantity to be controlled will depend on the task in hand.[32] In the case of feed-back control this is usually based on the composition of the product either at the top or at the bottom of the column, sometimes on the analysis of samples taken from an appropriate plate of the column. Feed-forward control is the most convenient for the concomitant control of both product tappings.[33] An important application of feed-forward control is the modification or adjustment

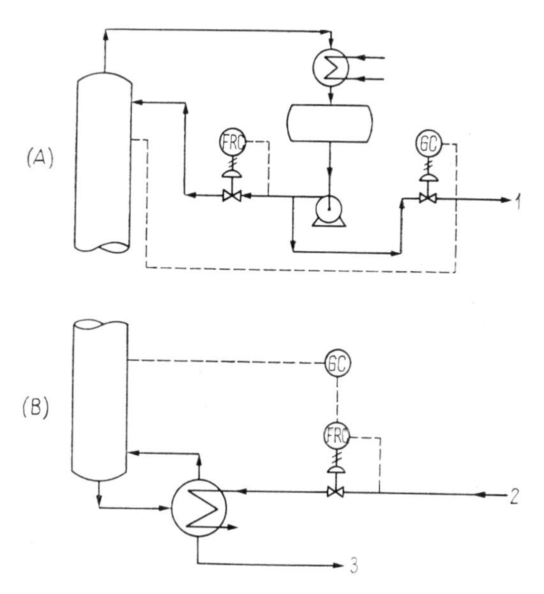

Fig. 10.10. Schematic diagram of the gas chromatographic control of a distillation column. (A) Direct control; (B) cascade control. GC–Gas chromatograph; FRC–flow rate control; 1. head product; 2. steam to reboiler; 3. bottom product

of the control points of the column according to the composition and quantity of the material fed into the column.

Other equipment performing similar separation operations, e.g. absorbers may also be automatically controlled.[3]

Gas chromatographic control is very useful to ensure the optimum operation of separation systems consisting of two or more units. Figure 10.11 shows a diagram of two distillation columns in series.[32] Chromatographs GC–1 and GC–2 forward the composition of the feed to the computer for feed-forward control. Chromatographs

GC–2 and GC–3 control the discharge of the bottom product on the basis of the concentration of the heavy key component in the head product (feed-back), thus chromatograph GC–2 serves a dual purpose. The process control chromatographs GC–4 and GC–5 serve the

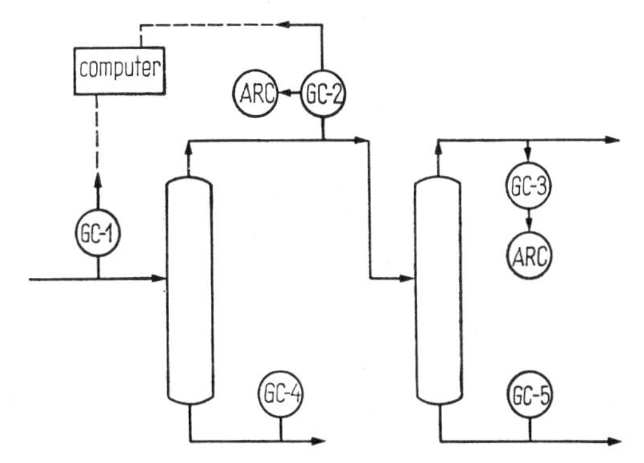

Fig. 10.11. Automatic control of two columns in series.[32] *GC–Gas chromatograph; ARC–automatic rate control*

continuous recording of the light key component in the bottom product, so that the entire process is easy to survey.

Besides the control of separating units gas chromatography is also used for the control and monitoring of the operation of various reactors, pyrolysis furnaces, catalytic cracking reactors etc., mainly coupled with computer control.[30]

10.4 Application of process chromatographs

Gas chromatographs can be used for process control in practically every branch of the chemical industry. They are used in their greatest number in petroleum and petrochemical plants; according to an estimate dating back to 1963 about 85 % of all process chromatographs are used in these industries.[34] Without an attempt at completeness a few examples will be quoted below as an illustration of the wide scope of process chromatographs.

In the *petroleum industry* process chromatographs are used mainly for the control and monitoring of equipment designed to separate light hydrocarbon mixtures, e.g. the distillation process in gasoline plants and refineries, the columns and products of the raw material preparation and product separation systems of alkylation plants,[3, 6, 32, 34–36] alkylation reactors reforming, butane isomerization, butane dehydrogenation, catalytic and thermal cracking plants, etc.[6, 30, 36, 37]

In the *petrochemical industry* process gas chromatographs are used in ethylene production to control the pyrolysis unit,[30] and the columns in which the gases are separated.[34, 36] A highly important application is the continuous monitoring of the ppm contaminations in polymerization grade ethylene and propylene.[5, 19, 34] Other fields are: acetylene production, analysis of the intermediaries and final products of butadiene production, analysis of the products of ethylene oxidation and methanol synthesis.[29]

In the *inorganic chemical industry* process gas chromatographs are used for the analysis of the recirculating gas of ammonia synthesis, detection of impurities in argon, monitoring in the production of rare gases, determination of acetylene traces in oxygen,[5] control and regulation of sulphur production,[38] analysis of chlorine for possible impurities, analysis of coke oven gases, etc.

In the *organic chemical industry* process chromatographs are used to analyze the intermediate and end products of various syntheses, separations, e.g. the analysis of fatty alcohols.[29] A very important application is the control of plants built for the recovery of solvents, where gas chromatographs may bring about considerable savings.[5]

In *metallurgy* the atmosphere of annealing furnaces may be controlled by gas chromatographs.[39] Other applications in this field are: analysis of the fuel and flue gases of furnaces and kilns, and of the contaminations in the oxygen used in steel manufacture.[34, 40]

Other fields where process gas chromatographs are used include the continuous control of the gases when petroleum wells are bored, and of the protective atmosphere of nuclear reactors.

A special type of gas chromatograph called the *Hydro-Chromatograph* serves the continuous detection of traces of moisture in hydrocarbon streams liable to polymerization (e.g. olefins). Water is here separated from the other components on a gas chromatographic column and the water only is led into the phosphorus pentoxide-filled hygrometer.[41]

References

1. B. O. Ayers, Chapter 1, ref. 59, p. 249
2. H. J. Noebels, *VDI-Berichte*, (97) 59 (1966)
3. M. M. Fourroux, F. W. Karasek and R. E. Wightman, *ISA Journal*, **7**, (5). 76 (1960)
4. A. Steppuhn *VDI-Berichte*, (97) 67 (1966)
5. R. Gittins *Instr. Pract.*, **18**, 235 (1964)
6. D. J. Fraade, *Petro. Chem. Eng.*, April. C-16 (1961)
7. D. W. Gillings, *Instr. Pract.*, **21**, 249 (1967)
8. Anon., *Chemical Processing*, Sept. 9. 24 (1963)
9. D. J. Fraade and H. Righi, *Z. anal. Chem.*, **189**, 106 (1962)
10. H. R. Karp, *Control Engineering*, June, 87 (1961)
11. C. C. Helms and H. N. Claudy, Chapter 1, ref. 59, p. 269
12. Carlo Erba, Process Chromatograph DT 51/1a E, Milan
13. C. S. Turner and R. Villalobos, Chapter 1, ref. 67, p. 363
14. E. L. Szonntagh, Chapter 1, ref. 68, p. 223
15. Greenbrier Instr. Inc. Bulletin No. 103
16. J. Hooimeijer, A. Kwantes and F. Van De Craats, Chapter 1, ref. 60, p. 288
17. R. F. Wall, W. J. Baker, T. L. Zinn and J. F. Combs, *Ann. N. Y. Acad. Sci.*, **72**, 739 (1959)
18. A. B. Littlewood, *J. Gas Chromatog.*, **2**, 186 (1964)
19. W. J. Baker and T. L. Zinn, *Control Eng.*, Jan. 77 (1961)
20. G. R. Nuss, Chapter 1, ref. 68, p. 119
21. H. Oster and E. Grimm, *Siemens-Zeitschrift*, **37**, (6) 3 (1963)
22. J. Fischer, *Z. für Messen, Steuern, Regeln*, **16** (1960)
23. Beckman Instr. Inc. Bulletin GC-4036
24. E. Grimm, Chapter 1, ref. 71, Nachtrag, p. 29
25. M. M. Fourroux, *Oil Gas J.*, Apr. 21, 114 (1958)
26. G. S. Turner, Chapter 1, ref. 64, p. 103
27. H. J. Noebels, *Erdöl u. Kohle*, **13**, 774 (1960)
28. H. L. Hoffman, *Hydrocarbon Proc. Petr. Ref.*, **42**, (2) 108 (1963)
29. E. Grimm, *Erdöl u. Kohle*, **18**, 791 (1965)
30. F. T. Ogle, *Chem. Engng. Progr.*, **61**, (10) 87 (1965)
31. R. Kaiser and H. Kienitz, Chapter 1, ref. 66, p. 234
32. M. W. Oglesby and J. W. Hobbs, *Oil Gas J.*, **64**, (2) 80 (1966)
33. D. E. Lupfer and H. L. Johnson, *ISA Trans.*, **3**, 165 April (1964)
34. A. Verdin, *Trans. Inst. Chem. Engrs.*, **41**, (3) CE111 (1963)
35. L. J. McGovern and L. J. Carlisle, *ISA Journal*, **6**, (5) 60 (1959)
36. H. J. Maier, *Control Eng.*, **8**, 89 (1961)
37. J. H. Williams Jr., R. L. Wilt, *Hydr. Pro. Petr. Ref.*, **40**, (5) 163 (1961)
38. R. L. Whitson and M. M. Fourroux, *ISA Journal*, **7**, (3) 40 (1960)
39. Anon., *Metal Treatment*, June/July 21 (1961)
40. C. Schoedler, *Chim. et Ind.*, **87**, (2) 231 (1962)
41. C. J. Penther and L. J. Notter, *Anal. Chem.*, **36**, 283 (1964)

11. Some Special Applications of Gas Chromatography

11.1 Simulated distillation by gas chromatography

Petroleum and petrochemical processes are in general characterized by the very high number of components with different boiling points in both the initial substances and in the end products, which are usually identified by their distillation curves. Engler distillation is a generally accepted standard method (ASTM D–86) in process control, though it provides only a rough approximation of the boiling point distribution with considerable inaccuracies mainly in the initial and final boiling range.

The boiling point curve may be plotted from the data of analytical fine fractionation known by the name of true boiling point (TBP) distillation. This is usually performed with a fine fractionation column of 100 theoretical plates and with the collection of very small fractions (1–2%). Distillation of this type requires a very long time, about 100 hours and even then the accurate determination of the initial and final boiling point is rather cumbersome. All distillation methods share the problem of a certain residue in the column, that is, of a recovery lower than the input.

With gas chromatography the boiling point curve can be determined rapidly and accurately. The method is based on the elution of components in the order of their boiling points from a column loaded with a non-polar liquid phase. Analysis is by programmed temperature, as wide boiling range mixtures have to be separated. The liquid phase is a non-polar liquid with low volatility, usually Apiezon grease L, or silicone grease SE–30 or SE–52. The boiling points of the components are determined from their retention times or from the temperatures at which they appear in the effluent. The weight composition of the sample is obtained from digital signals corresponding to the integration of the peak areas, so that the boiling point curve can be plotted from gas chromatographic data. In other respects the apparatus is the same as that used in programmed temperature gas chromatography with single or dual column arrangement and linear temperature program.

Eggertsen *et al.*[1] were the first to apply gas chromatography to the simulation of distillation and obtained boiling point curves corresponding to 10–20 theoretical plates. Detection was made by catalytic combustion and the CO_2 was measured, after adsorption of the water, by means of a thermal conductivity cell. Barras and Boyle[2] analyzed the boiling points of various petroleum fractions in a similar apparatus, and were able to determine the pour point of jet fuels from these data. By these methods boiling point curves similar to those obtained by the ASTM method but more accurate than the latter could be plotted. As an improvement of these procedures Petrocelli *et al.*[3] constructed a process gas chromatograph with dual column arrangement and flame ionization detector for the automatic analysis of boiling point curves. The accurate gas chromatographic data were transmitted after appropriate calibration in the form of the data of the less accurate ASTM distillation method. The process gas chromatograph operated with a minimum standstill reliably over a long period of time, thus with this method distillation could be controlled and monitored on stream, something quite impossible on the basis of the laboratory ASTM distillation.

Green *et al.*[4] introduced some further improvements into the method and apparatus of simulated distillation for the determination of the boiling point curve corresponding to TBP distillation. They obtained by dual column arrangement with flame ionization detectors and accurate temperature programming, from 1 μl samples boiling point curves corresponding to TBP distillation with 100 theoretical plates in about one hour, that is in a time which is about 1 % of the time requirement of TBP distillation. Comparison of simulated and TBP distillation is shown in Fig. 11.1, from which it appears that the gas chromatographic curve follows in fact the TBP curve, but is more accurate than the latter in the initial and final boiling point ranges. With the exception of sample injection the apparatus is automatic, and can be used for fractions with boiling points up to 450 °C after calibration with wide boiling range model mixtures of known compositions. The reproducibility of the boiling point curves was ± 1 °C.

These results led to the development of special commercial apparatus for simulated distillation[5] which can be used in the temperature range from -60 °C to $+300$ °C. Initial cooling is provided by the evaporation of liquid CO_2 and the apparatus is automatic but for the injection of the sample. The output response of the flame ionization detector is evaluated by a digital integrator which prints the results every 10 sec. A calibration curve of boiling point *vs.* integrator value is plotted from the analysis of a sample of known composition and from this curve the correlation boiling point *vs.* weight per cent can

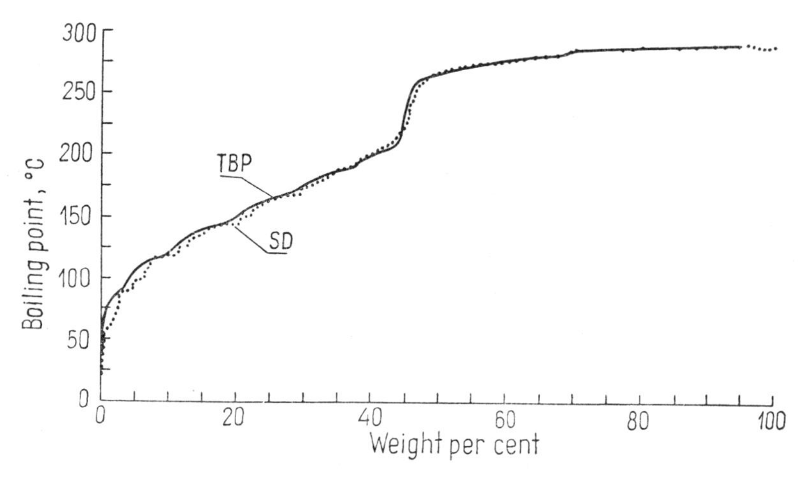

Fig. 11.1. Gas chromatographic determination of the boiling point curve (investigation of the pyrolysis products of hexadecane)[4]

be constructed. A computer program was worked out for the evaluation of the data and the plotting of the TBP curve which resulted in considerable time saving especially in the routine analysis of very large numbers of samples.[6]

With programmed temperature open tubular columns considerably higher resolutions can be achieved than with packed columns though quite obviously at the expense of analysis time. Thus for instance Diesel oil was analyzed by means of a 50 m long polyphenylether coated open tubular column in 160 min;[7] on the resulting chromatogram about 300 peaks were distinguished.

Gas chromatographic simulated distillation provides a very accurate and rapid method for the determination of the boiling point curves of wide boiling range samples. Analysis time will be between 30 and 60 min. The method is applicable for C_1 to C_{40} hydrocarbons. An added advantage is that analysis requires not more than 1–10 μl of the sample, and may replace not only analytical distillation but be applied to product control in distillation plants by providing data for computer processing in automatic plant control.

11.2 Elementary analysis

Elementary analysis is a special field of reaction gas chromatography which involves, like classical elementary analysis, oxidation and reduction of the initial substance to well-defined simple reaction products, e.g. carbon dioxide, water, nitrogen oxides, hydrogen, nitrogen, etc. These decomposition products are then separated on the gas chromatographic column, separation combined with the selective adsorption of certain products or their conversion to some other compound and the elementary components of the sample determined from the peaks on the chromatogram. This principle has been in use for some time with certain gas chromatographic detectors (see Section 4.5), but the first publications on the determination of the elementary composition of compounds appeared in 1960. In succeeding years parallel to the development of reaction gas chromatography the methods of elementary analysis also rapidly improved and today special methods for the determination of the more important elements and appropriate apparatus for this purpose are available.

One of the applications of elementary analysis is the qualitative identification of the gas chromatographically separated peaks by the determination of the elementary composition, that is, by the detection of the elements constituting the compound under investigation (see Section 6.1).

Beyond this application gas chromatographic elementary analysis has developed into an indispensable method of microanalysis for the determination of the elementary composition of various organic and inorganic substances. With the latest chromatographs elementary analysis can be performed much quicker than by any other method (e.g. CHN analysis in about 10 min) requiring only a very small sample (0.2–0.5 mg), while the accuracy of the method competes with that of any other.

The most important application of elementary analysis is the determination of the C, H and N content of organic compounds. The first gas chromatographic methods for the determination of *carbon* and *hydrogen* were described by Duswalt and Brandt[8] and Sundberg and

Maresh.[9] The first authors used for combustion an oxygen atmosphere and a microreactor filled with copper oxide, while Sundberg and Maresh[9] performed combustion in a copper oxide–copper filled microreactor under helium atmosphere. The water formed by the reaction was led through a tube filled with calcium carbide where it was converted to acetylene. In both methods combustion was slow (10–20 min) which necessitated the collection of the carbon dioxide and of acetylene in liquid nitrogen cooled traps from which they were swept by the carrier gas into the separating column. The introduction of rapid induction heating considerably improved this method as it enabled the direct introduction of the decomposition products into the column.[10]

Again other methods apply a combustion at 700–1000 °C with copper oxide, cobalt oxide or silver permanganate as the catalysts and helium as the carrier gas.[11]

Another possibility of carbon determination is the reduction of organic compounds to methane[12] over a Raney nickel catalyst at 400–500 °C, but this method is seldom used in elementary analysis.

In the analysis of organic compounds the carbon content is usually determined together with other elements, e.g. with hydrogen,[13] hydrogen and nitrogen,[10, 14] sulphur[15] and halogens.[16] This method can be applied to the determination of carbon contaminants in metals and inorganic compounds.[17]

Hydrogen can be determined by one of the oxidation methods described earlier. The water formed by the reaction can be directly analyzed on an appropriate column (Carbowax, Polymer gel). Other methods convert the water into acetylene by passing it through calcium carbide[9] or reduce it to hydrogen over an iron catalyst or calcium hydride and analyze it in this form.[18]

The determination of carbon and hydrogen is very important in the analysis of compounds containing labelled atoms.[13, 19, 20] The $^{14}CO_2$ formed by combustion and the tritium obtained after reduction of the water are sensed by a radioactive detector after the gas chromatographic detector.

Under the conditions of C and H determination *nitrogen* is oxidized to nitrogen oxide and can be analyzed[21] in the form of NO_2 or more generally as N_2 after reduction on metallic copper.[10] In the latter method part of the combustion tube is filled with metallic copper in which reduction takes place immediately after the oxidation process.

The *oxygen* content of organic compounds can be converted directly into carbon monoxide by passing the gas at high temperature (1100–1150 °C) through a quartz reactor packed with charcoal and analyzing the carbon monoxide by gas chromatography.[22] Analysis requires not more than 6 minutes and is at the same time highly accurate. Another possibility is to use a 1 : 1 Pt : charcoal packing at 920 °C. Free oxygen forms water with the hydrogen carrier gas on Pt catalyst and this water can then be determined by any of the methods described above.[23]

The *sulphur* in organic sulphur compounds is converted into SO_2 on Pt catalyst at 850 °C in an oxygen stream and after the adsorption of the water formed, CO_2 and SO_2 are separated on an appropriate column.[24] Sulphur dioxide may also be determined with a selective detector (e.g. electrical conductivity cell).[25] Another possibility is the reduction of the sulphur compound[26] on Pt catalyst to H_2S at 1000 °C.

Under conditions similar to the oxidation of the sulphur compounds, *halogen* containing compounds[24] can be converted into Cl_2, Br_2 and I_2 and analyzed by separation on a silicone grease loaded column.[16] Halogen compounds may also be reduced to hydrogen halides on a Pt catalyst at 800–1000 °C and selectively detected by micro-coulometry or the measurement of electrical conductivity.[25, 27]

Phosphorus is determined after reduction to phosphine.[27]

In respect to the selective detection of sulphur, halogen and phosphorus compounds we refer the reader to the operation of the specific detectors discussed in Section 4.5.

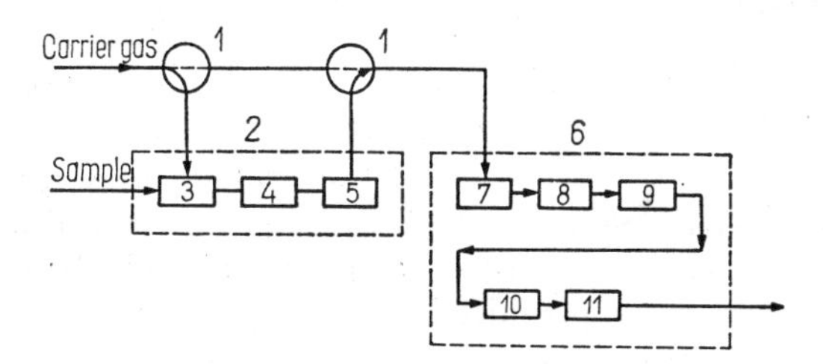

Fig. 11.2. Schematic diagram of a C–H–N analyzer. 1. Switch valve; 2. furnace; 3. sample transmitter; 4. combustion tube; 5. expansion chamber; 6. thermostat; 7. column No. I; 8. detector No. I; 9. dryer; 10. column No. II; 11. detector No. II

All these analytical procedures may be performed with appropriately modified laboratory gas chromatographs. Special apparatuses were however evolved for the determination of C, H and N with an accuracy required from elementary analysis and for routine analytical purposes. These types of apparatuses work with a considerably shorter analysis time (8–10 min) than the various forms of the Pregl and Dumas combustion methods, they provide satisfactory reproducibility within the generally required $\pm 3\%$ tolerance of the recorded results and give simultaneous determination of C, H and N.

Of these analytical apparatuses F. & M. Model 180 C–H–N analyzer is the best known, the schematic diagram of which is presented in Fig. 11.2. The apparatus is a modification of a dual column, dual detector gas chromatograph with a microbalance as an ancillary to weigh 0.2–2 mg samples with high accuracy. Analysis accuracy is primarily determined by the accuracy with which the samples are weighed.

The weighed sample in a small aluminium boat is transferred to the combustion zone packed with copper oxide and metallic copper. The combustion products are collected in the expansion chamber from where they are swept by the turn of a valve into the gas chromatographic section. The first column serves to separate the water which is sensed by the first detector. This water is then removed and the nitrogen and carbon dioxide are separated by the second column and sensed by the second detector. The quantities are determined from the peak heights or peak areas. By modification of the combustion section the apparatus is suitable for the rapid and accurate determination of the oxygen content of reducible metal oxides.[28] Fisher Scientific Co., Perkin–Elmer Corp., and Technikon Controls Inc.[11] have produced similar apparatuses for the same purpose which operate under somewhat modified conditions. The highest accuracies are: for C $\pm 0.08\%$, for H and N $\pm 0.2\%$. In addition to the already mentioned advantages gas chromatographic elementary analyzers are highly favourable, because, except for sample injection, they operate automatically and their handling is very simple.

11.3 Measurement of surface areas

Specific surface is one of the most important parameters of solids (adsorbents, catalysts) and is usually determined by the BET method.

An accurately weighed and appropriately prepared solid is placed in an evacuated static system where it is brought into contact with a known volume of a gas and after adsorption equilibrium has been reached the quantity of the adsorbed gas is determined from pressure measurement. By the addition of further known volumes of the gas the adsorbed gas quantities corresponding to different equilibrium pressures can be determined and the specific surface of the solid can be calculated from the Brunauer, Emmett, and Teller equation.[29, 30] The linearized form of the BET isotherm equation is:

$$\frac{p}{a(p^\circ - p)} = \frac{1}{a_m c} + \frac{c-1}{a_m c}\,\frac{p}{p^\circ} \qquad 11.1$$

where p is the partial pressure of the adsorbate, p° the vapour pressure of the component at the temperature of the measurement, a the adsorbed gas quantity, a_m the quantity of adsorbate necessary for the monomolecular coating of the adsorbent and c is the adsorption constant. When the experimental data are plotted in the $\frac{p}{a(p^\circ - p)}$ *vs.* $\frac{p}{p^\circ}$ co-ordinate system a straight line is obtained in the $p/p^\circ = 0.05$ to 0.35 relative pressure range from whose slope and intersection with the ordinate the value of a_m and from this the surface area can be calculated provided the space occupied by the adsorbate molecule is known (e.g. for N_2 about 17 Å^2).

The method is accurate but lengthy because of the time needed for equilibration and requires in addition fairly complicated vacuum equipment.

By means of the gas chromatographic method adsorption can be measured under dynamic conditions relatively rapidly and with a simple apparatus. Nelsen and Eggertsen were the first to evolve a dynamic method for the measurement of specific surface.[31]

The principle of the method can be summed up briefly in the following: The vessel containing the sample of the solid substance under investigation replaces the gas chromatographic column and is immersed in a liquid nitrogen bath which can be moved up and down. A helium–nitrogen stream flows through the apparatus, from which part of the nitrogen is adsorbed by the adsorbent. Changes in nitrogen concentration are sensed as a peak by the heat conductivity cell. The cooling bath is then removed, the adsorbent heated and the adsorbed nitrogen is desorbed; this temporary concentration rise is sensed by the detector as a peak of opposite sign. From the peak area the quantity of adsorbed nitrogen can then be calculated. Though in principle

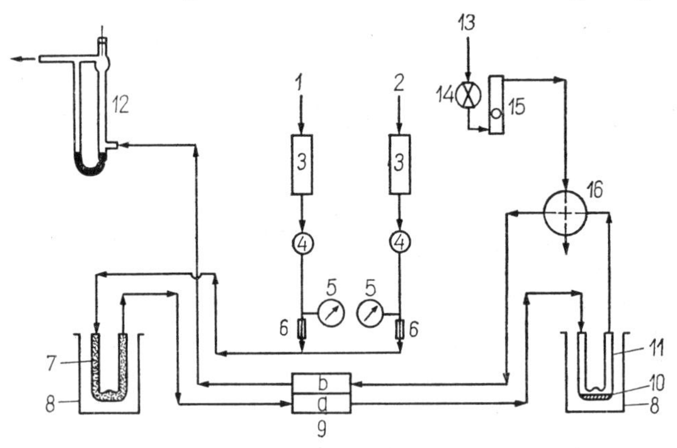

Fig. 11.3. Schematic diagram of the apparatus for determination of specific surface. 1. He inlet; 2. N_2 inlet; 3. filter and dryer; 4. pressure regulator; 5. manometer; 6. capillary restriction; 7. cooling trap (packed with glass wool); 8. dewar flask; 9. detector: a—reference branch; b—measuring branch; 10 solid sample; 11. sample holder; 12. soap film flowmeter; 13. N_2 for calibration; 14. needle valve; 15. rotameter, 16. sampling valve

the two peaks should be identical, in practice the desorption peak is used for the determination, being usually more symmetrical and easier to evaluate. The partial pressure of nitrogen is calculated from the composition of the nitrogen–helium mixture and from the barometric pressure. In this way one point of the BET isotherm can be plotted and the other points are obtained by repeating the measurement with different He : N_2 ratios. In general three points are sufficient to determine the specific surface from the linearized form of the BET equation.

The most important requirements which the measurements have to meet are: accurate control of the nitrogen and helium streams, their rapid and efficient mixing, constant pressure and uniform flow in the entire system and appropriate calibration.[32]

In principle, laboratory gas chromatographs are quite suitable for the determination of specific surfaces,[33] but it is more expedient to use for tasks of this type simpler special apparatuses which can be easily assembled in the laboratory. A commercial apparatus of this kind is the Perkin–Elmer Sorptometer.[34, 35] The schematic diagram of the apparatus is shown in Fig. 11.3. After drying and passage through a capillary restriction the helium and nitrogen are mixed, the mixture flows through the liquid nitrogen cooled trap, then trough the reference branch of the detector, the vessel holding the sample, the measuring branch of the detector and finally leaves the apparatus through a flow-meter. The apparatus is provided with a separate sampling system for calibration. By means of the sampling valve, different known quantities of nitrogen can be introduced into the gas stream before it enters the detector, so that the peak areas obtained with these quantities can be directly compared with the peak areas plotted during measurement. As the output responses of the detector for the adsorption and desorption peaks respectively have an opposite sign, a polarity switch has to be incorporated to the apparatus. The peak areas can be best evaluated by means of an integrator.

The dynamic method can be applied to substances with different specific surfaces, actually samples with specific surfaces ranging from 0.04 to 1300 m^2/g have been investigated. Depending on the surface, the sizes of the samples varied between 0.01 and 5 g. The sample is prepared quite simply by heating and purging with helium; depending on the nature of the sample, preparation may take a few hours at temperatures between 100 and 500 °C.

The main factor affecting the accuracy of the measurement is the constancy of cooling, that is the accurate measurement of the temperature of the cooling bath, as 1–2 °C changes in the temperature of the liquid nitrogen may result in about 15% changes in the value of the vapour pressure. The other factor to be considered is the deviation from linearity of the response of the thermal conductivity cell when the composition of the helium–nitrogen mixture changes, but this problem can be eliminated by accurate calibration.[32] Under adequate operating conditions the dynamic method furnishes reproducible results (σ = 2–4%) which are in good agreement with those obtained by the static method; the deviation between the results obtained by the two methods are of the order of 0.5–6%, depending on the specific surface.

The three points necessary for accurate specific surface determinations are obtained in about 40–50 min. For comparative measurements one point may be sufficient which is then joined to the origin and the slope of this line and from Eq. 11.1 the surface area is calculated with about 5% accuracy.[32]

When the specific surface is very low the adsorption on the wall of the sample holder may introduce a certain error which can be eliminated by carrying out a blank determination.

With modified sampling systems and a modification of procedure the apparatus may be used for the determination of the pore volume of solids and from the pore volumes measured at different relative pressures for the evaluation of the pore distribution on the capillary condensation section of the isotherm. For the measurement of pore volume in the adsorption step pure nitrogen, in the desorption step pure helium is passed through the sample holder and the quantity of nitrogen eluted by helium is measured.

In addition to the determination of specific surfaces the apparatus is quite obviously excellently suited for the plotting of adsorption isotherms.

A simplified version of this method senses both adsorption and desorption not from the concentration changes, but from the changes in the volume rate of the effluent gas with a soap film flowmeter.[36] This method is of course less sensitive and can be applied only to substances with specific surfaces over 10 m^2/g.

Dynamic specific surface determination is far simpler and more rapid than the static method, while its results are just as accurate as those obtained by the latter method.

References

1. F. T. EGGERTSEN, S. GROENNINGS and J. J. HOLST, *Anal. Chem.*, **32**, 904 (1960)
2. R. C. BARRAS and J. F. BOYLE, *Oil Gas J.*, **60**, (31) 167 (1962)
3. J. A. PETROCELLI, T. J. PUZNIAK and R. O. CLARK, *Anal. Chem.*, **36**, 1008 (1964)
4. L. E. GREEN, L. J. SCHMAUCH and J. C. WORMAN, *Anal. Chem.*, **36**, 1512 (1964)
5. Aerograph Research Notes, Spring Issue, 1965. Wilkens Instr. and Res. Inc.
6. T. H. GOUW, R. L. HINKINS, R. E. JENTOFT, *J. Chromatog.*, **28**, 219 (1967)
7. KEMMNER, B. KOLB and H. PAUSCHMANN, "*Instrumentation, Technique and Application of Gas Chromatography in Mineral Oil Analysis*", Bodenseewerk Perkin–Elmer GmbH (1963)

8. A. A. DUSWALT and W. W. BRANDT, *Anal. Chem.*, **32,** 272 (1960)
9. O. E. SUNDBERG and C. MARESH, *Anal. Chem.*, **32,** 274 (1960)
10. C. F. NIGHTINGALE and J. M. WALKER, *Anal. Chem.*, **34,** 1435 (1962)
11. K. J. FRANCIS JR., *Anal. Chem.*, **36,** (7) 31A (1964)
12. F. DRAWERT, Chapter 1, ref. 66, p. 347
13. F. CACACE, R. CIPOLLINI and G. PEREZ, *Anal. Chem.*, **35,** 1348 (1963)
14. W. WALISCH, *Chem. Ber.*, **94,** 2314 (1961)
15. W. K. STUCKEY and J. M. WALKER, *Anal. Chem.*, **35,** 2015 (1963)
16. J. C. MAMARIL and C. E. MELOAN, *J. Chromatog.*, **17,** 23 (1965)
17. A. DIJKSTRA *et al.*, *J. Gas Chromatog.*, **2,** 180 (1964)
18. A. T. JAMES and C. HITCHCOCK, *Kerntechnik*, **7,** 5 (1965)
19. A. T. JAMES and E. A. PIPER, *Anal. Chem.*, **35,** 515 (1963)
20. H. W. SCHARPENSEEL, *Angew. Chem.*, **73,** 615 (1961)
21. R. H. REITSEMA and N. L. ALLPHIN, *Anal. Chem.*, **33,** 355 (1961)
22. A. GÖTZ, *Z. anal. Chem.*, **181,** 92 (1961)
23. J. W. SWINNERTON, V. J. LINNENBOM and C. H. CHEEK, *Anal. Chem.*, **36,** 1669 (1964)
24. D. R. BEUERMAN and C. E. MELOAN, *Anal. Chem.*, **34,** 319 (1962)
25. D. M. COULSON, *J. Gas Chromatog.*, **3,** 134 (1965)
26. F. H. HUYTEN and G. W. RIJNDERS, *Z. anal. Chem.*, **205,** 244 (1964)
27. H. P. BURCHFIELD *et al.*, *J. Gas Chromatog.*, **3,** 28 (1965)
28. Hewlett–Packard F. & M. Sci. Div., *Facts and Methods*, **7,** (3) 1 (1966)
29. S. BRUNAUER, P. H. EMMETT and E. TELLER, *J. Amer. Chem. Soc.*, **60,** 309 (1938)
30. P. H. EMMETT, (Ed.) "*Catalysis*", Vol. I. Fundamental Principles (Part 1), p. 31. Reinhold, New York (1954)
31. F. M. NELSEN and F. T. EGGERTSEN, *Anal. Chem.*, **30,** 1387 (1958)
32. H. W. DAESCHNER and F. H. STROSS, *Anal. Chem.*, **34,** 1150 (1962)
33. J. F. ROTH and R. J. ELLWOOD, *Anal. Chem.*, **31,** 1738 (1959)
34. L. S. ETTRE and G. CAROTI, *La Chimica e l'Industria*, **42,** 864 (1960)
35. L. S. ETTRE, N. BRENNER and E. W. CIEPLINSKI, *Z. physik. Chem.*, **219,** 17 (1962)
36. H. BEYER and G. SCHAY, Chapter 1, ref. 71, p. 21

Index